LIGHT
&
COLOUR

photo—F S Smythe F R G S

The spectre of the Brocken

the nature of

LIGHT & COLOUR

in the open air

M. MINNAERT

translation H. M. Kremer-Priest
revision K. E. Brian Jay

Dover Publications Inc.

This Dover edition is an unabridged republication of the first English translation which was published under the title *Light and Colour in the Open Air*. This translation was originally published by G. Bell & Sons, Ltd. and this edition has been made available through a special arrangement with them.

Library of Congress Catalog Card Number: 54-10021

International Standard Book Number

ISBN-13: 978-0-486-20196-2
ISBN-10: 0-486-20196-1

Manufactured in the United States by Courier Corporation
20196123
www.doverpublications.com

Preface

A lover of Nature responds to her phenomena as naturally as he breathes and lives, driven by a deep innate force. Sun and rain, heat and cold are alike welcome to his observation; in towns and woods, on sandy tracts and on the sea he finds new objects of interest. Each moment he is struck by new and interesting occurrences. With buoyant step he wanders over the countryside, eyes and ears alert, sensitive to the subtle influences that surround him, inhaling deeply the scented air, aware of every change of temperature, here and there lightly touching a shrub to feel in closer contact with the things of the earth, a human being supremely conscious of the fullness of life.

It is indeed wrong to think that the poetry of Nature's moods in all their infinite variety is lost on one who observes them scientifically, for the habit of observation refines our sense of beauty and adds a brighter hue to the richly coloured background against which each separate fact is outlined. The connection between events, the relation of cause and effect in different parts of a landscape, unite harmoniously what would otherwise be merely a series of detached scenes.

The phenomena described in this book are partly things of everyday life, which it is fascinating to study from a scientific point of view, and partly things unfamiliar to you as yet, though they can be seen any moment, if only you will touch your eyes with the magic wand of 'knowing what to see.' And finally there are the rare, remarkable wonders of Nature, occurring only occasionally in a lifetime, so that even the most practised observer may wait year after year to see them. When he does see them he is filled to the very depths of his consciousness by the sense of their rareness, and by an ineffable feeling of pleasure.

However extraordinary it may seem, it remains a fact that the things one notices most are the things with which one is familiar; it is very difficult to see new things, even when they are before our very eyes. In ancient times and in the Middle Ages innumerable eclipses of the sun were observed; and yet the corona was scarcely noticed before 1842, although nowadays it is regarded as the most striking phenomenon of an eclipse and can be seen by anyone with the naked eye. In order to draw your attention to them, I have tried to collect in this book those things that in course of time have become known through the activities of many able students of Nature. No doubt there is far more yet to be seen in Nature; every year numbers of treatises are published concerning new phenomena; and it is strange to think how blind and deaf we must be to so many things around us that posterity is bound to notice.

The study of plants and animals is the usual meaning attached to 'observation of Nature,' as if the display of wind and weather and clouds, the thousands of sounds with which space is filled, the waves, the sun's rays, the rumblings of the earth were not part of Nature, too! And a textbook containing notes of all that is to be seen of particular interest to the student of physical science among the inanimate is just as necessary to him as a book on flora and fauna is to the biologist. We shall be led unavoidably into the domain of the meteorologist, and into the regions bordering on astronomy, geography and biology, but nevertheless I hope to have found a certain unity, in which the connection between the subjects treated can be perceived.

Since we are concerned with a simple, direct way of observing Nature we must systematically avoid: (i) Anything that can be found only with the help of instruments (concentrating instead on our senses, our chief helpers, whose characteristics, therefore, we ought to know). (ii) Anything deduced from long series of statistical observations.

(iii) Theoretical considerations not directly concerning what we see with our eyes.

We shall see that a surprising abundance of observations even then remain possible; indeed, there is hardly one branch of physics that is not applicable out of doors, and often on a scale exceeding any experiments in our laboratories. Bear in mind, therefore, that everything described in this book is within your own powers of understanding and observation. Everything is meant to be seen by you and done by you!

Where our explanations are perhaps too concise, we suggest that the reader should refresh his memory of fundamental physical theories by turning to some elementary textbook.

The importance of outdoor observations for the teaching of physics has not yet been sufficiently realised. They help us in our increasing efforts to adapt our education to the requirements of everyday life; they lead us by natural methods to ask a thousand questions, and, thanks to them, we find later on that what we learnt at school is to be found again and again beyond our school walls. And so the omnipresence of the laws of Nature is felt as a continually surprising and impressive reality.

Moreover, this book is written for all those who love Nature; for the young people going out into the wide world and gathering together round the camp-fire; for the painter who admires but does not understand the light and colour of a landscape; for those living in the country; for all who delight in travelling; and also for town-dwellers, for whom, even in the noise and clamour of our dark streets, the manifestations of Nature still remain. Even for the trained physicist we hope it will contain something new, for the field it covers is vast and often lies outside the ordinary course of science. So it will be understood why very simple as well as more complicated observations have been chosen, grouped according to their mutual relationship.

This book is very probably the first attempt of its kind, and as such is not perfect. I feel more and more overwhelmed by the beauty and extent of the material, and more and more conscious of my inability to explain it according to its merit. I have been experimenting systematically for twenty years and I have collected here the gist of some thousands of articles from every possible periodical, although only those articles have been quoted which give a comprehensive survey or throw light on very special points. But I am well aware how incomplete this collection still is. Many things already known are still unknown to me, and much remains a problem even to a professional. I shall, therefore, be all the more grateful to those who, through their own observations or by literary references, are willing to help me correct my mistakes or fill up any gaps.

M. M.

Contents

Plates

LIGHT
&
COLOUR

I

Sunlight and Shadows

1. *Sun-Pictures*

In the shade of a group of trees we see on the ground a number of spots of light, scattered irregularly, some large, some small, but all of them of a similar elliptical shape. Hold a pencil in front of one of them; the line connecting pencil and shadow indicates the direction from which the rays of light come that make the little patch on the ground. It is, of course, the sunlight piercing through an aperture in the crown of the tree; our eye sees a dazzling brightness here and there among the leaves.

The surprising thing is that all these spots have the same shape, and yet it is unlikely that all those chinks and openings should happen to be so nicely similar and round! Intercept one of these images by a piece of paper, held at right angles to the rays, and you will see that it is no longer elliptical, but circular. Raise the paper higher and the spot grows smaller and smaller. So we conclude that the pencils of light forming such a spot have the shape of a cone and the spots are elliptical only because the ground cuts this cone slantwise.

The origin of this phenomenon is to be found in the fact that the sun is not a mere point. Any very small opening P (Fig. 1) forms a small, well-defined image of the sun AB; another small opening P′ gives a somewhat shifted image A′B′ (dotted lines); a wider opening, which contains both P and P′, gives a less sharp but brighter image of the sun A′B. We can indeed see spots of light of every degree of brightness, and if there are two of equal size, the brighter one is at the same time the less sharp.

In confirmation of this, notice that when clouds pass before the sun you can see them glide over each patch of sunlight, but in the opposite direction; during a partial eclipse of the sun, all the sun-pictures are crescent-shaped. When there is a large sunspot it is visible on the sharpest images of the sun. You can make a very well-defined image of the sun by piercing a small, perfectly round hole in a sheet of

FIG. 1. Sun's rays penetrating dense foliage.

thin cardboard, and holding it up so that the sun's image falls on a well-shaded spot.

Examine the image of the sun formed at various distances by a square aperture.

The angle subtended by the sun's disc must be, therefore, the angle APB at the vertex of the cone forming the sun's image. Small angles of this kind we often measure in radians. We say, 'That angle is $\frac{1}{108}$ radian,' meaning that the sun seems to be as large as 1 inch at a distance of 108 inches, or as large as 10 inches at a distance of 1,080 inches (Fig. 2). Similarly, therefore, the diameter of a well-defined picture of the sun must be 108th part of that picture's distance from the opening; and for a hazy picture the size of the aperture in the foliage must be added. Intercept a weak, clearly defined sun-picture on a sheet of paper, hold it perpendicular to the rays of light, measure the diameter

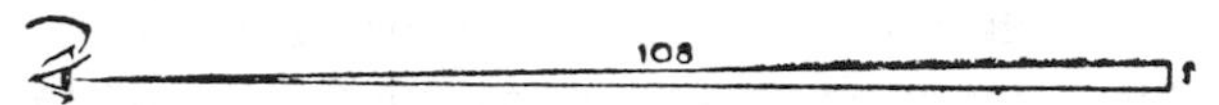

FIG. 2. We see the sun's disc at an angle of $\frac{1}{108}$ radian.

k of the light-spot, and determine by means of a piece of string the distance L from the paper to the opening in the foliage. Is k really equal to about $\frac{1}{108} \times L$?

If the sun-pictures formed on a level surface are ellipses, we measure the short axis k and the long axis b; the ratio between them equals that between the height H of the tree and the distance L. It follows, therefore, that $H = \frac{k}{b} \times L = 108 \frac{kk}{b}$. In this way the axes of a strikingly large sun-picture formed by the foliage of a beech-tree were found to measure 21 inches and 13 inches; the height of the aperture from the ground was, therefore, 870 inches or 72 feet 6 inches.

Observe that the sun-pictures are more oblong in the morning and evening, and rounder about noon.

Good sun-pictures are to be found in the shade of beeches, lime-trees and sycamores, but seldom in that of poplars, elms and plane-trees.

Look at the sun-pictures formed by the trees on the banks of shallow water; they can be seen very curiously defined on the *bed* of the water.

2. *Shadows*

Look at your own shadow on the ground; the shadow of your feet is sharp, the shadow of your head is not. The shadow of the bottom part of a tree-trunk or post is sharp, whilst the shadow of the higher parts becomes more and more hazy towards the top.

Hold your hand open in front of a piece of paper;

the shadow is sharp. Hold it further away; the umbra of each finger becomes narrower and narrower, while the penumbras grow larger until they merge into one another.

These peculiarities are likewise a consequence of the sun's not being a mere point, and correspond to what the sun-pictures showed us. Look at the shadow of a butterfly, of a bird (how seldom we usually notice such things !) and you will notice that it looks like a round spot; it is a 'sun-shadow-picture.'

The shadow of wire-netting used as a fence, consisting of rectangular meshes, struck me once as very odd, for only the shadows of the vertical wires were visible, and not those of the horizontal ones ! If a sheet of perforated paper is held in the rays of the sun, each hole in the paper is seen to form an elliptical light-spot on the ground; one can imagine the shadow of a wire to be due to a number of similar small ellipses, only dark this time, placed close together, which makes it fairly sharp when the wire lies in the direction of the longer axis, and indistinct in the direction of the shorter axis (Fig. 3).

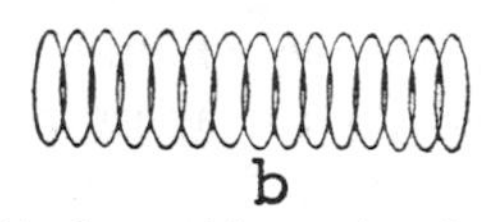

Fig. 3. Shadows of iron wires in slanting rays of the sun. (*a*) Distinct shadow. (*b*) Indistinct shadow.

Hold a piece of paper immediately behind the wire-netting, then take it farther and farther away, so that the gradual appearance of the remarkable shadows can be followed. Investigate similar cases where the sun's rays make different angles with the ground: examine also the shadows of slanting meshes, etc.

Shadows have played an important part in folklore. It used to be considered a terrible punishment for anyone to lose his shadow, and anyone possessing a headless shadow would die within a year ! Tales like these, which are told by all peoples at all times are interesting to us, too, as they prove how cautious one should be in believing the assertions of untrained observers, however numerous and unanimous they may be.

3. *Sun-Pictures and Shadows during Eclipses and at Sunset*

During an eclipse the dark moon is seen to glide in front of the sun's disc, so that, after a short time, only a crescent remains visible. It is worth while to notice at that moment the resemblance of small sun-pictures underneath the foliage to diminutive crescents, all lying in the same direction, large or small, bright or dim.

The shape of the shadows is influenced in a similar way. The shadows of our fingers, for instance, are peculiarly claw-shaped. Every small, dark object would at such a time cast a crescent-shaped shadow; the shadow of a small rod consists of a number of similar crescents, the curvature appearing at the end.

A good example of an isolated dark object of this kind is a balloon, and indeed it has been noticed that during eclipses of the sun the shadow of both balloon and basket are crescent-shaped. An aeroplane, if high enough, casts a curved shadow, too.

Eclipses of the sun, even partial ones, occur rarely; but similar shadow formations can also be seen while the sun is setting over the sea behind an open horizon, if you study the shadows of coins and discs of different sizes stuck on the window-pane, or hung on a piece of wire. The shape and the distribution of light vary according to the size of the coins, and also as the sun's disc sinks below the horizon.

4. *Doubled Shadows*

When the trees have lost their leaves, we often see the shadows of two parallel branches superposed upon each other. A branch quite near to us gives a sharp and dark shadow, one more distant gives a broader and more greyish shadow. The curious thing now is that, when they accidentally fall one upon the other, we see a bright line in the middle of the sharpest shadow, so that this looks double (Fig. 4). What can be the explanation ?

Let us suppose that the distant branch looks thicker, the nearer branch thinner. In order to find how strong is the illumination of the ground at different points of the shadow, and at points near to it, we imagine that we are looking successively from these several points towards the sun. Suppose first that our eye is looking from a point at some inches distance from the shadow; we should see the whole solar disc sending light towards us. Suppose now that we shift the position of our eye slightly, till we come, as shown at A in Fig. 4, into the penumbra of the distant branch. We now see that branch before the sun's disc; because it obliterates part of this disc, the illumination on the ground at the point where our eye is supposed to be is decreased. We shift our eye still more until it is at B; now the second branch comes before the sun, and both branches together intercept a considerable part of the light. But suppose that we shift our eye still further, till it reaches the position C where the two branches are seen superposed, then the intercepted fraction of the sun's disc is again *smaller*, the brightness on the ground, which we use as our screen, is again *greater*. Realising that when we look at the shadow on the ground we see simultaneously the various cases which we have discussed separately above, we understand why the central strip of the whole shadow is brighter than the adjacent parts either to the right or to the left.

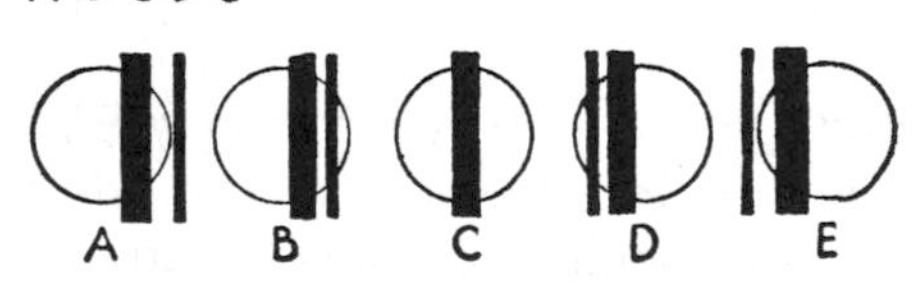

Fig. 4. How double shadows arise.

In Fig. 4 I have shown roughly how the eye will see the solar disc when looking successively from A, B, C, D, E, supposing as above that the distant branch appears thicker than the nearer one.

This phenomenon will apparently be visible whenever both branches subtend an angle smaller than the sun's disc.

'Some time ago I was walking along the shore. . . . It was an evening late in March. The sun was setting in the sea to the West, the moon was shining brightly to the East. For a long time it was the sunset that cast my shadow on the ground, so that it fell eastwards; *but then, for a short time I had no shadow at all*, until the brightness of the moon outshone the evening-glow and my shadow fell westwards.'[1]

Was this observed correctly?

(For shadows on a surface of water, *see* §§ 216, 217, and shadows on mist, *see* § 183; for contrast-edges along boundaries of shadows, *see* § 92.)

[1]From the Icelandic of S. Nordal: *Alfur of Windhael.*

II

Reflection of Light

5. *The Law of Reflection*

Look for a place where the moon is reflected in a very quiet surface of water. Compare the angle of the moon above the horizon and the angle of the reflected image below the horizon; they are both the same, within the errors of observation.

If the moon is not high in the sky you can hold your walking-stick vertically at arm's length, so that you can see the end of the stick opposite the disc of the moon, while your thumb covers the horizon. Then twist the stick upside down with your arm as axis, and see whether the end of it touches the reflection of the moon.

FIG. 5. Sunlight reflected from a deep-set window.

Similar measurements performed with the telescope on stellar images are the most accurate test of the law of reflection.

A window built far back in a wall catches the rays of the

sun when this is not too high (Fig. 5). The shadow AB shows the direction of the incident beam; the reflected light falls as a brighter spot of light in the direction BC. One can see now that both directions are symmetrical relative to the normal BN and that, therefore, $\angle ABN = \angle CBN$. This is not the same as the law of reflection, but a consequence of it. Prove this.

Why do the windows of distant houses reflect only the rising or the setting sun?

6. *Reflection by Wires*

A number of telephone wires shine in the sun; if you walk parallel to those wires, the light moves as fast as you do. In the same way we can see how at night the light from a street-lamp casts a bright line on the overhead tram-wire. What determines the exact spot of these reflections? Construct in your mind an ellipsoid which touches the wire, with your eye and the source of light as the foci (Fig. 6). The illuminated spot is the tangent point, for it is a well-known property of an ellipsoid, that the lines connecting any point with the foci, make equal angles with the tangent plane.

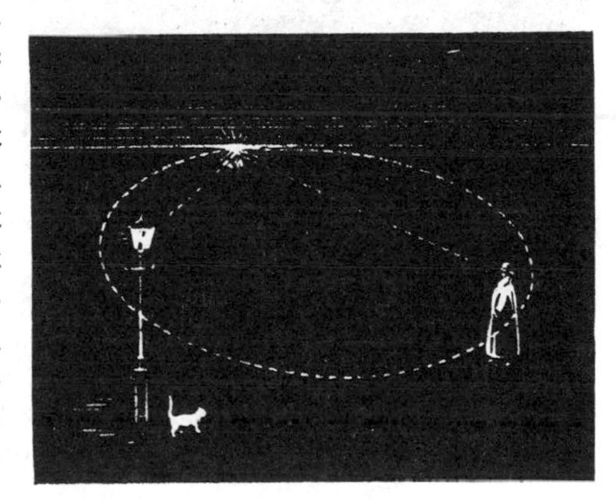

FIG. 6. The reflection of a street-lamp from telegraph wires.

7. *Differences between an Object and its Reflected Image*

Many people think that the reflection of a scene in a calm sheet of water looks exactly the same as the scene itself, upside down. Nothing is farther from the truth! Note how at night some groups of street-lamps are reflected! (Fig. 7A). The reflection of a dyke sloping down to the water appears shortened, and even disappears if we stand high enough above the water's surface (Fig. 7B). You can never see the reflection of the top of a stone lying in the water.

All these effects are a matter of course if you consider that the reflected image is really identical with the landscape itself, only the perspective is different because it is shifted. We see it as if we were looking at the object from a point as far below the surface of the water as our eye is above it. The differences grow less the further away the objects are (cf. §§ 5, 130).

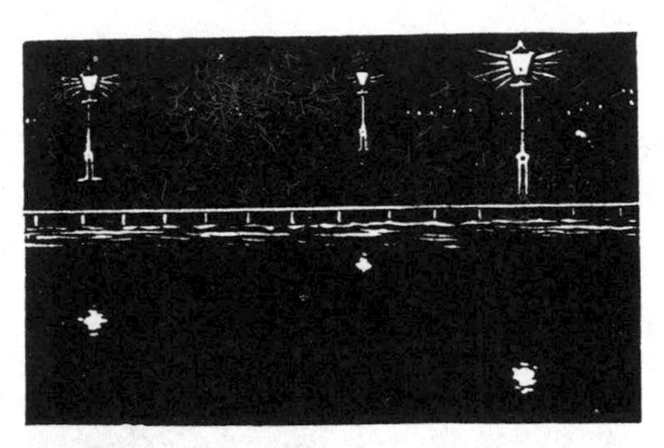

FIG. 7A

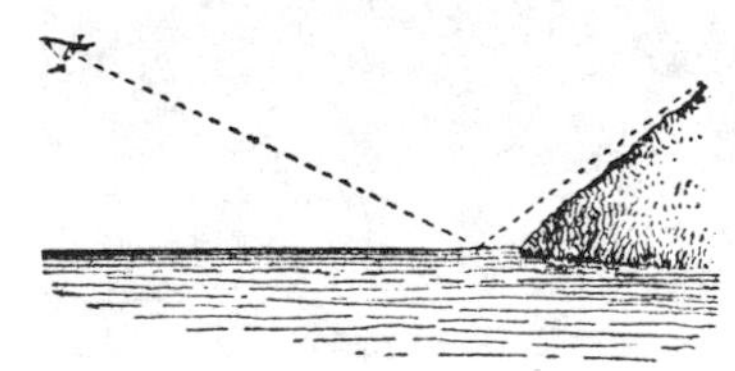

FIG. 7B.
An object can look different from its reflection.

There appears, however, to be still another point to be considered. The reflection of trees and shrubs in small ponds and in pools along the roadside often shows a clearness, a purity and warmth of colour, which seem greater than those of the object itself. We never see clouds so beautiful as those reflected in a mirror. A street reflected in the glass of a shop window with a dark curtain as a background is amazingly sharp.[1] These differences are due more to psychological than to physical causes. Some attribute them to the fact that a reflected scene arouses the same feelings as a picture lying in one plane (physically

[1] H. R. Mill, *Geogr. Journal*, **56**, 526, 1926. Vaughan Cornish, *ibid.*, p. 518.

speaking, the reflected images lie in various planes, just as the objects themselves do). Others say that the frame makes us feel uncertain as to the position of an object in space, and this gives rise to a stronger impression of relief.[1] To me, however, a more important reason seems to be that the eye is not dazzled by the vast bright field of the sky round the scene observed, that is the effect is of the same kind as one gets when looking through a tube (§ 71). Moreover, the decreased brightness of the reflection is in itself enough to make it easier for us to observe the sky and the clouds, which otherwise are too bright for our eyes.

8. *Freak Reflections*

A row of houses casts a dark streak of shadow across the street, but with unexpected spots of light here and there in the middle (Fig. 8). How does the light get there? Hold your hand before the spot of light and deduce, from the situation of the shadow, from which direction the rays fall. It appears that they are reflected by the windows of the houses on the other side.

FIG. 8. Patches of sunlight in a dark and narrow street.

Similarly, one can see spots of light shining on the surface of a canal which is itself in the shade. The houses on the other side throw back the light.

A row of houses along the waterside stand completely in the shade, and yet there is a play of light on them, regular, more or less parallel streaks of light moving forwards.

[1] *J.O.S.A.*, **10**, 141, 1925.

These are the reflections of the waves on the water (Fig. 9). The part AB of the waves acts as a concave mirror and produces a focal line at L; the part BC of the wave is less curved and unites the rays at a much greater distance. In this way, there is for every distance of the wall a part of the water's surface that produces a sharp line of light, while the other parts give the general light-effect. Similar effects can also be seen along the quays and on the underside of the arches of bridges (Plate IV, *a*; p. 32). We have here actually a model of the twinkling of the stars (cf. § 40).

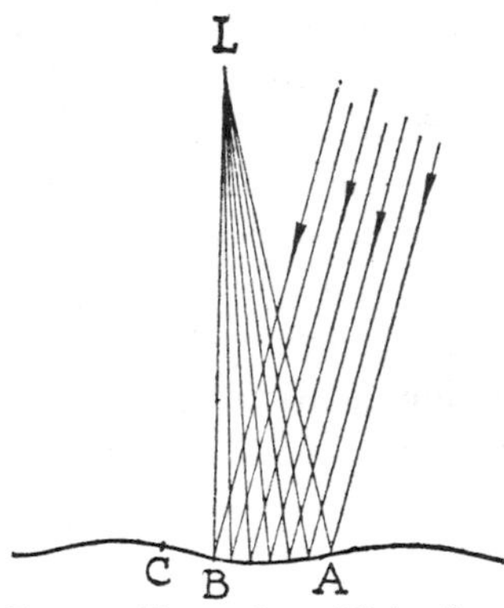

FIG. 9. Formation of light lines by reflection from slightly undulating water.

9. *Shooting at a Reflection*

Near Salzburg there is a lake, the Königsee, surrounded by high mountains, and therefore very calm. Shooting competitions are held there, at which the competitors aim at the reflection of the target in the water, the bullet rebounding from the surface to hit the mark. The chance of its actually hitting the mark seems at least as great as with a direct shot.

The curious thing is that the bullet does not rebound from the surface, but penetrates it and enters the water. Hydrodynamical theory shows that the effect of the movement of the liquid which surrounds the bullet is to force it towards the surface, so that it finally emerges at the same angle with the surface as that which it made on penetration. By hanging screens in the water it has been possible to follow the course of the bullet.[1]

10. *Gauss's Heliotrope*

Place a mirror in such a position that it reflects the sunlight. Close to the mirror the light-spot has the same shape

[1] Ramsauer, *Ann. d. Phys.*, **84**, 730, 1927.

as the mirror; farther away it becomes less distinct; still farther it becomes round; and at a great distance it is a true image of the sun. Now cover part of the mirror. The spot of light remains round, but becomes less bright. It will not be possible to follow the light-spot for more than 50 yards, but an observer at that distance will be able to see the mirror still shining brilliantly in the sun.

Fix the mirror in a clamp or between two stones, somewhere where the view is open, and in such a way that the reflected ray of the sun is perfectly horizontal. Walk backwards as far as you can while still seeing the light. It is rather difficult to keep in the beam, but fortunately its diameter increases the farther one recedes, as you will see if you move sideways in the beam and notice the width between its boundaries; at a distance of 100 yards, its width is one yard. Moreover, you must bear in mind that the sun in the meantime is moving across the firmament; for this reason it is advisable to carry out this test some time at midday, for then the reflected beam remains in the horizontal plane without need for much readjustment.

It is amazing how far such a tiny spot of light remains visible. Gauss, when triangulating, made very sharply defined light sources in this way, which could be observed through the telescopes of the measuring instruments at distances of sixty miles. A 'heliotrope' of this kind is fitted with special apparatus for directing the ray of light towards any spot one wishes. By covering and uncovering the light, Morse signals can be transmitted.

11. *Reflection in a Garden-globe*

The convex mirrors we are always taught about at school are small and only slightly curved. They correspond to that little portion AB of the garden-globe, turned straight towards us and in which we can see our own reflection (Fig. 10).

But the garden-globe as a whole is far more interesting, the most remarkable thing about it being that we can see the complete surface of the celestial sphere (more correctly, sky and earth) in it, confined within a circle. The garden-globe acts as an optical instrument with an ideally large aperture. Of course, this is only possible because the images are deformed; they are compressed in the direction of the radius, the more so the nearer they are situated to the surface of the garden-globe (Fig. 10). Suppose, for the sake of simplicity, that both the object and the observer are a fairly long distance from the globe (compared to the radius R); then an object in a direction making an angle α with CE will be imaged at a distance $r = R \sin \frac{1}{2}\alpha$ from the centre of the globe. One sees that as α increases to 180°, r increases to R, and so the whole of earth and sky is indeed imaged on the globe. The only part that is missing is the little piece lying exactly behind the globe, which gets smaller, the farther we stand away from the globe.

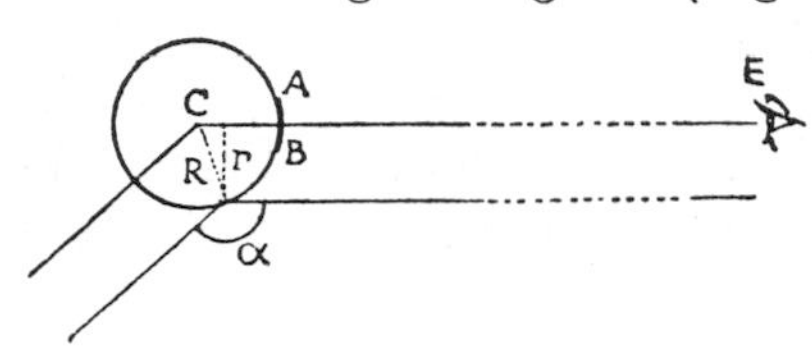

FIG. 10. How the universe is reflected in a small garden-globe.

Helmholtz once remarked that a landscape as deformed by a globe would turn out to be quite normal if the measuring-rod employed were deformed according to the same law. This statement is closely related to the principles of the theory of relativity.

The garden-globe can be used for very fine observations in the domain of meteorological optics, because it gives so good a survey of a considerable part of the sky. If you stand a few yards away from it, and in such a manner that the reflection of the sun is covered by your head, you will see with extraordinary clearness (*see* later): (*a*) rings, haloes, iridescent clouds, Bishop's ring, the twilight shades; (*b*) Haidinger's brush and the polarisation of the light from the sky. Because of the reduction of the image, the slowly changing tints are transformed into much steeper gradients, so that the differences of brightness and colour are more striking to the eye. In the shining surface of a convex bicycle mirror one can often see very delicate little clouds which one had not noticed by direct observation.

12. *Irregularities of the Surface of Water*

Imagine a pool of water in a hollow of the dunes where there is no wind to ruffle the surface. Here and there a stalk of grass or a reed sticks out of the water, and it is interesting to see how each stalk is surrounded by a patch of light just where it emerges from the water. The stalk acts as a capillary, so that the surface tension of the water causes it to heap up round the base of the stalk, and the mound of water so formed reflects the sunlight, thus making it visible a long way off. If one part of the pool reflects the dark slope of the dunes and another the bright sky, one can see close to the boundary line how all the tiny mounds of water show light and dark contrasts according to the direction in which one is looking.

In a similar way we can detect small eddies anywhere where a river has a current worth mentioning. Inside each whirlpool the pressure is somewhat smaller and the surface slightly scooped out. The order of magnitude of the hollow is 2 inches diameter and $\frac{1}{10}$ inch or so in depth. Close to the boundary between light and dark reflections even the feeblest disturbances of the surface can be seen very distinctly. Often a set of striae is very obvious.

It has been raining. Water is lying along the tram-rails, and now we can see a cross-line reflected in it horizontally, e.g. the suspension wire supporting the overhead cable. If we look along the vertical plane of the rail we see the reflected image altered in shape symmetrically (Fig. 11, *a*), which proves clearly that the water's surface is curved and forms a capillary meniscus. If we stand to the left of the rail, then the image is deformed as in Fig. 11, *b*, and to the right as in Fig. 11, *c*. Consider why the reflection assumes just this shape.

The images shown by a curved liquid surface can be studied from a steamer, because all along you are looking from the same position and in the same direction at the waves moving along with it. Note in particular how the shapes of the reflections are changed by the first swell caused by the bow. The images are strongly compressed; they are upright or inverted according as you look at a concave or convex part of the surface.

FIG. 11. Rain in a tram-rail forms a curved mirror.

13. *Ordinary Window Glass and Plate Glass*

You can tell at once by the reflection in the window panes of houses whether they are made of plate glass or window glass. If of plate glass the images are fairly regular; if of window glass they are so irregular that the lumps in the pane are visible at once.

It is remarkable how different this makes the aspect of houses in the poorer and in the well-to-do streets of a town. In the middle of a row of high-class houses with plate glass windows we distinguish at once an isolated house with window glass. Of two adjoining plate glass windows we note that they do not lie exactly in the same plane, for the reflections of a roofline are slightly shifted relative to one another. Elsewhere we notice that one pane of good plate glass has a weak spot, and that another is not quite flat.

14. *Irregular Reflection on Slightly Rippling Water*[1]

To me the long streaks of light of reflected lamps are inseparable from the quiet atmosphere of evening. I see the moon mirrored in the sea, casting over it a broad stream of light. Or I recall the houses and turrets of ancient Bruges, reflected in its quiet canals, every spot of light, every colour, extended to a vertical line, and all these lines, long or short, quivering in the changes of light and elusive brilliance.

When we see the moon or a lamp reflected close to us in

[1] See in particular: J. Picard, *Arch. Sc. Phys. Nat.*, **21**, 481, 1889; also G. Galle, *Ann. d. Phys.*, **49**, 255, 1840; A. Wigand and E. Everling, *Phys. Zs.*, **14**, 1156, 1913; E. O. Hulburt, *J.O.S.A.*, **24**, 35, 1934; W. Shoulejkin, *Nat.*, **114**, 498, 1924; K. Stuchtey, *Ann. d. Phys.*, **59**, 33, 1919.

water only faintly stirred, we notice that in reality each small wave gives a separate image. All these illuminated waves together form one figure, the average of which is an oblong spot with its long axis in the vertical plane containing the eye and light-source. This drawing-out of one point of light to a path-like patch of light directed towards our eyes, while the waves themselves are absolutely irregular and equally frequent in every direction, is the fundamental phenomenon to be explained. At the extremity lying in our direction we can see very clearly how the patch of light lengthens or shortens according to the formation of waves in the water, while at the other extremity, farther away from us, the light-spots flow closer together to form a certain mean.

To be correct, therefore, one ought to consider and compute the average distribution of the intensity of light over a path of this kind as a probability problem. This has never been done properly. We will, therefore, simplify things for ourselves by assuming that the slopes of the waves do not exceed a certain angle α, and find out only what are the boundaries of the patch of light formed in this way. Or, to express it in another way: if on each spot there are a large number of little waves, all sloping at an angle α, but in all directions of the compass, what then is the locus of the waves that will be illuminated? The problem becomes complicated enough, even when stated like this.

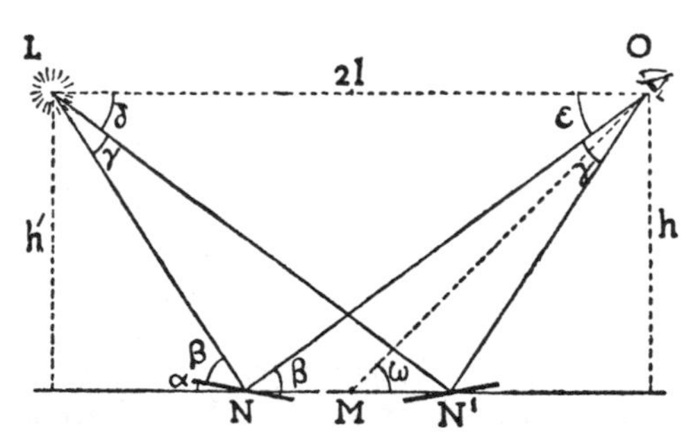

Fig. 12. Calculation of the long axis of a path of reflected light.

1. The simplest case: $h=h'$; observer and source of light are at equal heights above the water (Fig. 12).

A small horizontal mirror throws light in the eye of the observer O, when it is exactly half-way, at M, the place of regular reflection. A small mirror inclined at angle α must be shifted a little from the mid-point if it is to send light to the observer. How far?

For a shift in the vertical plane through eye and light-source this question is easily answered. Call N the required position if the mirror slants in the one direction, N′ if it slants in the other. For reasons of symmetry MN=MN′. Observe now the angles:

$$\beta+\alpha=\gamma+\delta$$
$$\beta-\alpha=\varepsilon=\delta$$

Therefore $\gamma=\alpha+\beta-(\beta-\alpha)=2\alpha$

This is an important result. *The angle subtended at the eye by the longest axis of the patch of light is equal to the angle between the two largest inclinations of the wavelets* (Fig. 13).

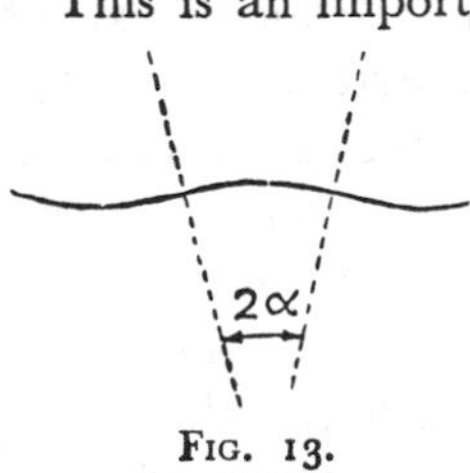

FIG. 13.

Let us now shift our mirror in the plane through M, perpendicular to the line connecting eye to light-source, and call P, P′ the points where the favourable reflection occurs (Fig. 14).

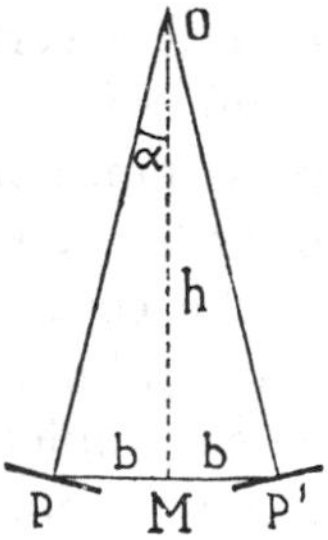

Obviously MP=MP′=$h \tan\alpha$. The width of the patch of light is therefore $2h\tan\alpha$, and the short axis subtends at the eye the angle $\dfrac{PP'}{OM}=\dfrac{2\,h\tan\alpha}{\sqrt{l^2+h^2}}$.

The ratio of the apparent axes of the patch of light is therefore:

$\dfrac{h\tan\alpha}{\alpha\sqrt{h^2+l^2}}$, or about $\dfrac{h}{\sqrt{h^2+l^2}}=\sin\omega$ if the patch is not too large. Therefore, when we look down at the water from a hill, the patch is only slightly oblong (ω large, $\sin\omega$ nearly 1). *The more obliquely we look across the water, the more oblong the patch.* If we let our glance graze the surface, it becomes infinitely narrow.

FIG. 14. Calculation of the short axis of a path of reflected light.

We must always distinguish the 'primary oval,' the curve that one can imagine as being drawn on the rippling water, indicating the boundary of the patch of light, from the 'secondary oval,' which arises from the former by projection on the plane at right angles to the direction of our gaze. The axes of the primary oval can indeed be simply calculated, but the entire figure is a complicated curve of the sixth degree, symmetrical with respect to M. The secondary oval becomes slightly asymmetrical; the greatest width lies in reality more towards us than the point M, at which we calculated the cross-axis. This asymmetry is particularly noticeable when looking at a small angle to the surface.

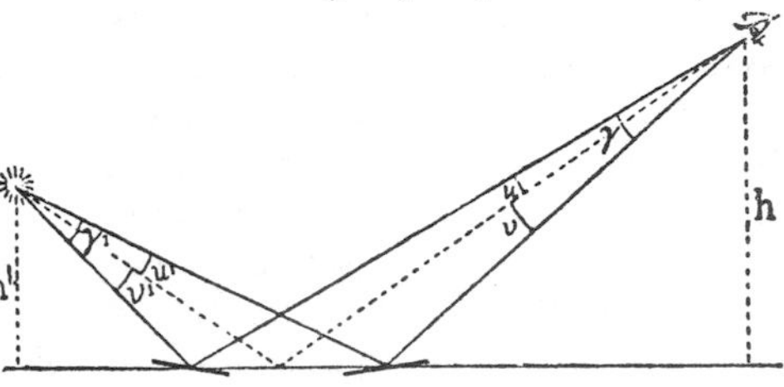

FIG. 15. Observing a patch of light from a level different from that of the source of light.

2. The general case: $h \neq h'$ (Fig. 15).

By similar arguments we can prove the two fundamental properties:

$$u + v' = 2\alpha$$
$$u' + v = 2\alpha$$
$$u + v + u' + v' = \gamma + \gamma' = 4\alpha$$

Further computation proves that the patch remains more or less elliptical in outline, but the results are complicated. Practically speaking, the difference in height between h and h' influences only the dimensions of the patch of light, not its proportions; approximately, of course

$$\frac{\gamma}{\gamma'} = \frac{h'}{h},$$

so that

$$\gamma = 4\alpha \cdot \frac{h'}{h + h'}$$

3. Special case: $h' = \infty$. This holds for the sun, the moon and very high lamps.

The formulae are now $\gamma = 4\alpha$; $PP' = 2h \tan 2\alpha$ (as can be proved). The axes of the oval subtend at the eye angles of about 4α and $4\alpha \sin \omega$. The ratio of the apparent length to the breadth of the light-patch is, therefore, $\sin \omega$, precisely the same as in Case 1, only all the dimensions are twice as large.

A general idea of the distribution of light in these reflections is obtained without computation by means of the following argument (Fig. 16). Imagine the reflecting surfaces on a very small scale to be close to the centre of a large sphere; the normal to the undisturbed surface of the water ends at N; the normals to the slanting sides of the wavelets end therefore in a small circle at angular distance α from N; the light-source at infinity is represented by the point L on the sphere. In order, now, to find how, for example, the surface with normal OS reflects the rays, it suffices to draw the arc of the large circle LS, and to extend it to S′, so that SS′=SL. By this means one sees at once that the rays reflected by all the wavelets form a cone with a very oval cross-section, which becomes more oblong the more obliquely we look at the water's surface. It is quite easy, too, to understand why the cone formed by the directions of the gaze of the observer—that is, from the eye to the boundaries of the light-patch—has the same shape.

Let us sum up the results of our calculations from the point of view of a practical observer: First, if we suppose ourselves to be at the same height above the water as the light-source, then the angle subtended by the longer axis of the patch is at the same time the angle 2α between the two steepest slopes of the wavelets (Fig. 13). In proportion to this, the transverse axis of the patch is smaller, the more slantwise we look at the surface of the water.

Secondly, if the source is higher above the water than our eye, all the dimensions of the light-patch become larger (in angular measure); they approach twice what they were originally if the source recedes to infinity; however, the ratio between the long and short axis remains about the same.

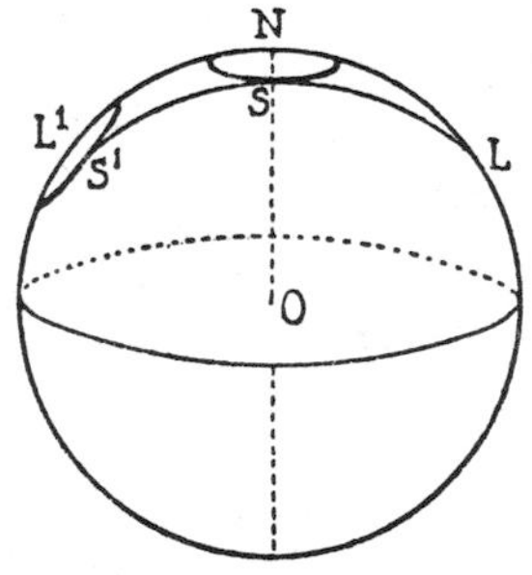

Fig. 16. Showing how path-like patches of light arise, by means of a construction on a sphere.

Compare the patch of light formed by the moon with that of a lamp whose reflection lies more or less in the same direction. The light-patches are generally larger the further away they are from the light-source. Objects quite close to the water give an almost point-like, not elongated, image. Compare the patches of light seen at different angles with the water's surface.

Determine angle 2α from the length (in angular measure) of the patches at various strengths of the wind.

Notice how beautifully long, regular and vertical the light-patches are when it rains; the waves, though small, slant sharply.

It is also worth while to watch the shapes of the reflections on each separate wavelet. Each wavelet bears a spot of light, spread out in a horizontal direction, which is reduced more and more to a small line the lower the position of the sun, and all these little lines together form the vertical column (Fig. 17 *left*).

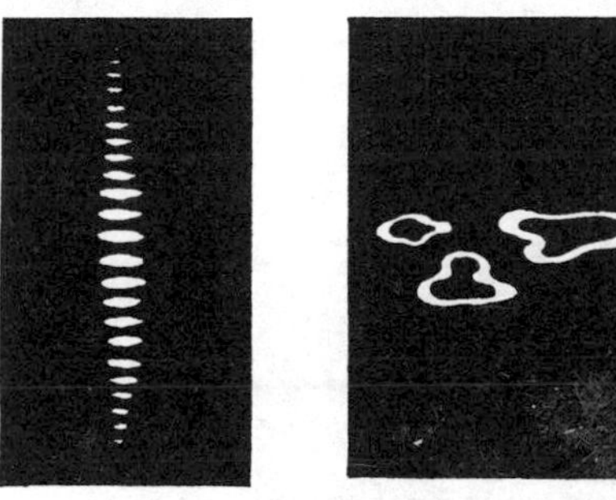

FIG. 17. *Left:* Patch of light on slightly rippling water. *Right:* Reflection from high source of light.

Remarkable is the appearance of closed coils of light (Fig. 17 *right*), when the light-source is high and has a large area (e.g. lighted neon-tube advertisements).

There is still one more peculiarity in perspective connected with these patches of light. Each patch always lies in the vertical plane through my eye and the light-source (for exceptions, *see* § 15). When I am drawing or painting, I project everything on to a vertical plane in front of me, and for this reason every patch of light is bound to run in a vertical direction, even when outside the centre of the scene. On a painting by Claude in the Uffizi the sun is close to the side of the canvas, and the painter has represented a column of light which falls obliquely from the sun to the middle of the foreground. But this is wrong ![1]

Focus your camera on the sea on which the sun is shining, and look on the ground-glass slide at the distribution of the light reflected by the waves; from this you can deduce the slope of the waves and their predominating direction; a general impression of the water's surface can be obtained at a glance, and can be fixed on the photographic plate.[2]

[1] Ruskin, *Modern Painters*, I, Pt. II.
[2] W. Shoulejkin, *loc. cit.*

15. *Reflection from the Rippled Surface of a Narrow Stretch of Water*

The patches of light frequently show distinct asymmetry: when one looks obliquely at a lamp across a canal, say towards the right, they no longer lie in the vertical plane through the eye and the source of light, but are inclined towards the direction of the canal, i.e. to the right (Fig. 18).

If you look obliquely at a lamp towards the left they are again inclined towards the direction of the canal, i.e. to the left.

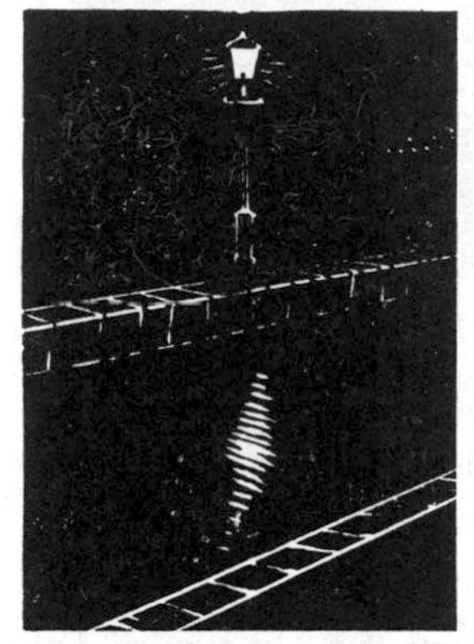

FIG. 18. A strange sight! The reflection does not lie in the vertical plane of eye and source of light.

And yet our theory is not wrong, for if it rains and there is no wind, the patches are perfectly vertical in whichever direction you look. The cause of the deviations is the wind, which shows a preference for blowing ripples across the direction of the canal, so that we can no longer take the ideally irregular wave-formation as our starting-point. The following observations may serve to prove this:

(*a*) In a very wide river the direction is much less systematic, the waves not showing a predominating direction at right angles to the banks.

(*b*) When the water is covered by a layer of ice, it appears that this layer has a lot of little lumps and gives a distinct patch of light, which, however, is vertical.

(*c*) On an asphalt road, wet after a shower, the same deviations are to be observed as those on a canal in windy weather, in the reflections of street-lamps as well as in those of the head-lights of cars and bicycle-lamps. In fact, irregularities are caused in the asphalt by the traffic (how they arise is in itself an interesting phenomenon!); if we examine the surface we can see these roughnesses at once, and notice that they are just like real waves, with their crests at right angles to the direction of the road.

The detailed treatment of this subject has not yet been given, but we can get some idea of its main features with the aid of our projection on the sphere, at least for the case of a light-source at infinite distance (Fig. 19). If the normals are distributed over the planes as represented by the curved line surrounding N, the mirrored rays will be directed towards the various points of the curved line round L′; the axis of the column lies, therefore, no longer in the plane LNL′, but deviates sideways.

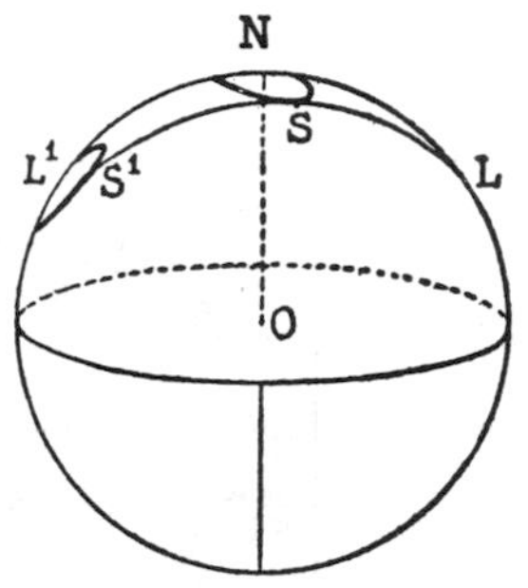

FIG. 19. How oblique patches of light arise when the waves have a definite direction.

16. *Reflection from large Rippled Surfaces of Water*[1]

Reflection in a mildly choppy sea is accompanied by a phenomenon which we shall call 'the shifting of reflected images towards the horizon' (Fig. 20). The reflection A′B′ of the boundary AB between cloud and blue sky lies much closer to the horizon than does the boundary itself in the sky. The first 25° or 35° of the sky above the horizon are, in reality, hardly visible in the reflection. All the images are naturally formed irregularly, but the effect is very clear all the same, and so striking that it dominates all the distribution of light on the sea. This explains why one never sees trees, dunes, etc., on the coast reflected in the sea; they are not high enough. Ships, too, are hardly ever seen in these circumstances, as the dark spot they ought to produce is forced, by this effect, close back to the ship.

The reflection of the sun in the waves is a single dazzlingly bright patch, which, as the sun sets, is more or less triangular in shape, showing thereby the shift towards the horizon (Fig. 21).

These phenomena are easily explained; at a great distance we can only see the sides of the waves turned towards

[1] E. O. Hulburt, *J.O.S.A.*, **24**, 35, 1934.

us. This makes it seem as if we saw all the objects in the sky reflected in a slanting mirror (Fig. 22).

This accounts for the shifting of the reflections towards the horizon. It follows from the disappearance of the lowest

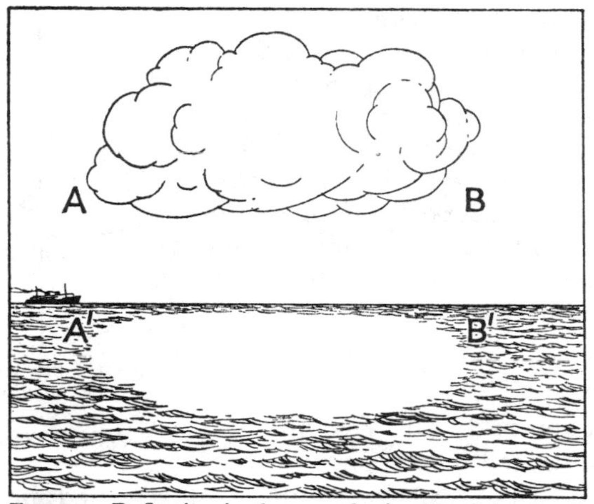

FIG. 20. Reflection in the sea: the image of the cloud is shifted towards the horizon.

FIG. 21. Sunlight on the sea.

30° in the reflection that the waves show average slopes of about 15° in each direction (if the sea is neither very calm nor very stormy).

Why was this phenomenon not mentioned in our theory in § 14? Because we were not considering the case where $\omega < 2\alpha$, that is, where we look very obliquely across the surface of the water. This case, for which our calculations do not hold, occurs whenever the surface of water is very wide, and especially where the sea is concerned. The calmer the surface, the more obliquely one has to look.

FIG. 22. Explanation of the shifted reflections. The angle of reflection is flatter than that of the incident ray.

One can see at once whether this condition is fulfilled on gazing at the sunlit sea; the path-like patch of light then reaches the horizon. We can no longer measure the inclinations of the waves from the length of the light-column, but must apply another method; if the inclinations of the

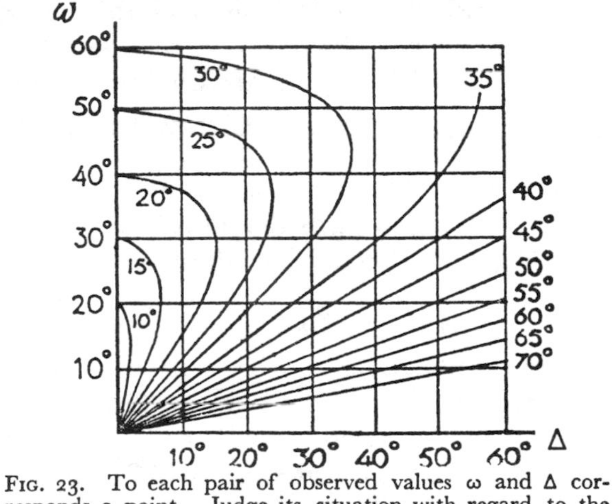

FIG. 23. To each pair of observed values ω and Δ corresponds a point. Judge its situation with regard to the curves each of which agrees with a certain value of α. (After E. O. Hulburt, *Journal of the Optical Society of America*)

waves get steeper, an increasingly broader part of the horizon is covered by sparkling light.

Measure this angle Δ, which is the breadth of the patch on the horizon; measure also the height of the sun ω and determine from this the inclination α of the waves with the help of the graph in Fig. 23, or with Spooner's formula, simplified for the sun's heights below 15°:

$$\alpha = \frac{\Delta}{2\omega} \text{ radians } (1 \text{ radian} = 57^\circ) \text{ (cf. Plate II).}$$

In a very calm sea the rising and the setting sun show an almost linear reflection, which merges into the fiery disc of the sun and forms a kind of Ω (Fig. 24). Sometimes,

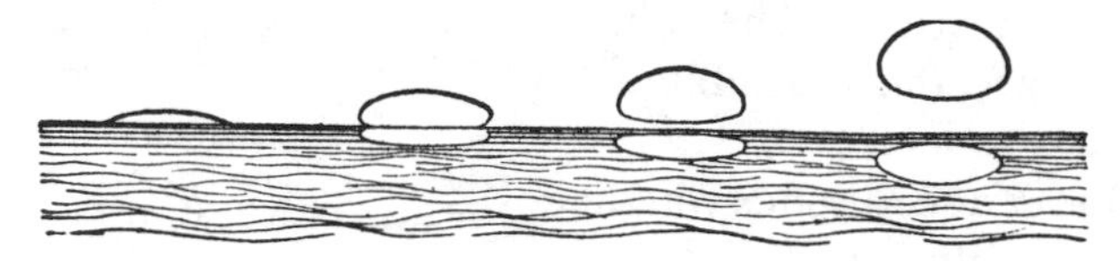

FIG. 24. Can one see, from the reflection of the rising sun in an extremely calm sea, the curvature of the earth?

when the sea is exceptionally calm, the elliptical reflection can still be seen when the sun is as much as 1° above the horizon, but usually the transition into the triangular spot of light mentioned above is very soon visible. In such cases the influence of the curvature of the earth's surface also comes into play; if there were no waves at all, one might say that the roundness of the earth was directly observable. In the most favourable case hitherto investigated, however, the observed shifting towards the horizon still remains twice as large as might be expected from the curvature of the Earth.

17. *Visibility of very Slight Undulations*

Very slight undulations can be seen better by looking at the wave crests at right angles than by observing them in a direction parallel to themselves. Therefore, in order to see how the wind makes the water ripple on a canal, one must as a rule look in the direction parallel to the canal. This explains, too, why the magnificent cross-waves behind a ship can be observed from the bridge, while they are practically invisible from the bank. The explanation of this is the same as that of the elongation of the image of a lamp into a patch of light. If you look at the waves at right angles, you can see, so to speak, the long axis of the patch of light; if you look at them in a parallel direction, then the short axis is seen. It amounts to this, that a wave causes a greater deviation in the direction at right angles to its crest line than in the direction parallel to it.

The sun reflected in the sea forms a column of light which is narrow or broad according to the height of the sun and the roughness of the sea. Note that the distant coast is not reflected. The column of light is always brightest near the horizon ('shifting of reflected images'; *see* § 16). (*From E. O. Hulburt, J.O.S.A.*, **24**, 35, 1934)

18. *Patches of Light on the Surface of Dirty Water*

Even although the surface of water is as smooth as a mirror, we often see columns of light round the reflections of street-lamps at night. These plumes of light do not show the lovely scintillations of patches of light on the waves; they are perfectly calm and motionless. They occur everywhere where the surface is not quite clean; apparently the small particles of dust on the water form so many minute irregularities on its surface that optically they act as wavelets. One would expect to see these columns grow thinner the more obliquely we look at the surface, and indeed this turns out to be so.

At more or less vertical incidence the light-patches can hardly be seen; at grazing incidence they are very noticeable and give a clear indication of the presence of dust on the surface. The difference in intensity is so striking, that there must be some special cause. The particles are so small that one is justified in speaking of the 'scattering' of light, and we shall see further on that the scattering by such particles is by far the strongest in the neighbourhood of the direction of the incident beam of light (§ 177). This explains, no doubt, why the scattering and the whole patch get stronger and stronger the more obliquely one looks.

19. *Patches of Light on Snow*

Sometimes the snow is covered with a layer of beautiful, small, flat discs and stars, all more or less horizontal. If you look for the reflection of the low-lying sun in the layer of snow, you will see a beautiful column of light which is to be ascribed to small irregular deviations of the snow discs from the horizontal plane. The sun must be low at the time because then the column of light contracts laterally and becomes more distinct.

The formation of patches of light is still more striking at night time when the street-lamps are burning and each light is reflected in the fresh snow.

20. *Patches of Light on Roadways*

Path-like patches similar to those seen on undulating water also appear on our roadways, most clearly when it has been raining and everything is wet and shining. They are splendid on our modern asphalt, but can also be seen on the old-fashioned cobbles, and even on gravel roads. Even without rain our roads usually reflect so well as to cause paths of light practically always, if only one looks at them obliquely enough (cf. § 15).

21. *Reflections during Rain*

Look at the reflection of a street-lamp in a pool at night when it is raining. It is surrounded by a lot of sparks arising wherever a drop of rain has fallen, and all of them look like *small lines of light radiating from the reflection* (Fig. 25). Forel noticed a similar phenomenon when looking through dark glass at the sun's image reflected in calm water, in which air-bubbles rose here and there.

FIG. 25. Spots of rain scatter fiery sparks round the reflection of a street-lamp.

The explanation is simple. Each drop makes a set of concentric wavelets, and the reflections from their sides must always lie on the line connecting the centre of the waves and the image of the light-source (Fig. 26). This can be seen at once when the source L and the eye E are at the same height above the surface of the water, and drop D falls at an equal distance from both of them. The points D_1 and D_2 lie on the line MD; if a wavelet expands in a circle round D, the reflected light travels over part of the line DM, and does this so rapidly as to make us think we see a line of light. The theory is equally obvious when the drop of rain falls in the plane EML, either in front of or behind M.

This phenomenon can be reproduced if, over a glass plate

III

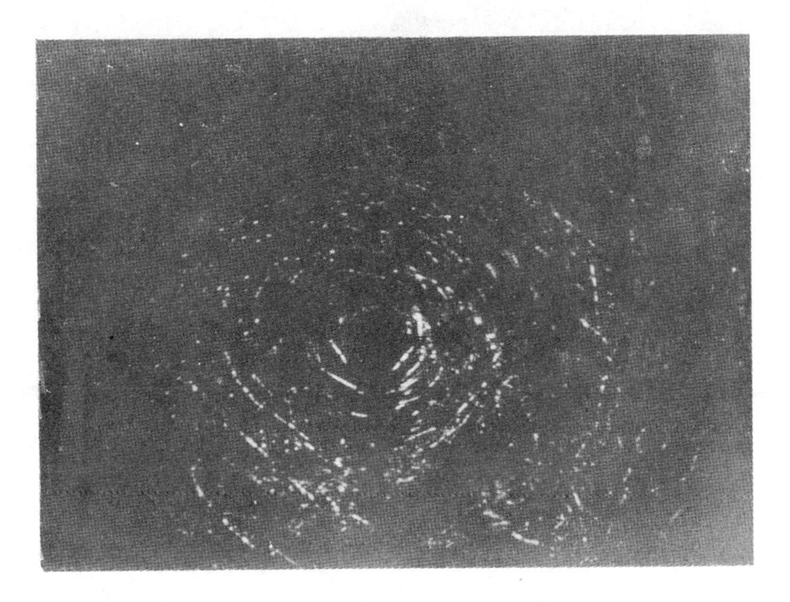

(*a*) When one looks at night through the crown of a tree at a street lamp, the shining branches appear as rings round the source of light

(*b*) The same tree by daylight. Each shining ring can be traced to a definite branch or twig. (*From photographs by Dr. In. A. J. Staring*)

in which a lamp is reflected, one moves an object that is concentrically ribbed, such as the lid of a sugar-basin or a disc of brass ground on a lathe.

Try to give a general proof of this.

22. *Circles of Light in Tree-tops*

One sees at night, when a street-lamp is shining just behind a tree, that the light is reflected here and there by the

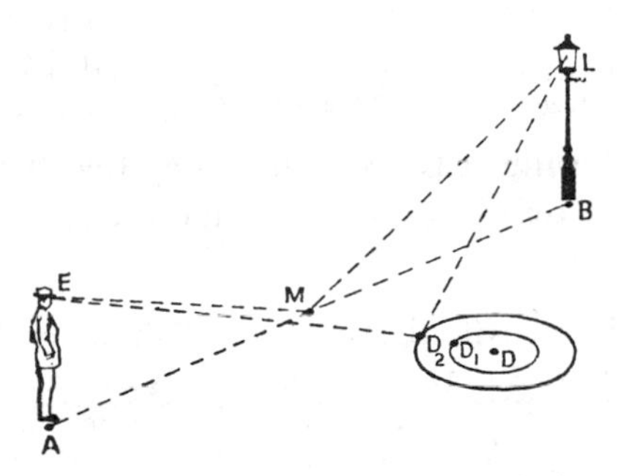

FIG. 26. How the sparks arise about the image.

twigs; these shining patches are, in reality, shorter or longer lines of light, and all these lines lie in concentric circles round the light-source (Plate III).

The best way to see this phenomenon is to stand in the shadow of the tree trunk, if the lamp is burning quite close to the tree. But it can be seen in sunlight, too, if, for instance, the branches are wet after rain, when the glistening twigs form a delicate pattern of dancing lines against a dark background. To prevent our being dazzled, the sun should be screened off by a wall or roof. The effect of glazed frost is also exceptionally beautiful.

This is explained as follows (Fig. 27): Consider a small plane V, reflecting the light of the lamp towards our eye. We shall see all the little branches in that plane glisten. But owing to perspective we see branches like AB greatly shortened, whereas those like CD show their full length. Since there are as many branches to be found in either direction, we shall see mainly light lines at right angles to the plane ELV. A similar statement is true for other small planes like V′, which we see above, to the right or left, of of the light-source; in this way we get the impression of concentric circles. It is easily seen that the effect of the direction is accentuated the smaller the angle our line of sight makes with the line EL, and that the effect will be slightly greater if the source is at infinity, like the sun, than if it is a lamp quite close to us.

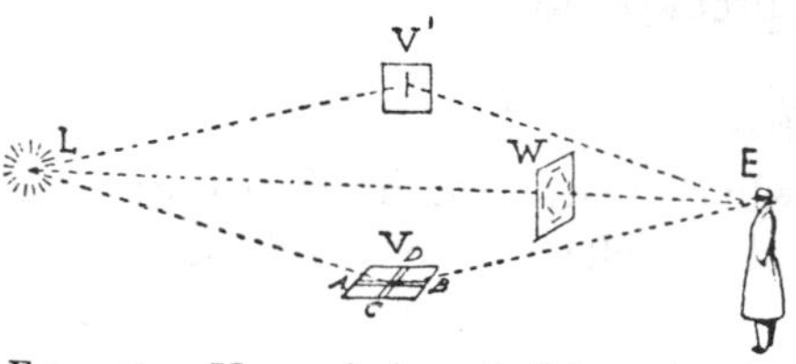

Fig. 27. How circles of light arise in tree-tops.

Compare this case with the patches of light on undulating water (Fig. 28). We must now imagine, as it were, that there are not branches everywhere in space, but only in one plane (the surface of the water). The only small lines lying in this plane and yet forming part approximately of concentric circles round EL lie, each of them, at right angles to the plane ESL, but together form a path of light in this plane. This is entirely analogous to the case of the water wavelets.

A similar phenomenon can also be observed when you see the setting sun shining on a cornfield, or in misty weather when the cobwebs are sprinkled with little drops of dew, and when you look at a street-lamp through one of these cobwebs. The scratches on the window of a compartment in the train show the same effect (cf. § 159). In all these cases, it is mainly the little lines at right angles to the plane of incidence of the light that glisten, so that you get the impression of concentric circles round the source of light.

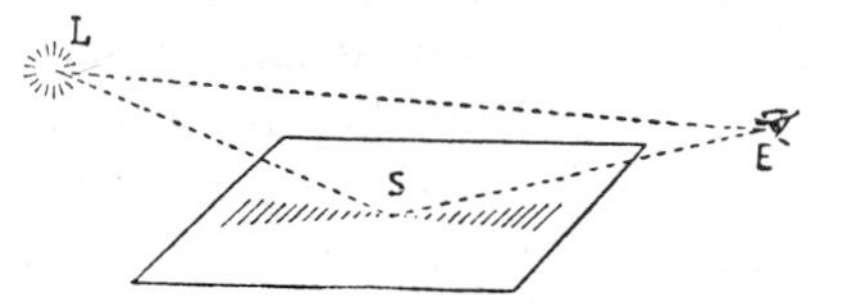

Fig. 28. Compare the circles of light in the tree-tops with the pillars of light on undulating water.

III

Refraction of Light

23. *Refraction of Light when it passes from Air into Water*

The pole used by a boatman to push his boat along looks as if it were broken just where it enters the water; this impression is caused by the fact that the rays of light bend when they go from the air into water, or *vice versâ*. Notice, however, that this 'broken stick' by no means represents the image of the broken ray of light, for the latter is bent in just the opposite direction. The connection between the two is seen in Fig. 29.

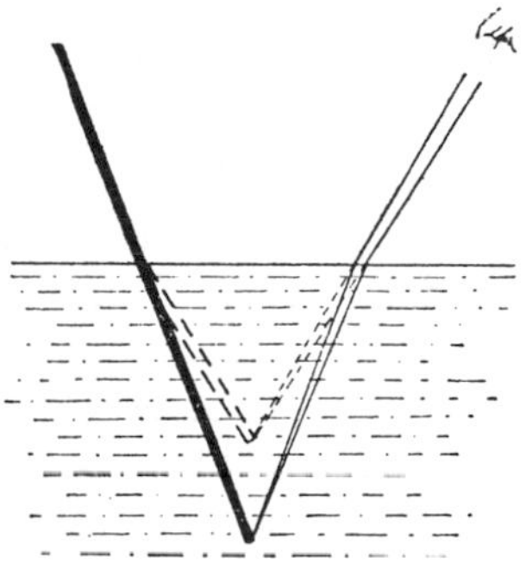

Fig. 29. Refraction of the rays of light makes the pole look bent.

Gauge with your eye the depth of an object under water and try to catch hold of it quickly. As a rule, you will miss it because, owing to the refraction of the light rays, the object seems to have been raised (cf. Fig. 29). It lies deeper than you thought. And yet the phenomenon is not so simple as to be correctly described by stating that the refraction of the light replaces, as it were, an object by an image lying in a higher plane. When, for instance, you are walking or cycling along the side of a ditch in which the water is clear, you will see that the positions of plants under water seem to undergo a peculiar change; their displaced image keeps moving and the more obliquely you look at it, the higher it is raised (Plate VII, *b*; p. 86).

The shadow of the floating leaves of the water-lily on the bottom of a clear pond looks remarkably indented, like the shadow of a palm leaf. The explanation is that the leaf curls up slightly at the edges, round which the water rises by capillary action, and in the tiny resultant prisms the rays of light are refracted and are thrown in irregular streaks into the shadow region.

The sun casts bright lines of light through the clear water on to the bottom of a shallow pool, or close to the banks of a river; the crests of the wavelets act as lenses and unite the rays of light in focal lines, which move on slowly with the waves (Fig. 30[1] and Plate IV, *b*). We have met with a similar phenomenon in reflected light (§ 8), and now find its counterpart in refraction. When the rays are incident obliquely the lines of light are edged with colours, blue towards the sun, reddish away from the sun, blue rays being more strongly refracted than red ones. This is the phenomenon of dispersion or colour shifting.

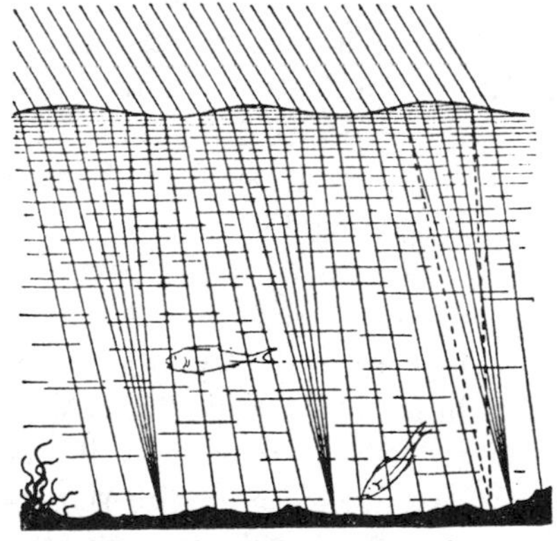

FIG. 30. Rays of sunlight penetrate the water and by the refraction of the waves are collected into lines of light. The blue rays (dotted) are more strongly refracted.

Throw a white pebble into deep, transparent water and then look at it from some distance; it will appear blue at the top and red underneath. This, too, is due to colour shifting.

24. *Refraction through Uneven Panes of Glass*

When looking through the windows of old-fashioned trains, you will often see that certain parts of the panes

[1] These phenomena can be observed still better by using a water-telescope (§ 209).

(*a*) The ruffled surface of the water in a canal reflects the sunlight in fantastic patterns on to the soffit of a bridge

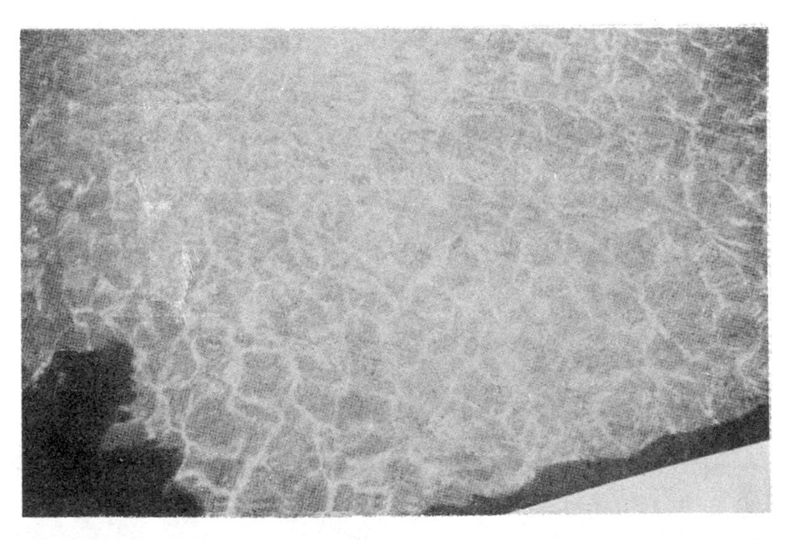

(*b*) Sunlight, refracted by slightly undulating and shallow water, is concentrated on the bottom in lines of light

distort everything completely. If the sun shines through one of these panes on to a sheet of paper, light and dark streaks will be cast on it by those parts. Hold it further away, and each streak will become a fairly sharp line of light.

The pane is evidently not a parallel plate, but has thinner and thicker parts which act as irregular lenses, spreading out or collecting the rays of light, and giving fantastic focal lines (cf. § 23).

25. *Double Images reflected by Plate Glass*

Look at a distant lamp or the image of the moon reflected in a window along the road. You will see two images, one moving irregularly in relation to the other according as the reflection falls on one or another part of the pane. A 'philosopher' stated not so very long ago that this was a case of effect without cause ![1] Physicists, however, must see if they cannot discover a cause !

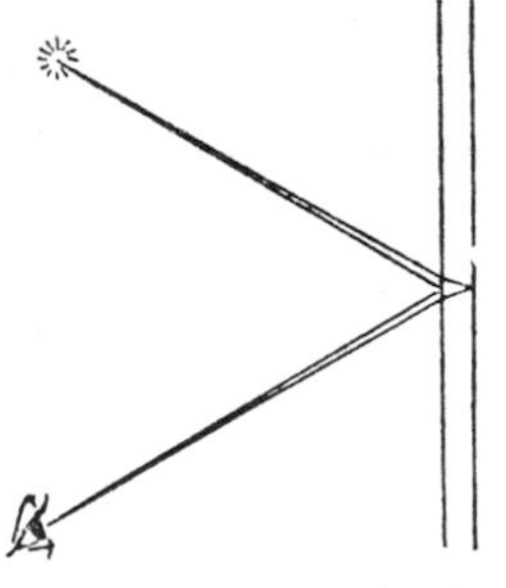

FIG. 31. An absolutely parallel plate-glass window pane produces double images, but they lie extremely close to one another.

We observe that the beautifully polished slabs of black glass adorning the parts of some shops and offices do not show double images. It is clear, then, that one image is reflected by the front surface of the plate glass, and that the other is formed by the rays that have penetrated the glass and are reflected by the back surface, reaching our eye through the glass. But in the case of black glass the rays of the second image are absorbed.

Refraction causes a slight deviation in the direction of one of the rays (Fig. 31). Can this be the cause of the double images ? No, because in that case (*a*) they would not draw so much closer to each other on some parts of the same pane than on others; (*b*) they would not lie farther apart than the thickness of the glass, which would hardly be

[1] E. Barthel, *Arch. for System. Philos.*, **19**, 355, 1913.

observable; (*c*) the shifting or displacement would be zero for very small and very large angles of incidence (with a maximum near 50°, as one can easily compute), while in the case of normal incidence we also observe double images; (*d*) for a source of light at infinity, such as the moon, the distance of the double images would always be zero.

The conclusion is that: *A plane parallel glass plate cannot produce double images of this kind. If, however, the pane of glass should be at all wedge-shaped, they may occur,* owing to the surfaces being slightly undulating. But before we can feel quite satisfied with this explanation, we must first calculate how large the angle must be between the front and the back surfaces in order to account for the distance observed between the double images, for it is not likely that, in good plate glass, the deviations from the parallel would be large.

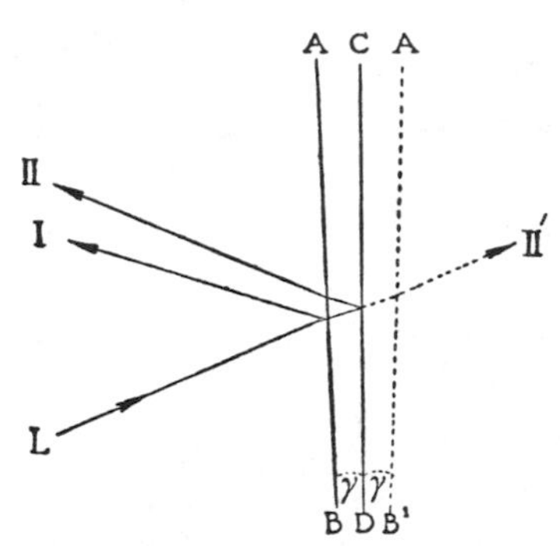

Fig. 32. How double images arise on a pane which is not of uniform thickness.

Suppose first that the planes are parallel, and let us follow one ray, after it is divided; the two reflected rays are still parallel, only slightly shifted relatively to each other.

Now let the face AB be inclined at a small angle γ (Fig. 32). Ray I will then have turned through an angle 2γ. To follow ray II on its path, we imagine CD to be a mirror, giving a reflected image of AB at A′B′, and an image of ray II along II′. We see, now, that the ray L II′ has passed through the small prism ABB′A′, with the small refracting angle 2γ; geometrical optics teaches us, that such a prism causes an angular deflection $(n-1)\,2\gamma$ in the path of the ray, provided the angle of incidence be not too large. The total angle between I and II is therefore, $2\gamma+(n-1)2\gamma=2n\gamma$; in the case of glass $n=1{\cdot}52$ so that the angle in question amounts to about 3γ. (n is the refractive index.)

Fig. 33 shows what follows from this when a person at E looks at the source L, when it is very far away; the two rays I and II, arising from that distant source, in, practically speaking, parallel directions, enter the observer's eye at an angle 3γ.[1]

So we conclude that *if we estimate the angular distance between the two reflections, the angle between the two glass surfaces is one-third of that amount.*

This estimate can be made, for example, by determining the distance a of the reflected images on the glass, dividing this by the distance R between eye and pane and multiplying by cos i.

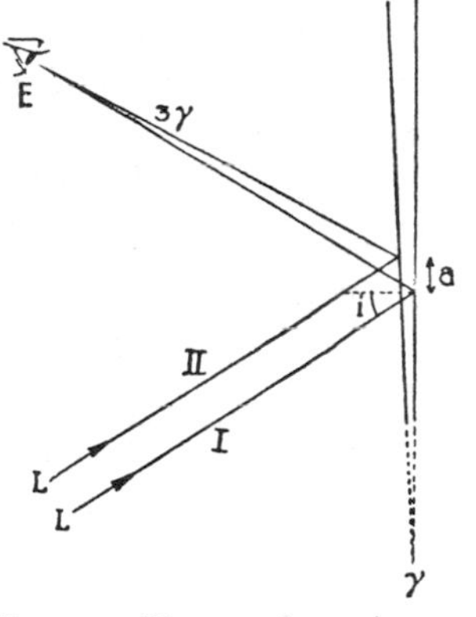

FIG. 33. How to determine from the angular distance γ of the two reflected images the extent to which front- and back-surfaces of a window-pane converge.

The angles obtained in this way amount, in ordinary plate glass, to a few thousandths of a radian,[2] or a few minutes of arc. Thus the thickness of the pane changes, over a length of say 5 inches, only $\frac{1}{100}$ inch. This is so slight, that, but for careful measurements of the thickness, we should not notice it at all. When these measurements were actually carried out the above estimate was confirmed.

Is it not splendid to be able to evaluate such extremely minute faults in the glass, without any further auxiliary means, simply as we walk along? And, moreover, we have seen now that our explanation of the double image is indeed correct. Whenever we are unable to find the cause of any natural phenomenon, it is our own ignorance which is to blame!

A more general and more accurate formula: the angular distance between the two images is $2m\gamma . \frac{R'}{R+R'}$, where R′ is distance from light-source to glass, R is distance from eye to glass and where $2m$ has the following values:

Angle of incidence $i =$	0°	20°	40°	60°	80°	90°
$2m =$	3·0	3·1	3·6	5·0	13·3	∞

[1] For a proof on different lines *see* § 26.

[2] For 'radian,' *see* § 1.

Ordinary window panes cannot be used for the investigation of multiple images, because they distort them very badly by their uneven surfaces; the method is too sensitive.

26. *Multiple Images shown by Plate Glass in Transmitted Light*[1]

One evening look sideways through a pane of good quality glass in a tram, train or motor-bus window at a distant lamp or the moon. You will see various images at pretty well equal distances apart, the first one quite clearly, the following ones fainter and fainter; the more obliquely you look through the window, the greater their distances become and the less they differ in brightness from one another.

It is clear that phenomena of this kind arise from repeated reflections from the front and back of the glass. They really resemble very closely the phenomenon of the doubly reflected images, and we have the same reasons for believing that the front and back surfaces are not parallel. But there is an additional reason; *in a parallel pane, the brightest image would necessarily lie always on the side nearest to the observer,* no matter whether we look through the pane in the directions from E or from E′; *experiment, however, teaches us that the brightest image lies invariably on the same side* (always to the right, or always to the left), so long as one looks through one definite point of the pane (Fig. 34). But in one and the same pane, parts are to be found where the brightest image lies to the right, and other parts, where it lies to the left: in the first case, we have a wedge-shaped region of which the greatest thickness is turned towards our eye; in the second case the greatest thickness is turned away from our eye.

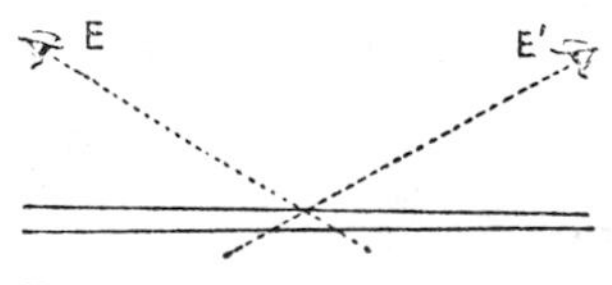

FIG. 34. The brightest of several images always lies on the same side as the observer.

Let us compute the angular distance in a way slightly different from that in § 25. From Fig. 35 we see that the angles at which the rays L_1, L_2, L_3 . . . are incident on the back surface are

[1] H. M. Reese, *J.O.S.A.*, **21**, 282, 1931.

$r+\gamma$, $r+3\gamma$, $r+5\gamma$ . . ., so if the angles of emergence of these rays are α_1, α_2, α_3 . . .

$$\sin \alpha_1 = n \sin (r+\gamma)$$

or since γ is a small angle

$$\sin \alpha_1 = n \sin r + \gamma n \cos r$$

Similarly $\sin \alpha_2 = n \sin r + 3\gamma n \cos r$

Subtracting, $\sin \alpha_2 - \sin \alpha_1 = 2\gamma n \cos r$

Now α increases only slightly, so we may put $(\sin \alpha_2 - \sin \alpha_1)$ equal to the differential of $\sin \alpha$, i.e.

$$\sin \alpha_2 - \sin \alpha_1 = \delta (\sin \alpha)$$
$$= \cos \alpha \,.\, \delta\alpha$$
$$= \cos \alpha \,.\, (\alpha_2 - \alpha_1)$$
$$\therefore \; \alpha_2 - \alpha_1 = \frac{2n \cos r}{\cos \alpha} \,.\, \gamma$$

Using Fig 32, a similar argument would also hold for the images formed by multiple reflections. The distances between the successive images are exactly the same, whether one observes them in reflected or in transmitted light; the factor by which γ is multiplied is in fact the same as the one denoted by $2m$ in § 25, where its values are given.

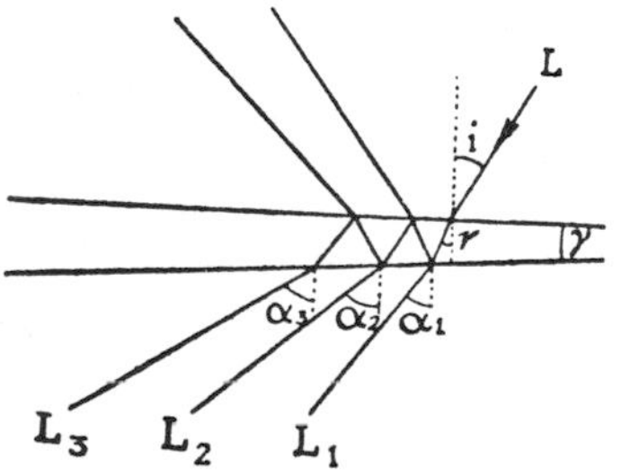

Fig. 35. Multiple images in transmitted light.

27. *Drops of Water as Lenses*

Raindrops on the windows of the compartment of a train produce very tiny images, just like a strong lens, only these images are, of course, deformed, since the raindrop is not in the least like a perfect lens in shape. They are upside-down, and whereas the scenery outside seems to move in *the opposite direction* to the train, the images are seen to move in *the same direction* as the train.

The image of a post is much thicker at the top than at the bottom, because the lens makes the images smaller according as its focal length is smaller, and therefore its curvature greater; the top part of our raindrop being much flatter than the lower part gives a larger image.

28. *Iridescence in Dewdrops and Crystals of Hoar-frost*

Who does not know the colourful gems of light in the morning dew? See how steadily and brilliantly they glitter on the short grass of the lawn, and how they twinkle like stars on long and waving blades.

Let us look more closely at the dew on a blade of grass. Don't pick it! Don't touch it! The tiny spherical drops do not wet it, they are quite close to it, but at most places there is still a layer of air between the dewdrop and the blade. The greyish aspect of the bedewed grass is due to the reflection of the rays of light in all the tiny drops, inside as well as outside; a great number of the rays do not even touch the blade of grass (cf. § 168). Large flattened drops have a beautiful silver sheen when seen at fairly large angles, because the rays are then totally reflected at the back surface.

Let us select one large drop and look at it with one eye, and we shall see colours appear as soon as we observe it in a direction making a sufficiently large angle with the direction of incidence. First we see blue, then green and then, particularly clearly, yellow, orange, red. This is, of course, the same phenomenon as that which we see on a large scale in any rainbow (§ 119).

Similar sparkling colours are seen in the crystals of hoar-frost and freshly fallen snow.

Compare §§ 129, 154.

'You must ask Professor Clifton to explain to you why it is that a drop of water, while it subdues the hue of a green leaf or blue flower into a soft grey, and shows itself therefore on the grass or the dock-leaf as a lustrous dimness, enhances the force of all warm colours, so that you can never see what the colour of a carnation or a wild rose really is till you get the dew on it.'

RUSKIN: *The Art and Pleasures of England.*

IV

The Curvature of Light Rays in the Atmosphere

29. *Terrestrial Ray Curvature*

The celestial bodies appear to us slightly higher above the horizon than they really are, and this displacement increases the nearer they approach the horizon. This accounts for the flattening of the sun and the moon on the horizon. At sunset the lower edge of the sun's disc appears, on an average, 35 minutes of arc higher than it actually is, but the top edge, which is further from the horizon, only 29. The flattening

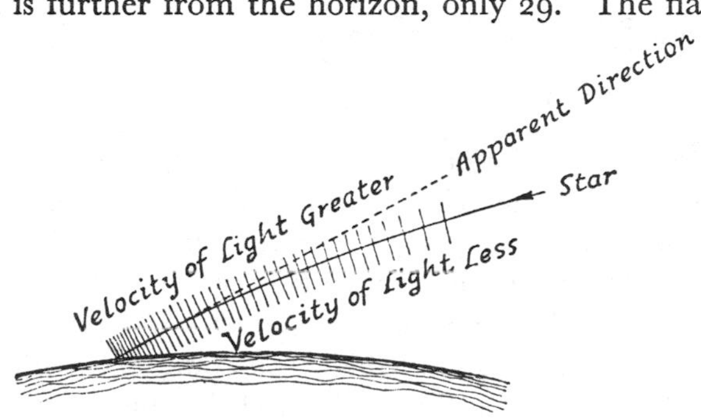

FIG. 36. Terrestrial ray curvature causes the celestial bodies to appear higher than they are in reality.

amounts, therefore, to 6 minutes of arc, or about $\frac{1}{5}$ of the sun's diameter.

This phenomenon, which shows us by direct observation how the apparent rise increases towards the horizon, is simply a consequence of the increase in the density of the atmosphere in the lower layers. According as the density becomes greater, the refractive index of the air increases and the velocity of the light decreases, so that, when the light waves emitted by a star penetrate our atmosphere, they move somewhat more slowly on the side nearest to the Earth and bend round gradually. The light rays, which indicate how the *wave-fronts* are propagated, curve too, therefore, and distant objects appear to be raised (Fig. 36).

The terrestrial curvature of rays changes from day to day, owing to the varying distribution of temperature in the atmosphere. It would be very interesting to note for a number of days the time the sun rises and sets, and to compare the results with the times calculated from almanacs and tables. One would have to aim at an accuracy of one

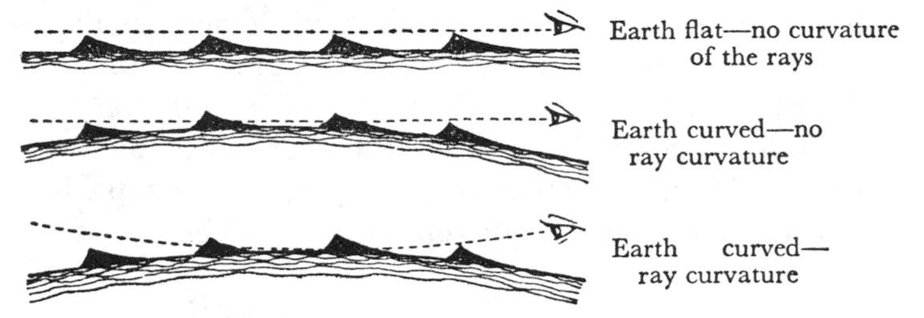

FIG. 37. Waves seen against the line of the horizon.

second, which is quite possible with the help of radio signals. It appears that differences in time of one minute, or even two minutes, can be expected. Anyone living at the seaside could carry out this experiment very well, as there the sunset can be observed above a clear and open horizon. An experiment of this kind might be combined with observations of the height of the horizon, of the shape of the sun's disc, and of the green ray (*see* §§ 30, 35, 36).

30. *Abnormal Curvature of Rays without Reflection*

Notice how often, when on the beach, we can see the waves in the distance standing out against the horizon, while waves of the same kind near to us do not reach the line of the horizon, although the line connecting equally high crests ought to be level and therefore meet the horizon too. This phenomenon can also be studied during a sea-voyage in stormy weather, if you keep watch on one of the lower decks, where you can see how the waves near to you do not quite reach the horizon, and compare them with the waves a long way off. It will be clear that our observation can only be explained by the curvature of the Earth, which we can see as a real fact here, before our very eyes (Fig. 37).

The phenomenon just described is, however, altered by the terrestrial curvature of the rays. This is extremely pronounced on some days when the horizon seems quite near, and the boats seem further off than usual and bigger, and it seems as if the curvature of the Earth were increased. On other days the calm sea resembles a huge concave dish. Objects normally beyond the range of vision become visible and seem near to us and smaller than might be expected. Distant ships which to the observer's eye should have been on the horizon or beyond it, still seem to sail in a valley of water. They look as if they were compressed more or less in a vertical direction, the line of the horizon running above their hull, while, as a matter of fact, our eye is lower than the top of the hull. The horizon seems abnormally far away.

The disappearance of distant objects; the water's surface seems convex. (In both drawings the curvature of the light-ray is exaggerated.)

Distant objects normally invisible become visible; the water's surface seems concave.

FIG. 38.

We can call these two characteristic conditions the *convex* and the *concave* surface of water[1] (Fig. 38). The first condition arises when the density decreases abnormally slowly from below upwards in the atmosphere or even (in the bottom air layers) increases, and the second condition arises when the density decreases with abnormal rapidity from below upwards. Anomalies of this kind are a consequence

[1] *Proc. R. Soc. Edinb.*, **32**, 175, 1912.

of exceptional temperature distribution. If the sea is warmer than the air, the lowest air layers become warmer than the upper layers and, therefore, optically more rarefied and less refracting, and the light rays curve away from the Earth. If the sea is colder, the curvature is the other way round. On such days it is desirable to measure the temperature at different heights to see whether this may account for the observations.

There is yet one more characteristic by which these two optical conditions may be distinguished—namely, the apparent height of the horizon. In order to measure this without instruments, we must choose a fixed point of reference, A, near the shore, and a variable point of reference B, on a post or a tree-trunk a few hundred yards inland (Fig. 39). We take B as our post of observation, and find the height at which we have to keep our eye, in order to see the horizon pass exactly through the point A. If the water is colder than the air, the horizon seems higher and B descends, but if the water is warmer than the air the horizon appears lower and B ascends. Differences of 6′, or even 9′ occur at times in one direction or the other, especially when there is no wind; if AB = 100 yards these differences correspond to about 7 and 11 inches. The use of field-glasses would make this method of observation more accurate.

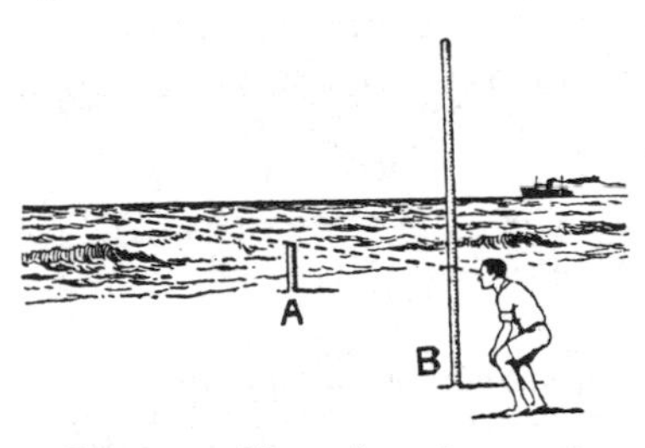

FIG. 39. Measuring changes in terrestial ray curvature.

In extremely rare cases the curvature of the rays is abnormally strong and gives rise to the most remarkable optical phenomena. There are days when everything can be seen with extraordinary clearness, and a far-away town or lighthouse suddenly becomes visible, which under ordinary circumstances it would be impossible to see at all, because it lies below the horizon. Very often it gives the impression of being surprisingly near. Two very striking cases of this kind were once observed along the Channel. Once the whole of the French coast opposite Hastings could be seen from the beach there with the naked eye, whereas in ordinary circumstances it cannot be seen even when the best field-glasses are used. Another time the whole of Dover Castle was seen from Ramsgate to appear from behind the hill which usually covers the greater part of it.

And, conversely, there are cases where distant objects, that usually project above the horizon, disappear as if they lay below it ('sinking'). These conditions, too, give a strong impression of proximity.

Observations of this kind should always be done in combination with measurements of the temperature of the sea's surface and of the air.

31. *Mirages on a Small Scale.* (Plate V)

The well-known mirage of the desert can be seen quite easily on a small scale. We find a long, even wall or stone parapet at least 10 yards long and facing south, on which the sun is shining. We then lay our heads flat against the wall and look along it sideways, while someone, as far away as possible, holds some bright object closer and closer to the wall, for instance an ordinary key, shining in the sun. As soon as the key is within a few inches of the wall, its image becomes strikingly deformed and a reflected image from the surface of the wall seems to approach the key. Often the whole hand holding the key can be seen reflected too. When this phenomenon has once been properly observed, it can be noticed with every distant object that can be seen by looking grazingly along the surface of the wall. This reflection can also be perceived when the wall is shorter, if the eye is placed quite close to the wall, which can be managed if there is enough room at the end of the wall to stand in.

If a very long wall is strongly heated, a second reflection can sometimes be seen as well as the first one, not inverted with respect to the object reflected but erect.[1] This is in agreement with a general law which states that the successive images of a mirage must be alternately upright and upside down (Plate V, *b*).

The reflection occurs in consequence of the air near the heated object being warmer and therefore more rarefied, so that its refractive index is diminished. This causes the rays of light to bend until they are parallel to the surface and afterwards to diverge from it (Fig. 40).

FIG. 40. Mirage on a sunlit wall (the vertical distances are much exaggerated for the sake of clarity).

This is sometimes called 'total reflection,' but this expression is wrong because the transition between the different layers is gradual everywhere; but on the other hand, one must bear in mind that the curving of the ray takes place almost entirely in close proximity to the heated object. Probably close along the wall there is a layer of air a small fraction of an inch deep, which has approximately the temperature of the wall itself; beyond this the temperature falls at first rapidly, and then more slowly.

It would be worth while to measure the temperature of the wall and of the neighbouring layers of air and to show how the observed curvatures of the rays can be explained quantitatively by these measurements.

Similar mirages on a small scale used to be sometimes noticed along the hot funnels of steamers.[2] The moon, Jupiter, the rising sun, were reflected as in a silver mirror; the mast of the ship, on the contrary, did not show this effect. But I do not think that the funnels of modern steamers are hot enough to give rise to this phenomenon.

If observed over the roof of a motorcar that has been standing for some time in the sun, the images of distant

[1] W. Hiller, *Phys. Zs.*, **14**, 713, 1913; **15**, 303, 1914.
[2] Ball, *Phil. Mag.*, **35**, 404, 1868.

(*a*) Inferior Mirage, Death Valley, California. (*By courtesy of the U.S. Weather Bureau*)

(*b*) Mirage on a long sunlit wall. One can see the reflected image of the boy (180 yards from the observer) and the beginning of an abnormal second reflection; the temperature of the wall was 4·5° C. higher than that of the air. (*From W. Hillers, Physikalische Zeitschrift*, **14**, 718, 1913)

objects are noticeably distorted, provided one looks closely along the heated surface.

When you look over a small board, not longer than 20 inches, lying in the sun you can often see every distant object, as it were, 'elongated' and attracted by it.

32. *Mirages on a Large Scale above Hot Surfaces* (*'inferior mirages'*) (Plate V, *a*)[1]

A flat surface, and observation at a long distance, are at least as essential as excessive heating of the ground, for the formation of a mirage. That is why a flat country like Holland is so especially suitable for observations of this kind: there the reflections in the air are often as fine as those shown by the scorched desert of the Sahara. Often these mirages can only be seen when stooping; by using opera- or field-glasses and scanning the horizon frequently, it is amazing how very much clearer they become, and how often they can be seen.

We shall now describe three cases in which this phenomenon occurs with extraordinary clearness and frequency.

First of all, it is to be seen on any sunny day *above flat asphalt roads*. The thermometer shows a fall in temperature of as much as 20° or 30° in the first half inch above the surface, after which the fall becomes a few degrees per inch.[2] My own experience is that the mirage is still finer above our modern straight concrete roads. It is true that the radiation of the sun is not absorbed so much by these as by the asphalt roads, but the re-emission of heat is less too. When the weather is sunny, this sort of road appears to be covered with pools of water, which grow larger and clearer if one stoops to look at them, and which appear to reflect the bright and coloured objects in the distance. What we take for water is nothing but the reflection of the clear sky

[1] Literature on the subject is extensive. *See* for instance: Pernter-Exner, *loc. cit.*; R. Meyer, *Met. Zs.*, **52**, 405, 1935; W. E. Schiele, *Veröff. Geophysik. Inst. Leipzig*, **7**, 101, 1935, with numerous references.

[2] H. Futi, *Geophys. Mag.*, **4**, 387, 1931. L. A. Ramdas and S. L. Malurkar, *Nat.*, **129**, 6, 1932.

in the distance. It is remarkable how this reflection remains undisturbed by the busy traffic, while paper, leaves and dust are whirled up by it! Observe accurately at what angle the mirage is visible, and calculate the temperature of the air which is in contact with the ground by the formula explained on p. 48.

Secondly, the mirage is a usual phenomenon in the *wide meadows of flat districts*, and can almost be called a characteristic of the landscape, at least in spring and summer, when the weather is bright, and there is not too much wind. Along the horizon you can see a white strip, above which towers and treetops in the distance seem to float apparently without any foundation. On bending you see the landscape nearer to you distorted, with large shining pools of water, reflecting the houses and clear sky in the background. This is particularly clear in the direction of the sun.

Towards midday, the curvature of the rays is often so strong that, even when you stand upright, it seems as if there are pools everywhere, and by stooping for a moment or by climbing a few yards higher, you will be surprised to see how the pools seem to expand or to shrink. Note how the images become distorted and elongated in a vertical direction, whenever the eye is just a little too high to see the reflection. If the eye is very low the base of objects in the distance is no longer visible, they are suspended in empty space. On the side away from the sun, the pools are less bright and, therefore, less noticeable, but the distortion and reflections of distant objects can be seen all the better.

It is interesting to note down a few temperatures in the lower air-layers, at heights, say, of 40, 20, 10, 4 and 0 inches. In the morning, if the sun is shining, the temperature will invariably be found to be highest close to the ground; if the difference between 40 and 0 inches amounts to 3°, there is little or no reflection. If it rises to 5°, the reflection is moderate, and at 8° the phenomenon is strongly marked. The greatest differences occur in the spring, on bright sunny days following chilly nights.

It was over the vast meadows near Bremen, that Busch, who was the first to make a scientific and thorough study of this phenomenon, was able to observe (in 1779) the mirage of the far-away city clearly.

The most beautiful and most evenly produced mirage is to be seen on the *beach, across firm and smooth sands*, when the weather is warm and there is no wind.[1] When we lie flat on the ground, our eyes as close as possible to the surface of the sand, we see no clearly reflected image, but if we raise

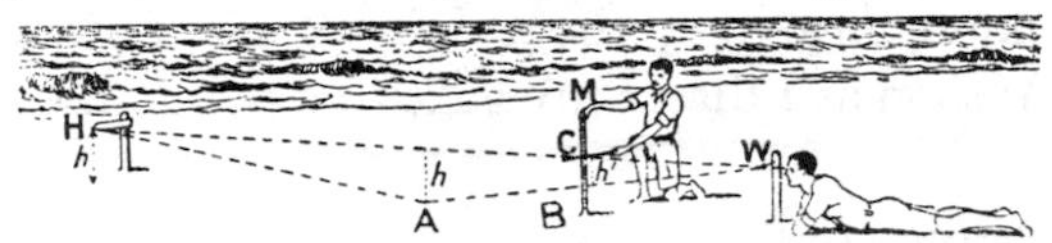

FIG. 41. How the path traversed by the rays forming a mirage can be determined. All horizontal dimensions are drawn on much too small a scale.

our heads a little it seems suddenly as if we were surrounded by a reflecting lake, and we can even see objects 10 to 5 inches high, 30 to 35 yards away, reflected in it. We pick out one clear, bright object H, and keep our eyes fixed at a definite point W, just as high from the ground as the object is, which can, for instance, be indicated by a twig or piece of stick.

We now determine by experiment the path of the ray of light by which we see the mirrored image. At C at a known distance, an assistant holds a small measuring rod M upright, and moves a little lath along it until it intercepts (*a*) the image in question at B, (*b*) the top of the object itself. We may assume the direct ray of light HW from H towards our eye to be a straight one, so that we are able to determine successively the height of each point of the deflected ray of light HAW, and therefore, point for point, the path of the ray itself. In this way, it turns out that close to the sand it must have suffered a rather sudden deflection. If this be true, we may expect $\frac{h}{AW} = \frac{h'}{BW}$ to be constant and equal

[1] L. G. Vedy, *Met. Mag.*, **63**, 249, 1928. Strikingly beautiful are the mirages across the 5 mile long stretches of sand along the Dutch North Sea islands.

to the angle between the surface of the sand and the ray traversing the longer path. This appears to be actually the case; and angles up to 1° are found. From this angle and the known refractive index of air for various temperatures, one deduces the difference in degrees between the temperatures of the air immediately next to the ground and at the height of the eyes: according to $\Delta t \text{ (centigrade)} = \frac{273}{29 \cdot 10^{-5}} \cdot \frac{1}{2} \left(\frac{h}{AW}\right)^2$. In practice, this may vary between 10° and 65°F.

In the foregoing case, the origin of the mirage is very simple. As soon as I direct my gaze to a point on the ground

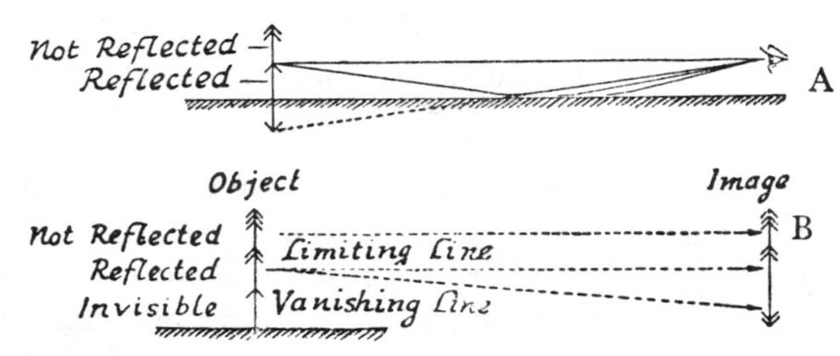

FIG. 42. A mirage shows only a reflection of part of the object. A, At short distances. B, At long distances.

beyond a certain limit, the visual ray impinges on the hot layers at a sufficiently inclined angle to suffer a sudden deviation. The effect is about the same as if a mirror were lying on the ground at that point. In this way distant objects are divided into two parts: the top part is seen single, and the bottom one shows a reversed reflection (Fig. 42A).

On mirages at long distances the curvature of the Earth and the ordinary curvature of the rays have a very marked influence. Below a certain 'vanishing-line' the foot of distant objects is invisible owing to the curvature of the Earth. Between this 'vanishing-line' and a still higher limiting-line lies that part of the object that is seen reflected, and its reflection is usually compressed in a vertical direction. Finally, above the limiting-line, we see those objects that have no reflection (Fig. 42B).

Instead of the rapid rise in temperature at the Earth's surface, we can imagine many much more complicated temperature distributions, each of which has its own peculiar optical consequences. In a very clear mirage above the beach it is possible by an experimental investigation, as described above, to find out the course of the vanishing line and the limiting-line, and from that to deduce the temperature distribution. Direct temperature measurements can be compared with this. But the possibility of the beach not being quite level makes investigations of this kind very difficult.

On every sea voyage, numbers of mirages are to be seen which can be explained by the preceding considerations (Figs. 43, 44). If the phenomenon is incompletely developed, as usually happens, the (inverted) reflected image becomes so flattened that it looks merely like a small horizontal line, merging into the base of the object itself. The only thing that strikes one now is the bright streak of light of the

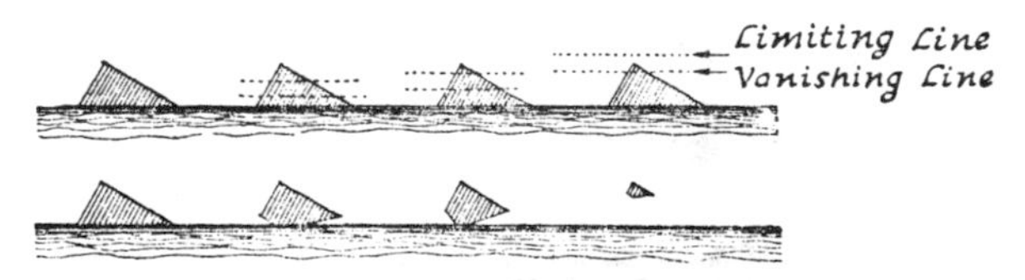

FIG. 43. An island seen at increasing distances where a mirage is present.

reflected sky, the fact of that being likewise compressed, being, naturally, not noticed. Far-away objects seem, therefore, to float, as it were, at a short distance above the horizon. This optical phenomenon, which is nothing but a slightly developed mirage, is to be seen almost every day at sea, especially when field-glasses are used. If the different parts of an island are at different distances from us, those farthest away are touched higher up by the limiting- and vanishing-line, and the condition in Fig. 44D is the result.

Measuring the height between the vanishing line and the apparent horizon is an easy way of expressing the 'intensity' of the mirage numerically, and can be carried out by one of the methods given in the appendix, § 235. Angles of a few minutes of arc are to be found.

There is another phenomenon that produces an effect which may sometimes be mistaken for this one, namely the formation of a layer of fine drops of water by the foam from surf. These drops float in the air above the sea, and cover the lowest parts of distant objects with a light layer of mist.

Mirages with their deformation and reflection of images have also been observed under the following circumstances:

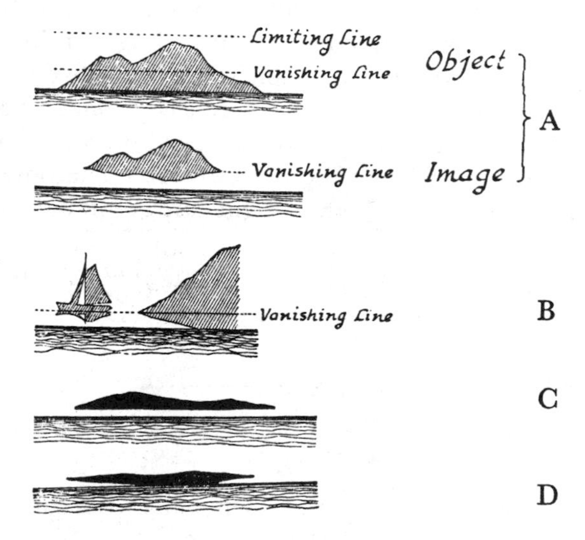

FIG. 44. Observations of a mirage during a voyage.

while bathing, when the water is hotter than the air; on large lakes under favourable atmospheric conditions; above railway lines, where, by stooping, one can see how completely distorted an engine looks in the distance; above flat sandy tracts or flat plough-land; along the slopes of dunes if one looks parallel to the slope; and along the streets of a town, paved with stones, especially if one can look very closely along the highest point of a rise in the street.

33. *Mirages above Cold Water ('superior mirage' or 'looming')*

Just as a downward reflection occurs mostly above heated land, upward reflection is to be seen chiefly over the sea, although far less often. This occurs when the sea is much colder than the air, so that the temperature in the lowest strata of air increases very rapidly with increase in height above the sea; this type of temperature distribution is called an 'inversion' of temperature by meteorologists (*see* Fig. 45).

Some classic observations of magnificent 'superior' mirages were carried out from the south coast of England,

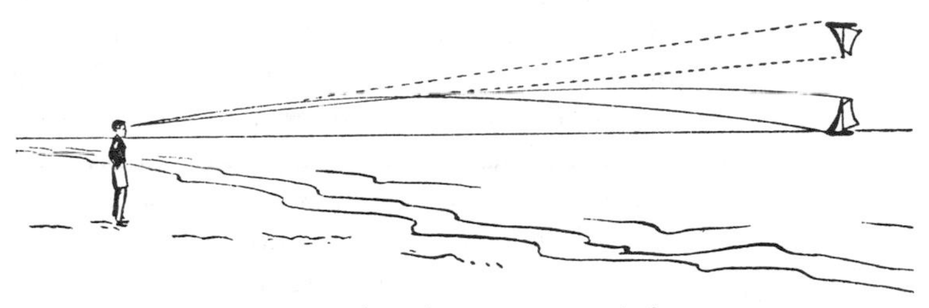

FIG. 45. The superior mirage, an unusual phenomenon.

by looking through a telescope across the Channel, sometimes in the evening after a very hot day, sometimes while a mist was lifting. Superior mirages are also seen under quite different conditions, namely, in the spring over the Baltic, when it has just thawed.

Such mirages can be seen above a frozen surface when it suddenly begins to thaw and the air close to the ice is colder than higher up, but, in order to see it, one has to stoop and look closely along the frozen surface.

Sometimes the bending of the rays upwards causes multiple reflections, free to develop now that there is nothing to cut off the light-rays (as the Earth does when the reflection is downwards), and strange images arise, upright and inverted, varying from one moment to the next, changing in accordance with the distance of the object and the distribution of temperature in the atmosphere.

34. *Castles in the Air*

In a few very rare cases most remarkable mirages have been seen by quite reliable observers, who describe them as landscapes with towns and towers and parapets, rising above the horizon, transforming, crumbling, fairy-like scenes, producing a deep sense of happiness and an endless longing—*fata morgana!* No wonder that these observations, already so beautiful in themselves, have been adorned by the fancies of poetry and folk-lore.

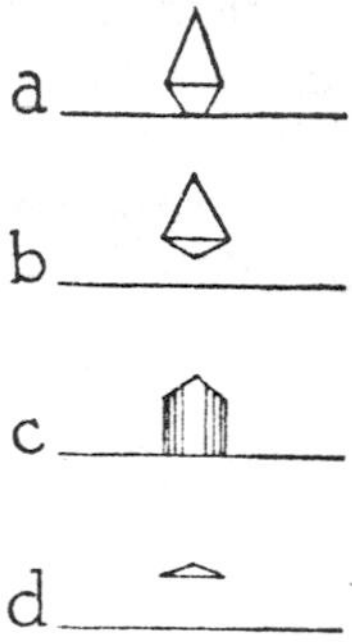

Forel observed simpler forms of this phenomenon time and again above the Lake of Geneva, and after fifty years of study described it in detail.[1] A calm surface of water ten to twenty miles across is essential, the eye must be 2 to 4½ yards above the water, and, this being very important, the exact height must be found by experiment. In the afternoon on bright days, when the water was warmer than the air, Forel saw four consecutive stages develop gradually along the opposite shore, the one succeeding the other, and remaining not longer than 10 to 20 minutes in the same place.

FIG. 46. How the *fata morgana* arises as a transition between the refraction of rays above warm and cold water.

These 4 stages were: (*a*) the mirage above warm water; reflection below the object; (*b*) the abnormal mirage above cold water; a very strange phenomenon, in which the object is seen quite normally, with its reflection below extremely compressed (probably a labile, temporary, transition shape); (*c*) the castles in the air; the distant coast line is distorted over a distance of 10° to 20° (in angular measure) and elongated vertically into a row of rectangles (the 'streaked zone'); and (*d*) the normal curvature of the rays above cold water; no reflection is visible, but the object itself is strongly compressed in a vertical direction.

The upper horizon in the stages *a* and *b* and the lower

[1] F. A. Forel, *Proc. R. Soc. Edinb.*, **32**, 175, 1912.

horizon in stage *d* are the boundaries between which the vertical shading of the streaked zone is developed (Fig. 47). The shifting of the castles in the air is a result of refraction of type *a* being gradually replaced by type *d*. The theory that the density of the air, in a transition region of this kind, is greatest in layers of average height, seems quite acceptable. The path of the rays, in that case, is that shown in Fig. 47, and, as will be seen, every point of light L is drawn out vertically into a line AB.

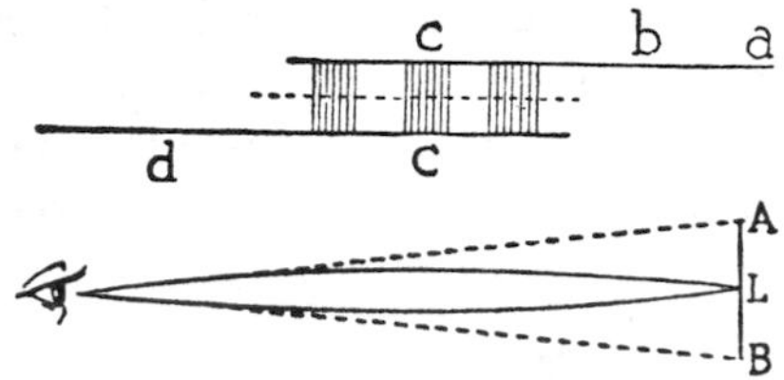

FIG. 47. How the *fata morgana* arises.

We may, perhaps, wonder whether there is any chance of our discovering the typical *fata morgana* in our own country. At least one splendid example is known to have been observed on the Dutch North Sea coast. On this one unique occasion practically all the characteristic features, mentioned by Forel, were noticed by the observer. He writes 'When I went down to the beach at Zandvoort, at 4.20 p.m. (summer time), I was at once struck by the unevenness of the horizon. North-West and West it was appreciably higher than in the South-West, in many places two horizons were visible, one above the other, merging on the one side into the higher level of West and North-West, and on the other side into the lower level of South-West. The distance between them was pretty well the same everywhere, about 7′ (0·08 inch at arm's length from the eye). Objects lying between the two levels suffered peculiar distortions, giving rise to all sorts of deceptive mock-images.' *See* Fig. 48.

35. *Distortions of the Rising and Setting Sun and Moon*[1] (Plate VI)

When the sun is low, the most curious distortions are to be seen at times. The corners of the visible segment are often rounded off, or the disc appears to consist of two pieces

[1] A. L. Colton, *Contrib. Lick Obs.*, **1**, 1895. *P.A.S.P.*, **45**, 270, 1933, etc.

joined together, or there is a strip of light below the sun, which rises as the sun's disc sinks. In other cases, the sun does not set exactly behind the horizon, but a few minutes of arc above it. These distortions seem to vary more in the evening than in the morning, and this is to be ascribed to meteorological factors (cf. § 193). On still, cloudless days the layers of different density are less disturbed during their

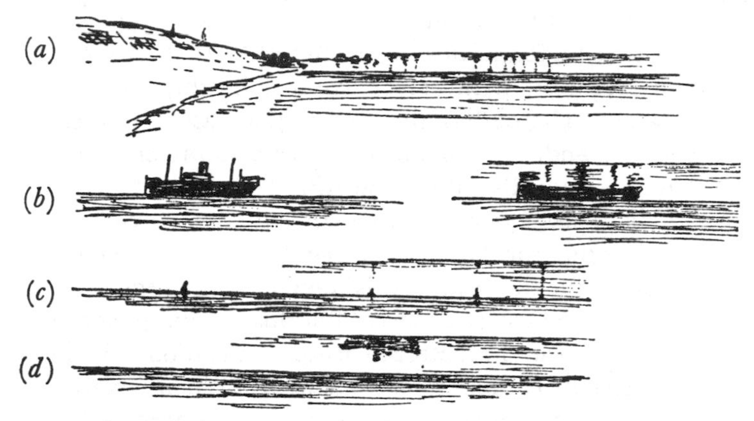

FIG. 48. Castles in the air, observed at Zandvoort, Netherlands.
(*a*) Noordwijk, Katwijk, Scheveningen, in the striated zone: a wood of palm-trees !
(*b*) An outward bound steamer left, no reflection: right, in the region of the *fata morgana.*
(*c*) Small sailing boats.
(*d*) Steamer itself invisible, behind the horizon; visible only in *fata morgana.* The inverted mirage hangs from the upper horizon.
(From J. Pinkhof, *Hemel en Dampkring,* **31,** 252, 1933. Block lent by the Royal Netherlands Meteorological Institute.)

formation, so that the distortions of the sun's edge may be taken to indicate a steady condition of the atmosphere, and are a sign of fine weather. If the sun is too blinding it is advisable to hold a sheet of silver-paper or ordinary paper, with a small round hole pricked in it, in front of one's eye, or to use a dark glass. A field glass is not essential, though it facilitates observation. A piece of blackened glass or a pin-hole diaphragm can in this case be held before one's eyes (not before the objective).

These phenomena enter their most interesting stage

VI

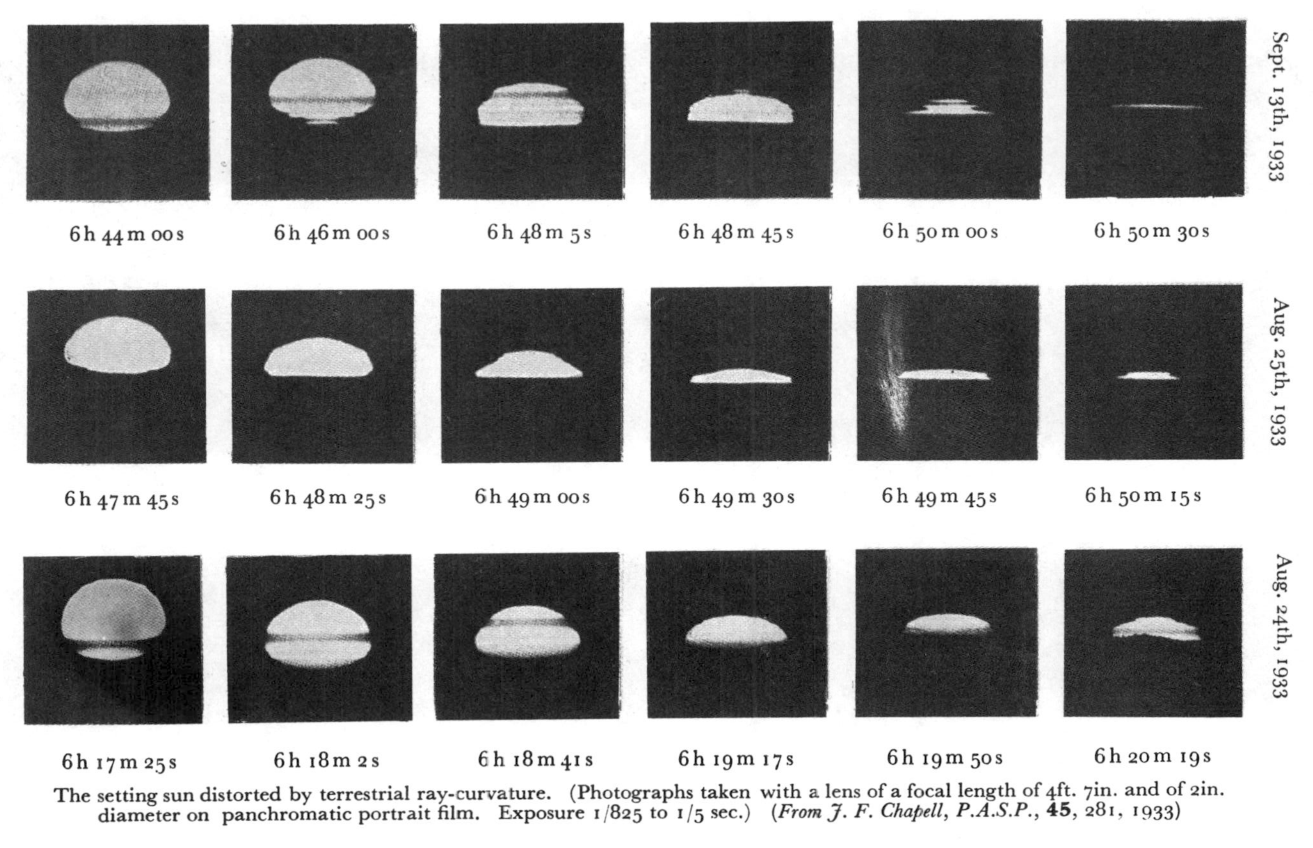

The setting sun distorted by terrestrial ray-curvature. (Photographs taken with a lens of a focal length of 4ft. 7in. and of 2in. diameter on panchromatic portrait film. Exposure 1/825 to 1/5 sec.) (*From J. F. Chapell, P.A.S.P.*, **45**, 281, 1933)

usually only 10 minutes before sunset (or last until 10 minutes after sunrise). Note, as well, the different shades of colour in the sun's disc, which is dark red on the side nearest the horizon, and in the upper part changes gradually into orange and yellow. Note, too, the large sunspots, present on the disc at times, which are drawn out into the shape of short rods.[1]

It would be interesting, although difficult, to take photographs; pictures of the sun made by an ordinary camera are much too small. Only with a telescope having a focal length of at least 30 inches and an aperture of from 1 to 4 inches can satisfactory photos be taken, requiring an exposure of less than one second, in which short space of time there will be no need to make the telescope follow the apparent motion of the sun. Use panchromatic plates and consult the literature on this subject.

These optical distortions are due to nothing but the ordinary mirage, and we must again distinguish between an upward mirage and a downward one. We obtain a fairly good approximation to the truth, if we assume (with Wegener) that there is a sudden bend in the light ray coming from the sun when it strikes a layer of discontinuity.

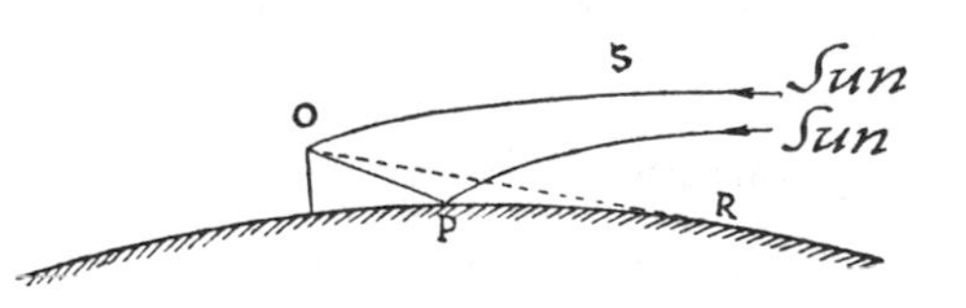

FIG. 49. Distortions due to a mirage according to Case A, during a sunset.

Case A. (Fig. 49). A thin stratum of warm air PR, rests on the earth, as assumed in Fig. 49. We see, therefore, the sun in the direction OS, and at the same time, underneath it, the reflected image in the direction OP, with the horizon OR lying between. At sunset a flattened 'counter-sun' rises

[1] Havinga, *Hemel en Dampkring,* **19**, 161, 1922.

out of the apparent horizon OP as the sun sets, and they unite at the spot where the real sun is about to disappear (OR). The two discs glide more and more into one another, while balloon-like shapes, etc. are formed.

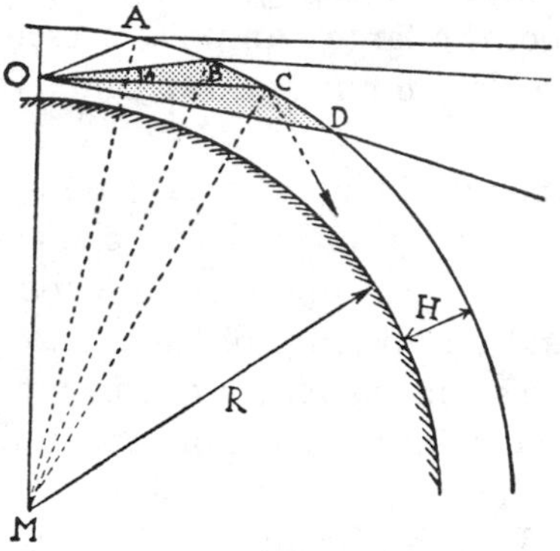

Fig. 50A.

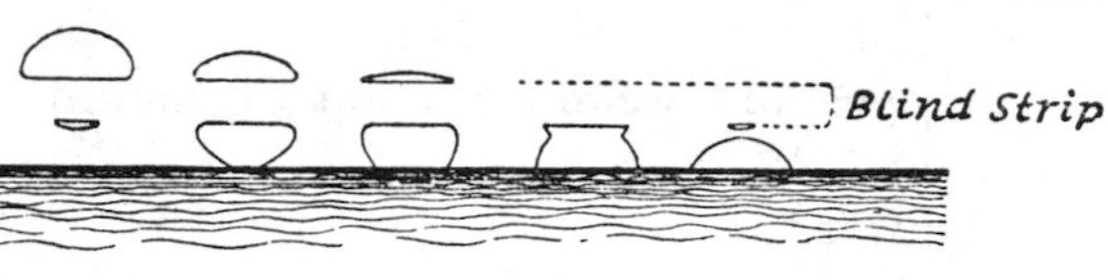

Fig. 50B.

Sunset with distortions caused by a mirage according to Case B.

Case B. (Fig. 50.) We now assume that the air near the ground is cold, while a warmer layer ABCD lies above it (inversion). Point M is the centre of the earth, round which two arcs are drawn to represent the level of the sea and of the layer of discontinuity. Imagine now observer O to be looking in directions nearer and nearer to the horizontal; in direction OA his gaze touches the upper rim of the sun; in direction OB he sees a point a little lower down, but his gaze is more inclined in relation to the discontinuity layer; in the horizontal direction OC it is incident on the layer at such a large angle that the visual ray is curved and can no longer leave the Earth. If the observer stands on a slight elevation above the earth's surface, he will also be able to look downwards at a small angle; thus if he looks in the direction OD the angle of incidence of his line of sight on the layer of discontinuity is decreased and is once more sufficiently small for the visual ray to be able to escape. Within the dotted angle on both sides of the horizontal direction, therefore, no rays from outside the earth reach the observer, he sees a 'blind strip' as high as $2h$. This is a consequence of the following theorem.

Of all the chords through O, the horizontal chord OC is the one that makes the smallest angle with the circle. Proof: in triangle MOB we have $\frac{\sin \widehat{OBM}}{OM} = \frac{\sin \widehat{MOB}}{MB}$, so that $\sin \widehat{OBM} = \frac{R}{R+H} \cdot \sin (90° + h) = \frac{R}{R+H} \cos h$. It is clear from this that $\widehat{OBM}$ attains its maximum value when $h=0$. In the limiting case of total reflection, $\sin \widehat{OBM} = \frac{1}{n}$, n denoting the refractive index of the one layer relatively to the other. Writing ε for $\frac{H}{R}$ and δ for $n-1$, and replacing $\cos h$ by its approximate expression $1 - \frac{1}{2}h^2$, the result for h will be:

$h = \pm \frac{\sqrt{2(\delta - \varepsilon)}}{n}$ or nearly $\pm \sqrt{2(\delta - \varepsilon)}$ as n is, practically speaking, equal to 1.

We see, therefore, that the blind strip extends just as far above as below the horizon (double sign). For H=55 yards, $\varepsilon = 78 \times 10^{-7}$; if one puts $\delta = 100 \times 10^{-7}$, then $h = \pm 0{\cdot}021$ radian $= \pm 7'$; the width of the blind strip is, therefore, 14′.

As a matter of fact we ought to take into account the ordinary terrestrial curvature of the rays as well, but we are only concerned with the main features of the phenomenon.

It is clear now, that, considering this structure of the atmosphere, the sun sets before it reaches the actual horizon, i.e., as soon as it enters the blind strip. If the observer is standing on the top of a hill or on the deck of a ship, he will probably be able to see the lower rim of the sun appear from behind the blind region. The images are of course distorted, that is, compressed above the blind region, and elongated below it.

In some cases the sun's image shows several small steps, indicating evidently the presence of more than one layer of discontinuity (Fig. 51). Occasionally one of the notches between these steps becomes so deeply indented, on both sides, that a strip is cut off as it were, from the top of the sun, remains for a moment afloat in the air, contracts, and vanishes, often with a magnificent display of the green ray phenomenon. This may be followed by another strip being cut off in the same way, and so on (Fig. 58).

FIG. 51. Distortion of the sun when several layers of discontinuity are present.

Distortions similar to those of the sun are shown by the moon as well, and can be observed particularly clearly in its slender crescent (Fig. 52) (cf., however, § 90).

36. *The Green Ray*[1]

According to an old Scotch legend, anyone who has seen the green ray will never err again where matters of sentiment are concerned. In the Isle of Man it is called 'living light.'

The green ray can be perceived more frequently than people used to think. During one sea-voyage from India to Holland, I observed it more than ten times. The best place to see it is undoubtedly over the sea, either from the deck of a ship or from the shore. It can, however, be seen above land as well, if the horizon is distant enough. It occurs, too, sometimes, when the sun is disappearing behind a sharply outlined bank of clouds. It seems that it is visible over mountains and clouds, provided these are not higher than about 3 degrees above the horizon. In a few cases the green ray has been seen at an amazingly short distance. Ricco

FIG. 52. Multiple moon crescents. (From *Onweders en Optische Verschijnselen in Nederland* and *Meteorologische Zeitschrift.*)

[1] Mulder, *The 'green ray' or 'green flash'* (The Hague, 1922). Feenstra Kuiper, *De Groene Straal* (Diss. Utrecht, 1926). In these treatises the extensive literature on these subjects is collected and reviewed. There are beautiful coloured plates in *Ann. Hydr.*, **65**, 489, 1937.

relates how he once stood at the edge of the shadow of a rock fairly close to him, and by moving his head a little more to one side or the other, he could see the green ray as often as he liked.[1] Whitnell and Nijland saw it along the top of a wall 330 yards off, but these are exceptional cases.

All who have observed it agree that the green ray is clearest on evenings when the sun shines brightly up to the moment of setting, whereas it is almost invisible when the sun is very red.

Opera- or field-glasses are generally a help, and a telescope still more so. However, care should be taken not to look through the glasses into the sun itself, except during the very last seconds before it sets, for fear of being dangerously dazzled. Nor should you look too soon into the last segment of the sun's disc with the naked eye, but turn your back to it until someone lets you know when it is the right moment for observation.

The phenomenon is very transitory, and lasts only a few seconds. Once, by running up the slope of a dyke, 6 yards high, I was able to watch the green ray for 20 seconds; it became at times bluer, and at times whiter, according as my own pace was too slow or too fast. It should be possible to see it occasionally from the different decks of a ship in turn. The movement of the ship enabled Nijland to see it several times in succession. In one very special case of abnormally strong ray curvature it has been seen for 10 seconds and longer. The Portuguese Gago Continho was able to observe it for any length of time in the light of a distant lighthouse.

During Byrd's expedition to the South Pole, the green ray was observed for 35 minutes, while the sun, rising for the first time at the close of the long polar night, was moving exactly along the horizon.

The phenomenon of the green ray can appear in the three following forms: (*a*) *The green rim*, which, as a matter of fact, can always be distinguished along the top of the sun's disc and becomes wider the nearer it descends towards the

[1] *Mem. Spettr. Ital.*, **31**, 36, 1902.

horizon; at the same time the lower part becomes red. (*b*) *The green segment* (Fig. 53). The last segment becomes green at the extremities, and this green colour shifts gradually towards the centre of the segment. This green segment is visible to the naked eye, often for a second or so, and with field-glasses sometimes for 4 or 5 seconds. (*c*) *The green ray itself* (Fig. 54). This phenomenon, visible to the naked eye, is seen extremely seldom. It is a green ray similar in appearance to a flame, which shoots up out of the horizon just as the sun is disappearing.

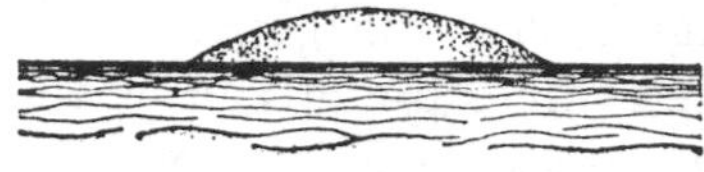

Fig. 53. The green segment.

In all these three forms its colour is mostly emerald, seldom yellow. Sometimes it is blue or even violet. The colour was once seen to run from green through blue into violet in the course of the few seconds the phenomenon lasted.

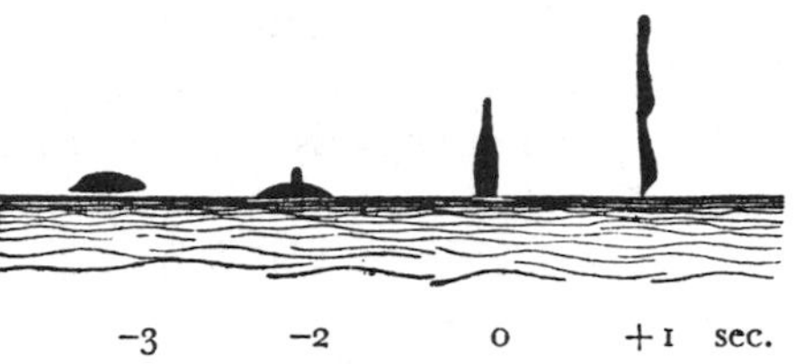

Fig. 54. The green ray proper. The times are calculated from the moment of the sun's setting. (After D. P.Lagaaij.)

There can nc longer be any doubt as to the explanation of the green ray. The sun is low, so that its white rays have a long way to travel through the atmosphere. A great part of its yellow and orange light is absorbed by the water vapour, the absorption bands of which lie in this spectral region. Its violet light is considerably weakened by scattering (cf. § 172), and there remain, therefore, red and green-blue, as can also be seen by direct observation[1] (Fig. 55).

Now the atmosphere is denser below than above, so that the rays of light on their way through the air are bent (cf. § 29); and this bending is somewhat slighter for red light, and somewhat stronger for the more refrangible blue-green rays. This causes us to see two sun discs partially covering one

[1] With very strong scattering the green-blue disappears also; that is why the green ray is invisible if the sun is deep red when it sets.

another, the blue-green one a little higher, the red one a little lower, which accounts for the red rim underneath and the green rim at the top (Fig. 56). One can now understand why the extremities of the segment are green when the sun is low, and why the white part disappears gradually behind the horizon while the green covers the whole of the remaining segment. In many circumstances, however, refraction is abnormally strong near the horizon and the green segment is more clearly visible for a longer time. Should mirages arise it may even be extended into a kind of flame or ray.

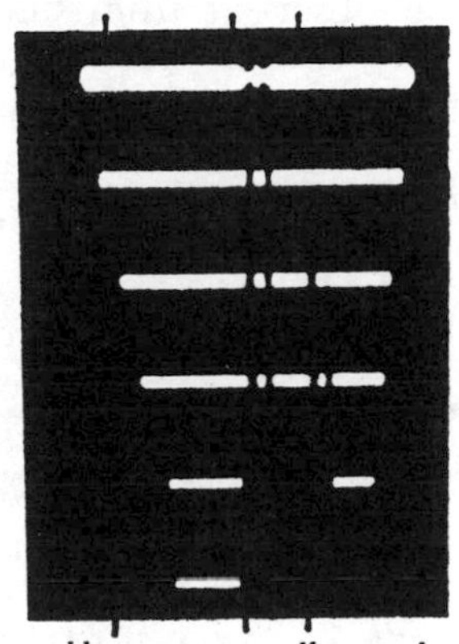

Fig. 55. Spectrum of the setting sun. Observed by N. Dijkwel, *Hemel en Dampkring*, **34**, 261, 1936.

This view would be supported if it should turn out that the green segment and the green ray are absent when the sea is warmer than the air, the decrease in density and the ray curvature then being particularly slight. There are indeed indications that this is so.[1]

It is said, too, that the green segment can be seen exceptionally well when the characteristics of a mirage are present

Fig. 56. How the green ray is caused.

underneath; that is when the lower edge is not straight but bent upwards at the corners.[2]

When the sun's disc is indented at the sides by layers of discontinuity, one can see how a strip at the top is detached now and then and disappears in a green glare, a very wonderful

[1] R. W. Wood, *Nat.*, **121**, 501, 1928.

[2] *Nat.*, **111**, 13, 1923.

sight! (Fig. 58; cf. Fig. 51, § 35). Here is yet another fact that speaks strongly for the considerable influence of *abnormal* refraction; on two occasions the green ray could be observed from one deck of a steamer, and not from another, which shows that it depended on the height at which the observer was standing.[1] And yet, on the other hand, there are competent observers of Nature who insist that the *ordinary* terrestrial ray curvature is quite sufficient to produce the green ray.[2]

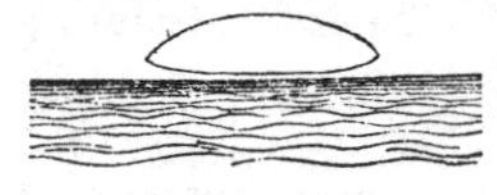

FIG. 57. The last segment shows upturned corners. There is a chance of the green ray occurring!

The chief problem still to be solved in regard to the green ray is, therefore, *how strong must the refraction be to cause a given intensity of the phenomenon?* In order to solve this, it would suffice if someone on the shore could determine for a number of days exactly what time the sun sets, observing the green ray phenomenon at the same time. The difference between the observed and the computed times is a good indication of the deviation of ray curvature from the normal (cf. § 29).

It used to be thought that the green ray might be a physiological after-image in the complementary colour of the last vestige of the red setting sun (§ 88). This supposition is sufficiently refuted by the fact that the green ray can also be seen when the sun is rising, though it is then more difficult to know where to look in anticpation of the appearance of the light. One must find the brightest point or take as an indication the crepuscular rays or Haidinger's brush (§§ 191, 182). Another argument is that the green ray can only be seen when the distance to the horizon is sufficiently great; though this would not affect the after-image in any way, it is naturally very important as

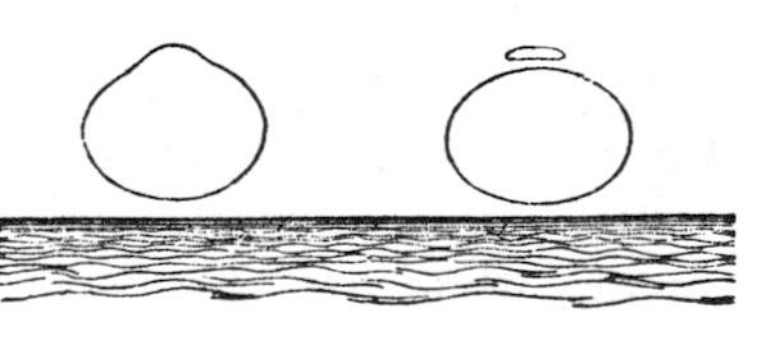

FIG. 58. How the green ray is caused by the detachment of the top portion of the setting sun.

[1] This observation would be worth repeating, preferably by the same observer standing on each of the decks in turn.

[2] *Proc. R. Soc.*, **126**, 311, 1930.

regards curvature of the rays. At the expense of much trouble the green ray has been successfully photographed on an autochrome plate.

The green ray has been observed too on rare occasions in connection with the Moon and Venus, and another time with Jupiter. One observer describes how he saw the reflection of Venus rise towards the planet and how, the moment they met, the colour suddenly turned from dull red to green.

37. *The Green Surf*

On the coast of Sumatra it was perceived that on the distant horizon the white-crested breakers seemed green and that this applied only to the lower ones, the higher breakers being white as usual. The sea was grey and the line of the horizon dipped clearly.

This phenomenon seems to be identical with the green ray, the gleaming white of the lower waves corresponding to the extreme edge of the setting sun.

38. *The Red Ray*[1]

It follows from the explanation of the green ray that there must be a red ray also, which would occur, for instance, when the sun has gone down behind a heavy, sharply outlined, bank of clouds near the horizon and the very lowest edge of it comes peeping out below. This has been observed at times, but very seldom, and it appears to last for a still shorter time than the green ray.

Whitnell, watching the green ray through an opening in a wall 330 yards away, was able to see the red ray on the same occasion.

39. *Scintillation of Terrestrial Sources of Light*

The phenomenon known as scintillation or twinkling is to be seen in its most intense form above the braziers or stoves used to melt asphalt for the surface of our streets. The

[1] *Nat.*, **94**, 61, 1914. For a most delightful account of an observation of the red ray with glasses during the disappearance of large sunspots at sunset, *see* W. M. Lindley, *J.B.A.A.*, **47**, 298, 1937.

objects in the distance seem to quiver and ripple so much as to be hardly distinguishable, and the air itself appears to be no longer transparent. Also one can see how everything in the distance quivers above the boiler of a railway-engine or above a sheet-iron roof exposed to the sun. A field of stubble, or a stretch of sand, heated by the sun suffices to bring about this effect.

This scintillation phenomenon is shown most clearly by bright and shining objects, the trunks of silver birches, white posts, patches of white sand, garden-globes, or distant windows lit up by the sun. In the summer or on cold days in spring one can see railway lines twinkling in the distance; they no longer appear to be straight, but twist and turn. If one lays one's head on the ground, the twinkling becomes greater, one sees air 'striae' borne along by the wind. These 'waves' can be higher than the waves of the sea. When using glasses one can never see the objects in the distance really distinctly while the sun is shining. (Observe this especially in directions away from the sun.) In the winter a practised eye can see by the quivering vibration of the images of distant objects the warm air ascending above the roofs of the houses (Oudemans).

'For the Air through which we look upon the Stars, is in a perpetual Tremor; as may be seen by *the tremulous Motion of Shadows cast from high Towers*, and by the twinkling of the fix'd Stars' (Newton's *Opticks*, 4th Edition, p. 110). Who of my readers has ever observed this?

All these phenomena can be explained by the curvature of the light rays in the currents of warm air, which rise like small fountains from the heated earth. At a height of not more than 2 yards they have already mingled considerably with the cold air, and the 'striae' have become smaller.

On a smooth white wall, illuminated by the sun, the ascending air striae can often be seen dancing above a window-sill, and casting shadows as delicate as very thin smoke. The parallelism of the light rays is disturbed by these striae, light becomes more localised in some parts, and less so in others. It is an effect similar to that caused in a much stronger degree by an undulating surface of water or by an uneven window-pane (§§ 23, 24).

It is evident that the scintillation will be more intense the greater the distance we look through the unevenly heated stratum of air. Lights a few miles away scintillate at night, and as one draws nearer to them the twinkling becomes less and finally disappears. A stationary motorcar in the road reflects the sun with a fierce brilliance which, at a distance of 500 yards, is one mass of scintillation; at 200 yards it is much more steady, and as I approach still closer the scintillation vanishes completely.

It has been observed that those parts of the light path nearest to the eye contribute most to scintillation. In the same way a pair of spectacles is most effective when close to the eyes: if you put them on the printed page you are reading, you will see that they do not change the size of the letters at all, but on moving them towards your eye the letters are magnified or diminished, the change being greater the nearer the lenses are to your eye. Similarly most of the scintillation arises from temperature variations in the air near the observer. This is confirmed by the fact that if the sun's radiation is obstructed for a short time by a thick cloud, so that the light-path in the immediate neighbourhood of the observer is overshadowed, the scintillation ceases almost immediately: conversely it reappears as the cloud passes away. Apparently the surface temperature of the ground follows any change in the sun's radiation very quickly.

By observing scintillation repeatedly from the same place one can soon discover how it varies under different weather conditions. It is always much less pronounced when the sky is clouded (general clouds that is, so that the *whole* light-path is more or less in shadow). Before sunrise it is rather feeble, becomes strong fairly soon after the sun is up, reaches a maximum about midday and is much less pronounced towards 4 or 5 o'clock. On some days, however, the development is quite different.

Scintillation can be seen not only above sand or soil or houses, but also above a surface of water, above snow, and above the foliage in a wood, which shows that the temperature of all these things can be so affected by radiation as to differ greatly from the temperature of the air. The rows of street-lamps along the distant boulevards of seaside towns are a fine sight seen from a ship entering a port, or steaming down the Channel or through the Straits of Messina.

Terrestrial sources of light sometimes show colour phenomena during scintillation, but only when they are a long way off. On one exceptional occasion, distinct colour changes were seen in the light from lamps not more than 3 miles away.

40. *Scintillation of the Stars*[1]

Note how Sirius, or any other bright star, twinkles when close to the horizon. When looking through a telescope one notices slight changes of position. When looking with the naked eye you will see variations in the brightness and also changes of colour.

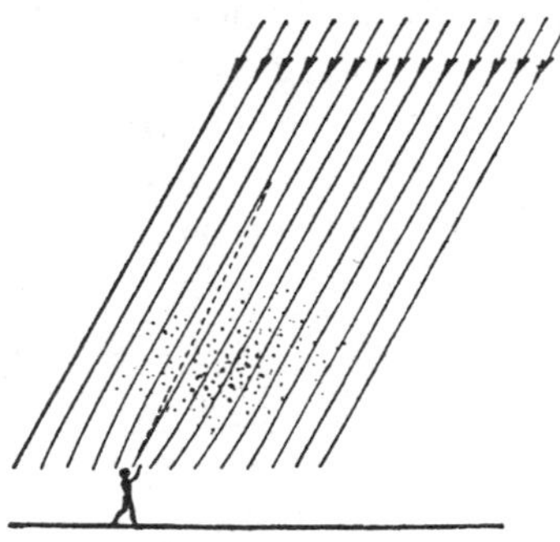

FIG. 59. How irregularities in the atmosphere cause the light-rays of a star to bend and produce scintillation. The observer here sees the stars displaced upwards and intensified.

Needless to say, this flickering is not a phenomenon taking place on the star itself, but is explained in the same way as the scintillation of terrestrial light-sources (§ 39). The changes in position are caused by curvature of the rays in the striae of hot and cold air, both of which are always present in the atmosphere, and especially where a warm layer of air passes over a cold layer and air waves with eddies are formed (Fig. 59). The changes in brightness arise from the fact that at the surface of the earth the irregularly deviated rays of light are concentrated at some places and sparsely distributed

[1] *See* extensive treatment by Pernter-Exner.

at others. If the continually changing system producing this is borne along bodily by the wind the observer will stand, now in a brightly illuminated region, now in one of less brightness. The colour changes are to be ascribed to slight dispersion of the *normal terrestrial ray curvature*, so that the rays from the star travel along slightly different paths in the atmosphere, according to their colour. For a star at a height of 10° above the horizon we compute the distance between the violet and red rays to be as much as 11 inches at a height of 1·25 miles, and 23 inches at 3 miles. The air striae are, on an average, fairly small, so that it may often happen that the violet ray passes through a striation and is deflected, whereas the red ray passes on without deviation (Fig. 60). The moments when the light of a star becomes brighter or feebler as a result of scintillation are, therefore, different for the different colours.

FIG. 60. How colours appear in the scintillation of the stars.

Scintillation is least near the zenith; there, when the atmosphere is calm, one can only just see, now and then, a twinkling of the bright stars. The closer the stars are to the horizon, the more they scintillate, simply because we are then looking through a thicker layer of air, and therefore through more striae (Fig. 63). Colour changes never occur, apparently, at altitudes of more than 50°, but frequently below 35°. The most beautiful scintillation of all is that of the bright star Sirius, which is visible in the winter months rather low in the sky.

Scintillation is so rapid that we cannot see what actually takes place, but anyone wearing glasses for short-sightedness can make a splendid study of scintillation by holding his eye-glasses in his hand and moving them slightly to and fro in their own plane before his eyes. This causes the image of the stars to be drawn out into a short line. It is even better if one moves the glasses in a circle, which can be done easily without jerks after a little practice (3 to 4 revolutions per second). As a consequence of the persistence of visual impressions (§ 80), one can now see, distributed along the circumference, all the variations of brightness and colour shown in succession by a star—a marvellous sight when the scintillation is strong! Sometimes dark spots occur in the band of light, which shows that there are moments when we receive hardly any light from the star. One can estimate how many different colours are to be seen along the circumference and calculate from that the number of colour variations in a second. This method of observation is based on the fact that an eye-glass acts not only as a lens but also as a weak prism if we do not look through the centre of it.

There are other means of analysing this scintillation phenomenon:[1] (*a*) Anyone with normal sight can use a weak concave eye-glass in the way indicated, but he will have to accommodate his eye as if the star were nearer. (*b*) By looking through opera-glasses while tapping them gently. (*c*) By looking at the reflection of the star in a pocket mirror, while rotating it through small angles. (*d*) Simply by letting your gaze move across the star (this can only be done after much practice, cf. § 82).

There is a simple method of observation which gives you a direct estimate of the dimensions of air striae.[2] Look at a brightly scintillating star with your eyes *slightly converged*, that is, focus your eyes on some object at, say, a distance of 5 or 6 feet, and more or less in line with the star. You will now see not one, but two images of the star, and *these two images do not scintillate in step*, because the eyes are so far apart, that a striation, while passing before one eye, has as yet no effect on the other eye. A large proportion of air striae must, therefore, be smaller than 3 inches, the space between the eyes.

A very beautiful scintillation is that of the Pleiades whose

[1] *Phil. Mag.*, **13**, 301, 1857.

[2] R. W. Wood, *Physical Optics* (New York, 1905), p. 76. For the carrying out of this interesting experiment all that is required is practice in controlling the eyes!

stars lie so close together that the mutual connection in the twinkling as a whole, makes it possible for us to distinguish the separate air striae as they pass by.

41. *How can the Scintillation of the Stars be measured?*

1. If one does not know how to measure a phenomenon one can always begin, by way of introduction, with an arbitrary qualitative scale: for a *non*-twinkling star I use the symbol 0; the strongest scintillation I have ever seen near the horizon I call 10; and the steps between these two I distinguish by the other numbers. It is remarkable how useful preliminary scales like this have been in the study of all natural sciences. One gets accustomed to the significance of each number of the scale sooner than one would expect, and there very soon comes a time when one finds a means of calibrating this qualitative scale quantitatively.

2. Another simple standard for the turbulence in the air is the altitude above the horizon at which colours disappear, or the altitude at which scintillation becomes practically imperceptible.

3. The number of changes of the light per second, determined by the rotating eye-glass, provides, too, a rough criterion for the nature of the scintillation (cf. § 40).

42. *When do Stars Scintillate most Strongly?*[1]

Strong scintillation really proves only this, that the atmosphere is not homogeneous, and that layers of different densities are intermingled. Because, however, this inhomogeneous atmosphere is usually accompanied by certain meteorological conditions, it would seem as if scintillation were a consequence of a special kind of weather.

In general scintillation increases with low barometric pressure, low temperature, intense humidity, strong curvature of the isobars and great change in pressure with altitude, and it is stronger when the wind is of normal strength than when the wind is either slight or very strong. It is, therefore, clear that atmospheric rest or motion depends on

[1] Dufour, *Phil. Mag.*, **19**, 216, 1860. Bigourdan, *C.R.*, **160**, 579ff, 1915.

so many complicated factors, that, for the present, scintillation of stars could not be made use of for weather forecasts.

It is interesting to note that scintillation becomes stronger *in the vicinity of clouds*, which proves that layers of different temperature occur there.

It is also said to increase *in the dusk*, which must either be a physiological optical illusion or a consequence of peculiar atmospheric conditions about that hour. It is even said that Northern Lights promote scintillation, but that is difficult to understand, considering the great height (60 miles) at which the Northern Lights are usually formed in the atmosphere.

Scintillation is strongest in the Northern sky, as could be explained by slightly more complicated considerations.

43. *Scintillation of Planets*

Planets scintillate far less than stars. This seems so strange because in other respects they seem quite alike to the naked eye. The cause of this difference lies in the fact that the disc of the stars on account of the tremendous distance appears as a mere point even in the largest telescopes (0.05″ at the most), while the planets show an apparent diameter, e.g. 10″ to 68″ (Venus), 31″ to 51″ (Jupiter). In the case, therefore, of planets, there will pass through any small, flat area AB, high up in the atmosphere, a cone of light rays, a few of which will enter our eye. A striation, which, as we know, deviates a light ray through only a few seconds of arc, will cause rays entering our eye to be replaced by other rays of the same cone, so that the brightness is not altered at all. We shall only notice a variation in brightness if it so happens that a bundle of rays, originally falling just beyond our eye, is now made to enter it. But the variation will be only slight, owing to the fact that there are many striae, some of which bend the rays towards our eye, while others bend them away from it. In the case of Jupiter, for instance, at 30° above the horizon, the pencil from our eye towards the planet has, at the height of 2,200 yards, a diameter of 27 to 40 inches.

We can understand now that the scintillation of a planet will become noticeable as soon as the changes in direction suffered by its light rays are of the same order of magnitude as its apparent diameter.

That is why Venus and Mercury, which at times are observed as fairly narrow crescents, do occasionally scintillate quite appreciably, and why Venus can even show changes of colour when very close to the horizon. When the disturbance in the air is very pronounced and the planets are low in the sky, one will almost invariably notice some changes of intensity.

In this way, therefore, scintillation provides us with a means of estimating the size of mere specks of light, which to the naked eye show no trace of a disc-like shape. It has even been said that in this manner it would be possible to estimate the diameter of the fixed stars, but for the time being this would seem to be too optimistic.[1]

44. *Shadow bands*[2]

Scintillation of stars is caused, therefore, by the irregular fluctuations of density in the ocean of air, at the bottom of which we inhabitants of the earth move and have our being. It is, properly speaking, the same phenomenon as the localised gathering and spreading of the sun's rays in gently undulating water (§ 23): to the fishes, the sun twinkles just as the stars do to us (Fig. 30), with only this difference, that for fluctuations in the *thickness* of the water-layers are substituted fluctuations in the *density* of the air-layers. The latter are so much less effective that we can see the scintillation of only the sharpest point-like sources of light.

In the same way as we have shown concentrations of light in clear water, so can we make air striae directly visible!

At night, in a very dark room, with only a small window opened so as to let in the light of Venus, a wispy cloudiness can be seen to pass over the smooth background formed by a

[1] It is not quite clear why red stars seem to scintillate more than white ones.
[2] Cl. Rozet, *C.R.*, **142**, 913, 1906; **146**, 325, 1906.

wall or a white cardboard screen. These are 'shadow bands.' They are to be seen clearly only when the planet is situated close to the horizon. Each time it twinkles, with only a slight increase in brilliance, a bright band is seen to pass over the screen, and conversely each decrease of brightness has a darker band to correspond (cf. Fig. 59). That which one observation shows us subjectively, is shown by another objectively. These air striae have no preferred direction and they move the same way as the wind prevailing at the time in the layer of air where they originate.

Jupiter, Mars, Sirius, Betelgeuse, Procyon, Capella, Vega and Arcturus are likewise suitable for this kind of observation, though it may be difficult on account of the intensity of the light being weaker. Air striae can be seen much better when light from a searchlight a long way off, at a distance of, for instance, 15 miles, happens to fall on a wall near to you.[1]

Very remarkable shadow bands can be seen immediately preceding or immediately following the totality of a solar eclipse, on a white wall or sheet. They remind one of the folds of a gigantic curtain. These, too, are air striae, made visible in the light of a linear light source, namely the last crescent of the sun before it disappears entirely. This causes the phenomenon to be more complicated than with a point-like source of light, each spot being drawn out into a little arc (§§ 1, 3) and the cloud-like striae seeming to consist of bands, all of which are parallel to the sun's crescent (at its brightest point). The bands are moved by the wind, but we only see the component of their motion at right angles to their own direction. Sometimes this phenomenon lasts only a few seconds, often a minute or longer. The distances of the bands give us an idea of the average thickness of air striae; 4 to 16 inches is most usually given.

However, it is not necessary to wait for total eclipses of the sun, which are very few and far between, in order to see shadow bands. We can carry out observations in the way

[1] *Nat.*, **37**, 224, 1888.

described, at sunrise (or sunset), during those brief moments when only a narrow segment of the sun stands out above the horizon. The bands are then horizontal and move up or down according to the direction of the wind. Their velocity is 1 to 8 yards per second, according to the force of the wind, the space between them is from 1 to 4 inches. They are generally visible for not longer than 3 or 4 seconds, on account of the sun's segment very soon becoming too broad.

V

The Measurement of Intensity and Brightness of Light

45. *The Stars as Sources of Light of known Intensity*

The stars form a natural series of sources of light of every intensity. By means of photometers these intensities have been measured with great accuracy and graded in a scale of magnitudes. This scale of 'magnitudes' has, however, nothing whatever to do with the actual dimensions of a star, but refers only to its brightness or luminous intensity.

m=magnitude	i=light-intensity in arbitrary measure	m	i
−1	251		
0	100	0	100
1	39·8	0·1	91
2	15·8	0·2	83
3	6·31	0·3	76
4	2·51	0·4	69
5	1·00	0·5	63
6	0·40	0·6	58
7	0·16	0·7	53
		0·8	48
		0·9	44

Each class is 2·51 times weaker than the one preceding it. Apart from a constant factor we have $i=10^{-0\cdot 4m}$.

In Fig. 61 magnitudes are given for the stars in the neighbourhood of the Great Bear, which are visible the whole year round. In Fig. 62 the magnitudes are given for the brilliant winter constellation, Orion. The following are for bright and very well-known stars:

Sirius=α Great Dog ..	−1·3	Altair=α Eagle ..	1·1
Vega=α Lyra.. ..	0·3	Aldebaran=α Bull ..	1·1
Capella=α Charioteer	0·3	Pollux=β Twins ..	1·3
Arcturus=α Boötes ..	0·2	Regulus=α Lion ..	1·6
Procyon=α Little Dog	0·6	Castor=α Twins ..	1·7

For other stars an atlas should be consulted.

Most people can observe as far as the 6th magnitude at least, on bright nights, and beyond the lights of the towns.

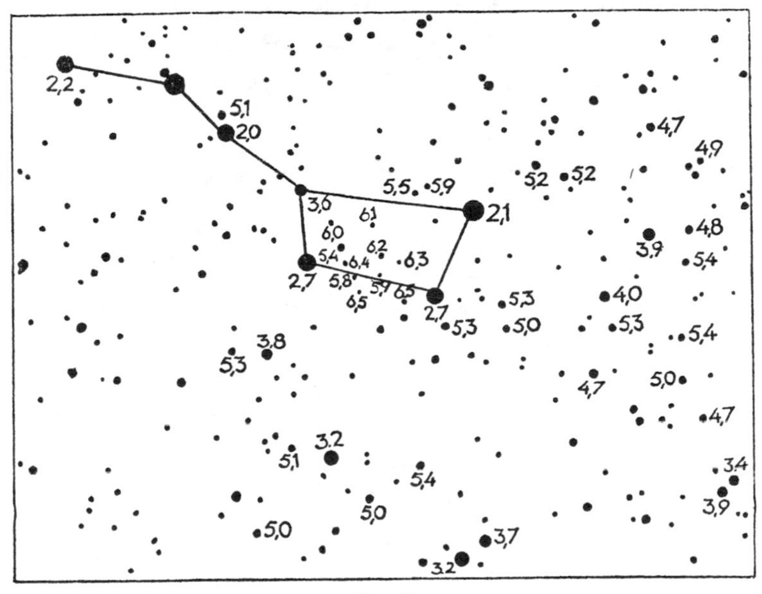

Fig. 61.

46. *Extinction of Light by the Atmosphere*

Close to the horizon we usually see only very few stars owing to the absorption of the rays of light on their way through the air. Rays running practically horizontal have traversed a much longer path than slanting rays and have, therefore, undergone a greater diminution in brightness.

We will now determine this diminution, if possible, with

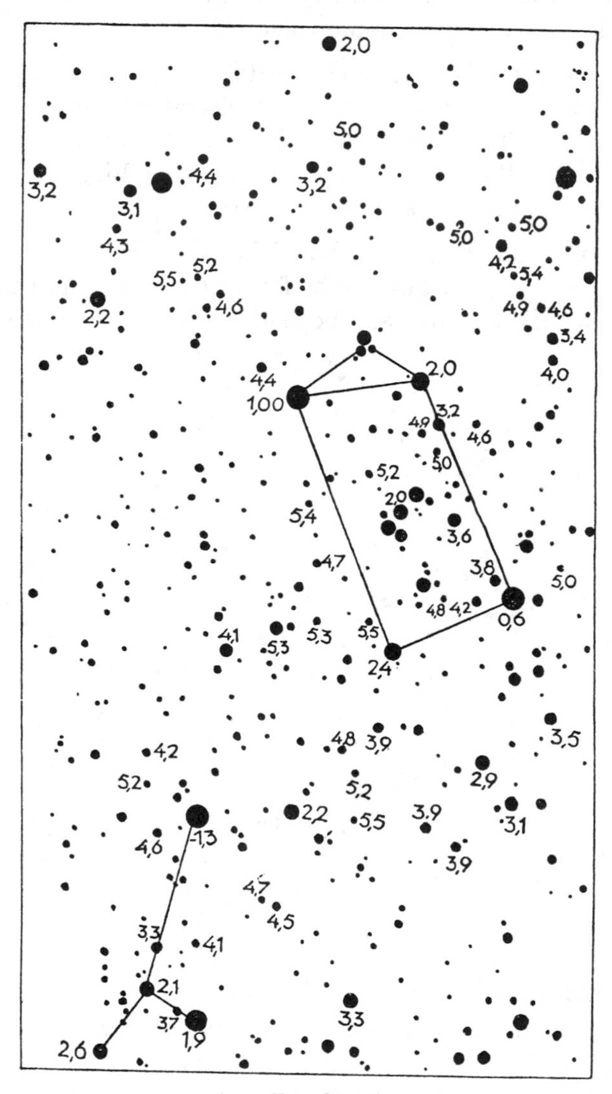
2,0
5,0
3,2
4,4
3,2
3,1
5,0
5,0
4,3
5,2
5,5
4,2
5,4
2,2
4,6
4,9
4,6
3,4
2,0
4,4
4,0
1,00
3,2
4,9
4,6
5,0
5,2
2,0
5,4
3,6
4,7
3,8
5,0
4,8
4,2
0,6
4,1
5,3
5,3
5,5
2,4
4,8
3,9
3,5
4,2
2,9
5,2
5,2
2,2
5,5
3,9
-1,3
3,1
4,6
3,9
4,7
4,5
3,3
4,1
2,1
3,7
1,9
3,3
2,6

Fig. 62.

the aid of a star map, supplemented by a list of magnitudes, though as a matter of fact, our own tables in § 45 are sufficient when Orion is low and the Great Bear high in the sky.

h	Δ	Z	sec Z
90°	0	0°	1
45°	0·09	45°	1·41
30°	0·23	60°	2·00
20°	0·45	70°	2·92
10°	0·98	80°	5·73
5°	1·67	85°	11·4
2°	3·10	88°	—

The magnitudes indicated in them refer to when the position of the stars is high in the sky. We find star A, which is not far from the horizon, and compare its brightness with that of the stars round the zenith (stars higher than 45° are practically unweakened). As far as we are able, we find stars exactly equal to A, or between which A lies. The difference between the *apparent* magnitude of A and the *true* magnitude as given in the tables is noted and called, say, Δ, the altitude of star A being determined at the same time (§ 235).

If we carry out this process for different stars at various distances, h, above the horizon (10 is sufficient to obtain a first impression), our table will resemble, more or less, that shown above.

The numbers in the second column, which represent the extinction caused by the atmosphere, are the average values for our part of the world and for a very clear sky, but they vary from place to place and still more so from night to night.

The distance from the zenith is also included, $Z=90°-h$, and sec Z, which is proportional to the length of the

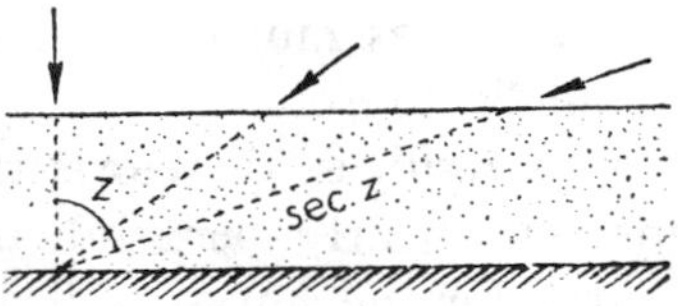

Fig. 63. The more oblique the light ray, the longer its path through the atmosphere.

path traversed by the light through the atmosphere (Fig. 63).

Now plot Δ against sec Z, and you will find a collection of points, lying more or less close to a straight line, which is drawn so as to fit the various points as well as possible (Fig. 64). From this drawing, therefore, we can conclude

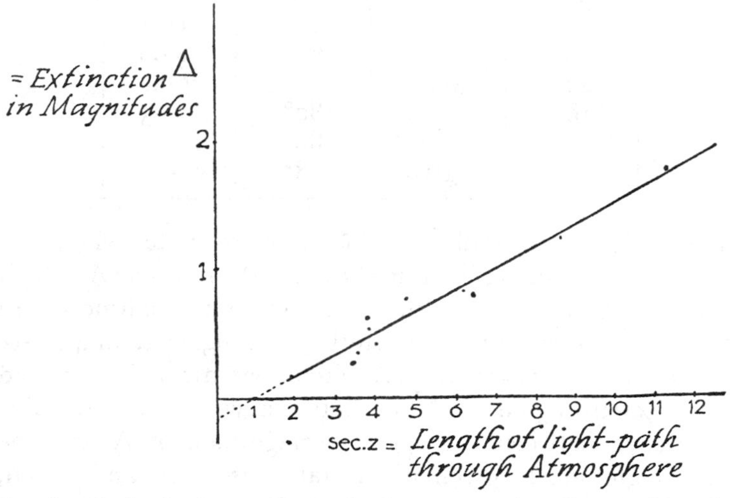

Fig. 64. Extinction in magnitudes Δ of a star at various distances from the zenith.

how many magnitudes a star is weakened as the length of the path travelled through the atmosphere increases. An extraordinarily interesting feature of this figure is that, by producing the line, one can find how much brighter the stars would seem to shine if we could rise above the atmosphere surrounding the Earth; that is, higher than the stratosphere. A star near the zenith would increase in brightness as much as 0·2 magnitudes, which means from about 83 to 100.

Our result is, therefore, that about $\frac{1}{6}$ of the light of rays incident nearly perpendicularly is extinguished and this is true for the sun as well as for the stars. This extinction is not absorption of light, but *scattering*, to which the blue colour of the sky is also due (cf. § 172).

47. *Comparing a Star with a Candle*

We choose at night-time an open stretch of country, and compare there the intensity of a candle with that of a bright star, e.g. Capella. It is surprising at what a great distance we have to stand from the candle in order to see its light reduced to the same brightness as that of the star: about 1,000 yards or 900 metres. We get, therefore, for Capella an intensity of illumination of $\frac{1}{(900)^2} = \frac{1}{810000}$ lux or 'Metre Candle.'

A pocket lamp can also be used for this experiment, but this requires still greater distances. Fix the lamp on the roof of a house, or outside the window of a high tower.

Note the difference in colour.

48. *Comparing Two Street lamps with one another*

On our evening walks we perceive that, now and then, we have two shadows, namely, whenever we happen to be between two lamps. The nearer we draw to one of them, the darker one of the shadows becomes. When both shadows are equally dark, the illumination from both lamps is equally strong; it follows from their distances, a and b, that the ratio of their intensities of illumination is $\frac{A}{B} = \frac{a^2}{b^2}$.

There is a striking difference of colour in the shadow cast by the light of incandescent and the light of electric lamps.

49. *Comparing the Moon with a Street-lamp*

Find once more two shadows, cast by these sources of light. The one opposite the moon is reddish, that opposite the lamp, dark blue (cf. § 96). We move away from the lamp, and the moon-shadow remains just as strong, whereas the lamp-shadow grows fainter. Let us assume that both shadows are equal at a distance of 20 yards from the lamp.

I estimate the strength of the lamp, which is an ordinary, not very strong, electric bulb, to be 50 candles, at 20 metres distance its intensity of illumination is $\frac{50}{20^2} = 0 \cdot 1\dot{3}$ lux.

This, then, must also be the illumination produced by the full moon, when we carried out this experiment.

Let us now repeat it during the first or third quarter of the moon. The intensity of illumination is now *far less than half* the other, for a considerable portion of the surface of the moon is darkened by the oblique shadows of the lunar mountains (cf. § 168).

The exact values are: for the full moon 0·20 lux, and for the first or third quarter, 0·02 lux.

50. *Brightness of the Moon's Disc*

When Herschel set out on his voyage to South Africa, and his boat arrived at Cape Town, he saw the moon, at the time nearly full, rise above Table Mountain which was illuminated by the setting sun. It struck him that the moon was less bright than the rocks, which made him conclude that the surface of the moon must be formed of dark rock.

A similar observation can be made in our own surroundings, if we compare the full moon, rising at about 6 o'clock in the evening, with a white wall illuminated by the setting sun. The distances between sun and moon, and sun and earth, are practically the same. If moon and wall were made of the same material their brightness would be the same, *however much their distances from our eye might differ* (a splendid application of a classical photometric theorem !). The difference observed must be ascribed to the fact that the moon consists of dark rock (volcanic ashes ?).

For this observation to be absolutely accurate, the sun and moon should be at the same height above the horizon, so that the reduction in intensity of their light by the atmosphere may be the same for both.

51. *A few Ratios of Brightness in the Landscape*[1]

The sun's brightness = 300,000 × brightness of the blue sky.

The brightness of a white cloud = 10 × brightness of the blue sky.

On a normally sunny day with a blue sky, 80 per cent. of the light comes directly from the sun, 20 per cent. from the sky.[1]

Illumination of a horizontal plane after sunset, with a cloudless sky:[2]

Sun's position:	−1°	−2°	−3°	−4°	−5°	−6°
	250	113	40	13	5	2 lux.

The eye adapts itself to every illumination intensity so well and so rapidly, that we never sufficiently realise how tremendous are the brightness ratios around us. Let us compare a landscape, illuminated by the sun at its height, with another illuminated by the moon:

the sun's disc	100,000 × 40,000	the moon's disc	20,000
pure white object	50,000	pure white object	0·25
dull black	1,000	dull black	0·005

This shows that in one and the same landscape the greatest brightness-ratio is not higher than 50 : 1, yet as regards the absolute value, the illumination changes enormously. Dull-black in the light of the sun is as much as 4,000 times as bright as white paper in moonlight.

52. *Reflecting Power*

Have you ever seen stars reflected in water? In town this is hardly possible, in the country only sometimes—in a pool or lake when there is no wind, and on dark nights, it is very striking.

Bright stars of the 1st magnitude, near the zenith, give a feeble reflection, about equal to stars of the 5th magnitude. A difference of 4 magnitudes corresponds to a ratio 40, approximately, of light intensities, so that water reflects

[1] Nela Research Lab., Postma, *Daylight Measurements in Utrecht* (Diss., 1936).
[2] K. Kähler, *Met. Zs.*, **44**, 212, 1927.

only 2·5 per cent. of the vertically incident rays. Stars situated lower are better reflected.

The reflecting power is connected with the index of refraction by Fresnel's formula. For perpendicular incidence it is

$$\left(\frac{n-1}{n+1}\right)^2$$

In the following table, the values of the reflecting power of glass and water are given for various angles of incidence.

Angle of Incidence	Reflecting Power	
	of water	of glass ($n=1·52$)
0	0·020	0·043
10	0·020	0·043
20	0·021	0·043
30	0·022	0·043
40	0·024	0·049
50	0·034	0·061
60	0·060	0·091
70	0·135	0·175
75	0·220	0·257
80	0·350	0·388
85	0·580	0·615
90	1·000	1·000

We now understand why we can never see stars reflected in a town; the sky is not dark enough, stars of the 3rd magnitude are hardly visible, and, moreover, the surface of the water is illuminated too much. Only planets are visible at all by reflection and then only when they are much brighter than the 1st magnitude.

The brightness by day of the reflected blue sky, the houses, and the trees, seems much greater than 2 per cent.; in some paintings the difference between the brightness of the objects, and of their reflection can hardly be seen. This is simply an optical illusion.

There is a superstition that stars are never reflected in *deep* waters. This is, of course, without any foundation whatever.

A pane of glass reflects 0·043 of the light at each surface, that is, 0·086 altogether. In many small glass buildings, such as telephone-booths, etc., lighted by a hanging electric bulb, reflections can be seen repeated between two parallel and opposite windows; an ordinary ground-glass bulb gives as many as four visible reflections at each side. The first is formed by once reflected light, the second by three times, the third by five times, and the fourth by seven times reflected rays. The brightness of the last reflection is, therefore, only $(0{\cdot}086)^7$ times, that is less than one ten-millionth

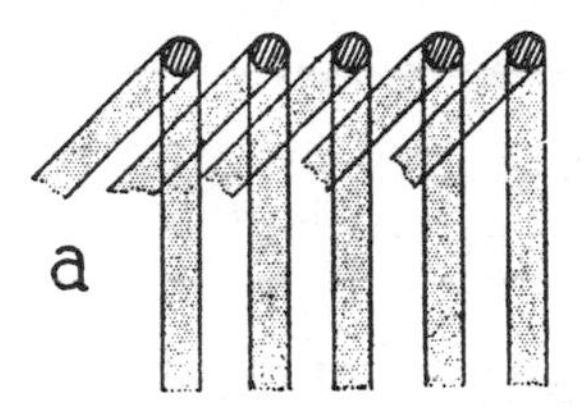

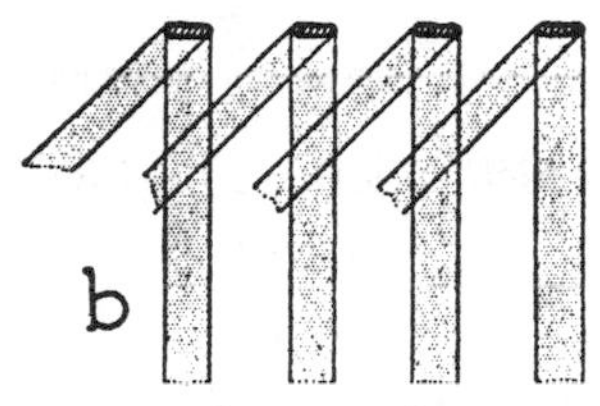

FIG. 65. Light intercepted by wire-netting seen from two directions: (*a*) when the wires have a round cross section and (*b*) when the netting consists of flat strips.

part of the original brightness. This very simple computation is a very good example of the tremendously wide range of brightnesses to which our eye can react.

53. *Transmission through Wire Netting*

Illuminated advertising signs on roofs are often fixed on wire netting attached to metal frames.

At a distance the separate wires are no longer distinguishable, and the netting resembles a sheet of uniformly grey glass. It is interesting to look at the netting at an increasingly acute angle, and to see how it gets darker and darker against the sky. This proves that the wires of which it is made have a *round* cross-section, for if it were built up of small, flat bands it would remain equally dark at every angle (Fig. 65).

54. *The Degree of Opaqueness of a Wood*

Looking through a narrow strip of woodland, we can see between the tree-trunks the light sky beyond. There must obviously be some expression indicating what part of the light is allowed to pass unobstructed, if we assume that the distribution of the trees is accidental, that N trees occur per square yard, and that they have a diameter D at a height on a level with our eyes.

Let us consider a beam of light of a width b, which has already travelled a distance l through the wood (Fig. 66). Let i be the amount of light that still remains of the original amount i_0. When the rays of light penetrate a little further over a short distance dl, an amount of light di is removed; we have

$$\frac{di}{i} = -\frac{\mathrm{N}\,\mathrm{D}\,b\,dl}{b} = -\,dl\,\mathrm{N}\mathrm{D}$$

from which it follows by integration that

$$i = i_0 \,.\, e^{-\mathrm{ND}l} = i_0 \,.\, 10^{-0{\cdot}43\,\mathrm{ND}l}$$

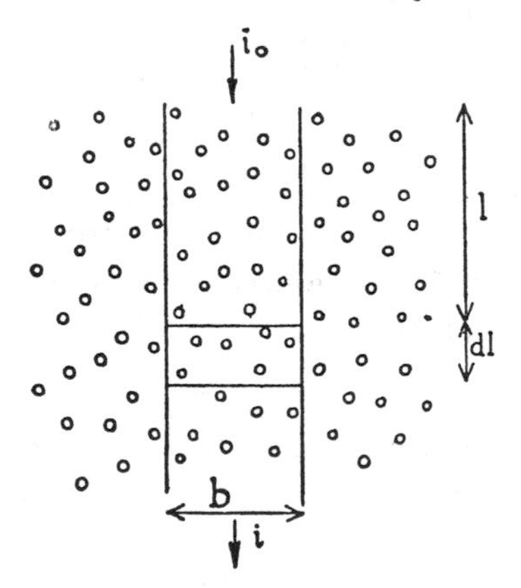

Fig. 66. How to calculate how much light can be seen between the tree-trunks in a wood.

The amount of light let through will, therefore, become less and less according as the wood is more extended in the direction of incidence of the light, in exactly the same way as the light transmitted by a dark liquid is less, according as the layer increases in thickness. Let us put, in the case of a wood of fir-trees, N=1 per sq. yard and D=0·10 yard. We obtain then approximately:

l=10 yards .... $\frac{i}{i_0}$ =0·37

l=25 yards .... =0·10

l=50 yards .... =0·01

l=70 yards .... =0·001

The rapidity of the increase in opacity is very striking. From a rough estimate of the fraction of the horizon which is as yet not intercepted by the trees, one can gather the depth of the wood.

How much is ND in the case of a beech wood, and in the case of a wood of young and of full-grown fir-trees?

55. *Beats between Two Sets of Railings* (Plate VII, *a*)

Whenever one can see the posts of one set of railings between the posts of another set, one perceives broad light and dark bands in the intensity of the light, which move when one moves. These are due to the fact that the apparent distance between the posts of the two sets of railings differs more or less, either because the one has wider spaces than the other, or because they are at different distances from our eye. In certain directions the posts seem to coincide, and in others the posts of the first railing fill exactly the space between the posts of the second, so that a difference arises in the average brightness. We can say that they are 'in step' or 'out of step.'

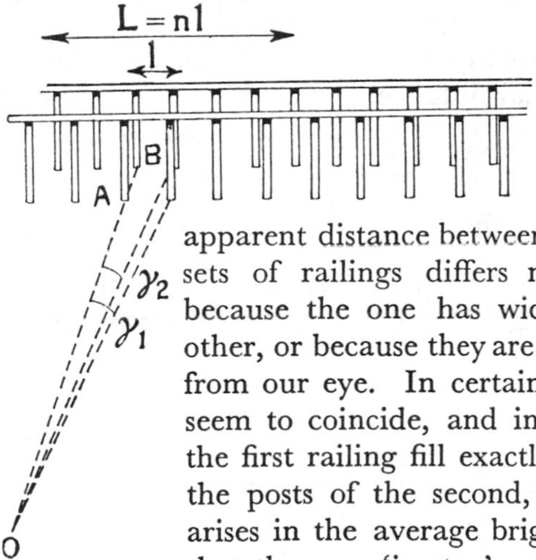

FIG. 67. Beats between two railings.

When one has once noticed these beats, one sees them in all sorts of places. Every bridge with a parapet in the form of a railing on both sides shows these undulations in intensity when seen from a certain distance. They appear, too, when one sees the shadow of a railing between its own posts, in which case the period is the same, but the distance to our eye is different.

In some stations a goods-lift is surrounded by wire-netting, and the combination of the side nearest to us and the side farthest away forms a kind of *moiré*, such as one sees when one lays two pieces of wire-gauze on one another, or two combs with unequal distances between the teeth.

Let us consider in more detail the simple case (Fig. 67) of two equal sets of railings seen at the unequal distances x_1=OA and x_2=OB from our eye; let l be the distance between two successive posts, which subtends the angles $\gamma_1=\frac{l}{x_1}$ and $\gamma_2=\frac{l}{x_2}$ at our eye. The length of a beat will contain n posts, where n is given by $n=\frac{\gamma_1}{\gamma_1-\gamma_2}=\frac{x_2}{x_2-x_1}$; that is, they increase in number as we move away from the railings. On the other hand, the angular distance θ covered by a beat, as seen by us, will remain the same, for $\theta=n\gamma_2=\frac{l}{x_2-x_1}$. We can determine the true length $L=nl=\frac{lx_2}{x_2-x_1}$ of a beat, by moving parallel to the railings; the beats will move with the same speed as we are moving. Now measure the distance you must

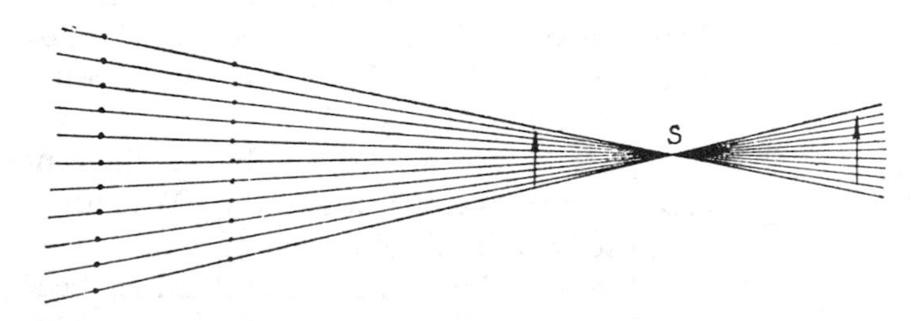

FIG. 68. Beats between two sets of railings of different periods.

walk so as to see a beat occupy exactly the same place as the one before. Test the validity of the various formulae. Or, conversely, on determining n, θ, and L, one can solve for x_2, x_2-x_1 and l. In this way it is possible to obtain all dimensions of the railings at a distance without any further means.

If the periods of the two sets are different, the beats will be seen to move in the most remarkable way, whenever our eye moves; now in the same direction as we are moving, now in the opposite one, according as we ourselves are in

(*a*) Beats between the boards on opposite walls of a shed

(*b*) The pole of a man punting appears to him 'broken,' and the bottom of the river 'lifted.' *From 'The Universe of Light' (G. Bell & Sons, Ltd.) by permission of Sir William Bragg, O.M.*

front of or behind the radiating point S (Fig. 68); in other words, according as $\gamma_1 < \gamma_2$ or $\gamma_1 > \gamma_2$. The beats will move faster and faster as we approach S.

When a vertical fence casts its shadow on level ground, the beats look somewhat different (Fig. 69); at the top they lie closer together than at the bottom, and also a slight curvature is noticeable. But this is in accordance with the above considerations, for the distance between the two interfering grid-like systems is greatest at the top. Therefore, the

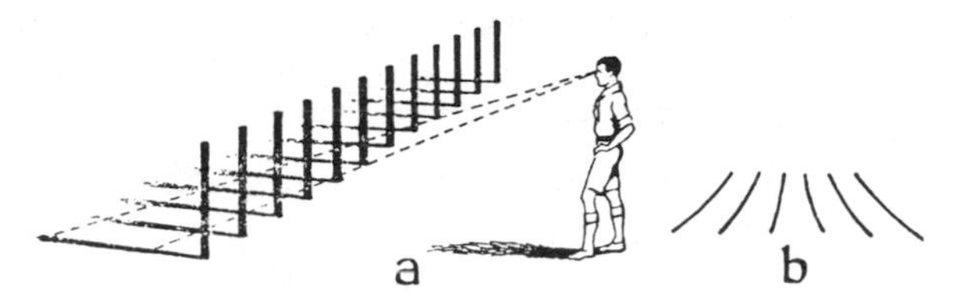

FIG. 69. Beats between a railing and its shadow. (*a*) Conditions during observation; (*b*) shape of the beat-waves.

angular distances seen by our eye between successive bars differ considerably, which means that the beats lie close together. At the bottom it is just the other way round.

56. *Photographic Photometry*[1]

Every photographic shop sells 'daylight paper,' which turns brown-purple very quickly in the sun. Roughly speaking, the time required by the paper to acquire a certain colour is inversely proportional to the intensity of light falling on the paper (the law of Bunsen and Roscoe). If, therefore, one always uses the same kind of paper, and has chosen one piece of normal brown-purple paper as the 'normal tint,' it will be easy to determine light intensities anywhere, simply by noting the length of time required for sensitive paper to acquire this normal tint. The normal paper must be exposed to the light as little as possible or it will fade.

The normal tint should be chosen with great care. We hold a strip of daylight paper in the sun and cover up little

[1] J. Wiesner, *Der Lichtgenuss der Pflanzen* (Leipzig, 1907).

steps or bars which have successively undergone exposures of respectively 10, 20, 40, 80, 160, 320 and 640 seconds. We examine this strip of paper in a dim light, and find that the first and the last steps show but little contrast, and the ones in the middle are the best. Select as norm a sheet of paper (a book-cover or a poster), very uniform in colour and as far as possible of exactly the same tint as one of the middle steps. Should the tint not be absolutely the same, while making a comparison, you must pay more attention to the *brightness*, judging it with half-closed eyes. Remember that our day-light paper need not be developed nor fixed, and the strips, once used, cannot be kept.

In this way Wiesner carried out a number of investigations in connection with the 'light-climate' required for the development of various plants. This method may be rough and ready, but it is an excellent way of estimating in many places and under different circumstances a value of which we had not the slightest idea.

Study the illumination of a horizontal plane when the sun is at various heights.

Compare the light on a horizontal plane when the sun is shining: (*a*) when a shadow is thrown on it by a screen and (*b*) without a screen, and compare in this way the light coming directly from the sun and that emitted by the blue sky.

Compare the illumination on the upper and under sides of a horizontal piece of paper. The ratio above water is 6, above gravel 12, above grass 25.

Compare the luminosity of the blue sky in various directions by fixing photographic paper at the bottom of equal-sized tubes pointing at various angles. Usually the sky turns out to be at its darkest 90° away from the sun (cf. § 176).

Compare the light inside and outside a wood. ('Outside' means at least 7 yards from its edge.)

Compare the illumination in a beech wood (*a*) in the middle of April; (*b*) while the first leaves are unfolding; (*c*) at the beginning of June. In one case the illumination was found to be $\frac{1}{11}$, $\frac{1}{30}$, $\frac{1}{64}$ respectively of the illumination outside the wood.

Measure the intensity of the illumination in the places where the following plants grow:

Greater plantain (*Plantago major*)	1
Ivy (*Hedera helix*)	1 to 0·22 (when in bloom) 1 to 0·02 (with sterile branches)
Heather (*Calluna vulgaris*) ..	1 to 0·10
Bracken (*Pteridum aquilinum*)	as low as 0·02

Determine the intensity of illumination inside densely growing crowns of trees, which is about the smallest illumination in which twigs can develop at all. For solitary trees the following values were found: larch, 0·20; birch, 0·11; ordinary pine, 0·10; fir, 0·03; beech, 0·01 (expressed as fractions of the illumination outside the tree).

VI

The Eye[1]

The study of Nature must necessarily involve the study of the human senses as well. In order to be accurate in our observations of light and colour in the landscape, we must first of all be familiar with the instrument we use continually—namely, the human eye. It is very enlightening to learn to distinguish between what Nature actually shows us and what our organ of vision adds to or subtracts from it. And no surroundings are more favourable for studying the peculiarities of the eye than those out of doors, to which we are adapted by Nature.

57. *Seeing under Water*

Have you ever tried to keep your eyes open under water? A little courage, and it will be easy enough! But every object we look at now is extraordinarily indistinct and hazy, even in a swimming-bath with very clear water. For in the air it is the outer surface of the eye, the cornea, which collects the rays of light and causes the formation of the images on the retina, helped only slightly by the crystalline lens. But under water the action of the cornea is neutralised owing to the fact that the refractive indices of the fluid in the eye and in the water outside it are nearly equal, so that the rays go straight on at the bounding surface of the cornea (Fig. 70). We have now an excellent means of judging how insufficient the working of the crystalline lens would be, if that alone were responsible for the formation of images. We are at this moment so hopelessly far-sighted

[1] When reading this and the three following chapters it is advisable to refer to the classic work of Helmholtz, *Physiologische Optik* (second or preferably third edition).

that focussing is practically of no use, and a point of light remains equally hazy at whatever distance it may be. The only possible way to distinguish an object at all is to hold it so close to our eye that it subtends a large angle there, while the unavoidable haziness of the outlines is not too much of a disadvantage.

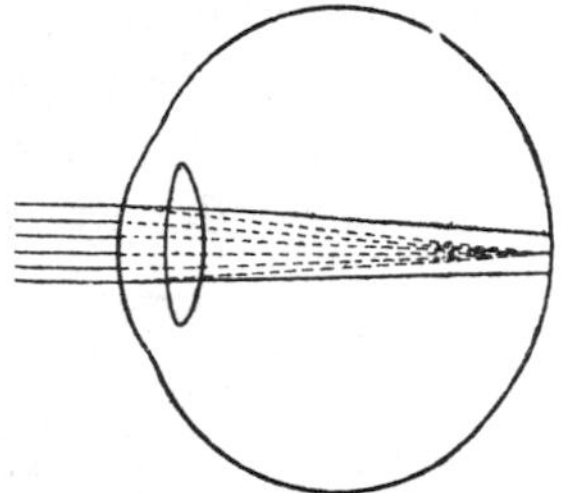

FIG. 70. When seeing under water there is no formation of images in our eye. *Thick lines:* the path of light rays when seeing under water. *Dotted lines:* light rays when seeing in the air.

In clear water a farthing becomes visible at arm's length (25 inches), and a piece of iron wire is not visible at any distance at all. On the other hand, anyone swimming past can be observed as far as 10 yards off, for such a large object is bound to be noticed. Roughly speaking, the presence of an object of length v, is observable at a distance of $30v$ at the most; its shape can, more or less, be ascertained at a distance of $5v$, and one can only speak of seeing it properly when it has come within a distance equal to its own dimensions.

In order to make our power of vision at all normal, we shall require a pair of very strong spectacles, but unfortunately spectacles are four times less effective under water than in air. And to make matters worse, such powerful glasses lose their full effect when held a few millimetres from the eye! Taking all this into consideration it will be necessary to use a lens of strength 100, that is, with a focal length of $\frac{1}{2}$ inch! The lens of a thread-counter as used for examining linen would be suitable.

Notice how difficult it is to estimate distances, as well with water-spectacles as without them. The objects look shadowy and almost ghostly.

One should also look upwards while submerged in water. Rays of light from outside when entering the water make an angle with the vertical smaller than 45° at most, so you will see a large disc of light above your head, and if you look sideways the ray from your eye will be completely reflected at the surface, and the mirrored image of the feebly illuminated ground will be all it shows (Fig. 71). This is what the world looks like to fishes!

One way of obtaining a very good impression of the view from under water is to hold a slanting mirror beneath

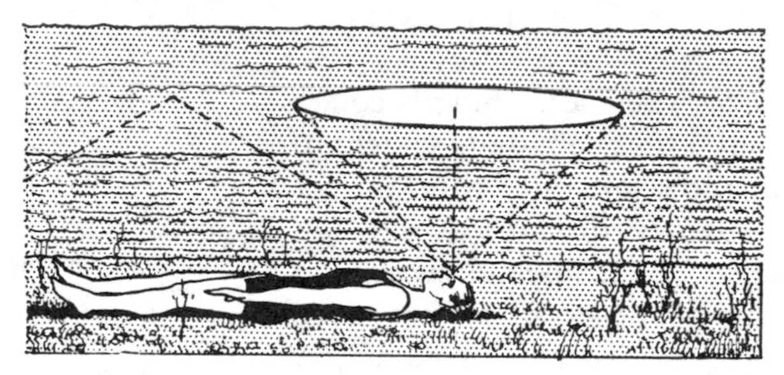

FIG. 71. We look at the view for a moment in the same way as fishes do!

the surface, standing upright in the water and taking special care not to cause ripples. Note then how all the objects out of the water appear to be strongly compressed in a vertical direction, the more so the nearer they are to the horizon, and how everything has a beautiful fringe of colour.

58. *How the Interior Part of Our Eye may be made Visible*

A practised observer can see the yellow spot of his own eye (the central, most sensitive point of the retina), surrounded by a darker ring in which there are no blood-vessels.[1] In the evening, when you have already been outside for a time, you should look at the vast, cloudless sky just when the first stars are making their appearance. Close your eyes for a few seconds, and open them again quickly, looking in the direction of the sky. The darkness will disappear first of all at the circumference of the field of vision, and contract rapidly towards the centre where the yellow spot, with its dark edge, becomes just visible and sometimes lights up for a moment.

[1] H. von Helmholtz, *Physiologische Optik*. Personally I could not succeed in carrying out this experiment.

If you walk beside a high fence, with the bright sun shining through it, the sunlight will flash several times a second in your eyes. If you keep looking straight in front of you, and don't turn your eyes in the direction of the sun, you will be surprised to see that each flash of light is accompanied by an indistinct figure of irregular spots and network, and side-lines, bright on a dark background.[1] It is possible that these are certain parts of the retina made visible to us by this unusual illumination.

59. *Imperfect Images formed by the Eye*

Stars do not appear to us as perfect spots, but as small irregularly shaped figures, often as a point of light from which rays diverge. The usual representation of five rays is not according to reality. For this experiment the brightest stars of all should be taken, preferably Sirius, or, better still, the planets Venus or Jupiter, because the disc they show us is so small that it is practically a point, and their brightness exceeds that of the brightest stars.

Hold your head on one side, first towards the right, then towards the left, and the shape slopes accordingly. This is different for each person, and also for each of his eyes, but if you cover one eye with your hand, and look with the other eye at various stars, you will always see the same shape.

This shows that it is not the stars that look so irregular, but our eyes which are at fault and do not reproduce a point exactly as a point.

The shape of the rays becomes larger and more irregular when the eye is in dark surroundings and the pupil is wide open. It becomes smaller in well-lighted surroundings, when the pupil is contracted to a small hole. And indeed Gullstrand has proved that the crystalline lens of our eye is distorted mostly at the edges by the muscle to which it is attached, so that the distinctness of the images diminishes when light passes near these edges.

Take a sheet of paper and prick a hole in it 1 millimetre in

[1] This observation probably agrees with the light-shadow figure of Purkinje (Helmholtz, *Physiologische Optik*).

diameter and hold this before the pupil. After searching for a while you will be sure to find Sirius or some planet, and you will see that the image is perfectly round. Now move this aperture to the edge of the pupil, and the point of light becomes irregularly distorted; in my case it stretches out into a line of light in the direction of the radius of the pupil.

FIG. 72. A star or a distant lamp seen by a short-sighted person, without glasses.

Many persons see the cusps of the moon's crescent multiplied. These deviations from the distinct image are to be ascribed mainly to small deformations of the surface of the cornea. Similar deformations appear to anyone, who is short-sighted, when he removes his spectacles (Fig. 72): every lamp in the distance becomes a disc of light, in which, however, the brightness is very unevenly distributed. Should it happen to be raining you will see, now and then, a small round spot appear suddenly in the little light disc; a portion of the cornea is covered by a raindrop (Fig. 73). You will see that it keeps its shape for a good ten seconds, that is, of course, if you can refrain from blinking for so long!

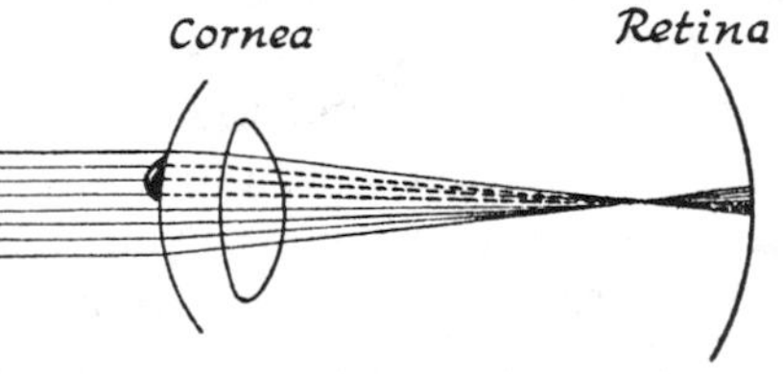

FIG. 73. A myopic eye without glasses sees distant lamps as small irregular discs; a drop of rain on the cornea is projected in the shape of a dark spot.

When the glaring lights of a motorcar a long way off shine towards us, we can see the entire field of view round the intense point of light covered by a haze of light, which is speckled, and sometimes striped radially. This structure is caused by diffraction or refraction of light at a great number of irregularities in the eye. Sodium lamps in the shape of long, narrow tubes also give a diffuse glow round the light-source, but this shows fine hatching, the lines of which run exactly parallel to the source of light, for each diffracting grain has produced a line of light instead of a point.

60. *Bundles of Rays which Appear to be Emitted by Bright Sources*

Distant lamps seem at times to cast long straight rays towards our eye, especially when we look at them with half-closed eyelids; along the edge of each eyelid lachrymal moisture forms a small meniscus by which the light rays are refracted.[1] Fig. 74, *a* shows that the rays are refracted at the upper eyelid in such a way that they seem to come from below; the source gets a downward tail, and the lower eyelid gives, in the same way, an upward tail. The formation of these tails can be followed very well by holding one

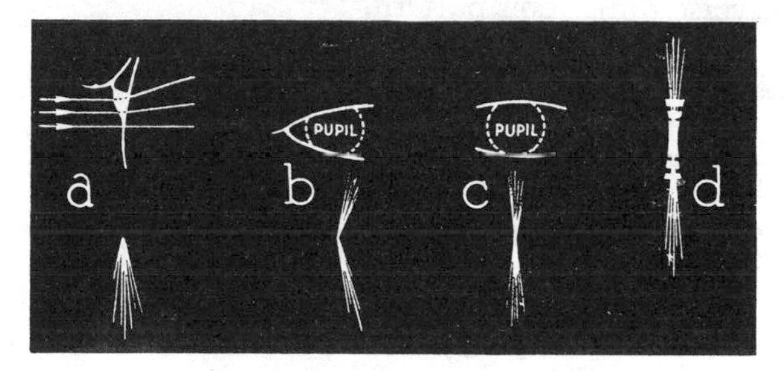

Fig. 74. How light rays arise round lamps in the distance.

eyelid fast, and closing the other slowly, or by holding the head up or down, while half-closing the eyes. The rays appear at the very moment the eyelid begins to cover the pupil; to a short-sighted observer this is easily visible, for the source of light, which he sees as a broadened disc, is at that moment partly screened off.

The rays are not quite parallel, nor are those from one eye. Look at a source of light in front of you, turn your head a little to the right and then move your eyes until you see the source again. The rays are now slanting (Fig. 74, *b*). The reason for this is, apparently, that the edges of the eyelids where they cross the pupil are no longer horizontal, and each bundle of rays is at right angles to the edge of the eyelid that causes it; the observed directions fit in exactly with this explanation. We can understand now why the

[1] H. Meyer, *Pogg. Ann.*, **89**, 429, 1853.

rays are not parallel when we look straight in front of us, for the curvature of the eyelids is already perceptible even within the breadth of the pupil. Hold your finger against the right-hand edge of the pupil, and the left-hand rays of the bundle will disappear exactly as they should do.

Besides the long tails (Fig. 74, *c*) there are short very luminous ones, caused by the reflection against the edges of the eyelids (Fig. 74, *d*). Convince yourself by experiment that this time it is the upper eyelid that causes the short upper tail, and *vice versâ*. These reflected rays usually show a transverse diffraction pattern.

61. *Phenomena caused by Eyeglasses*

Lines are distorted by ordinary eyeglasses when one looks through them slantwise. This distortion is 'barrel' when the glasses are concave and 'pincushion' when they are convex (Fig. 75). If one wishes to judge whether a line in a landscape is perfectly straight or vertical this deformation is particularly annoying. Astigmatism arises at the outer boundaries of the field so pronouncedly that every kind of minor detail becomes obliterated. These faults in the formation of images are more marked according as the glasses are more concave or more convex. In the case of meniscus glasses they are much slighter.

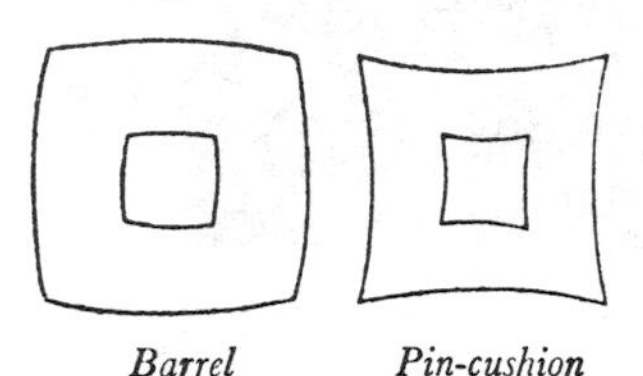

FIG. 75. Formation of images through eyeglasses.

Anyone looking through his glasses at a lighted lamp in the evening will see somewhere in its vicinity a floating disc of light. It is not very distinct, and if we keep staring at it, the accommodation of our eye changes automatically and we see the disc grow or diminish. If we remove our glasses and hold them away from our eyes a little way, we see the disc change into a point of light which, apparently, is a much reduced image of the lamp itself. If you look at a group of three lamps you will see that the image is upright. This is explained as follows: the light disc is due to a double reflection on the surfaces of the glasses or at the cornea of the eye. Actually three discs should be seen,

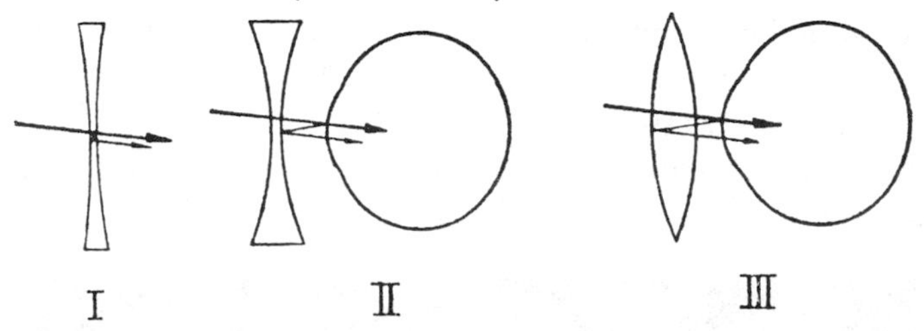

FIG. 76. How double reflections arise when one looks through a pair of eyeglasses. I Weak glasses. II Concave glasses stronger than −5. III Convex glasses stronger than +3.

but one can only see these when they are not too indistinct. Practically only one kind of double reflection occurs with a given pair of spectacles (Fig. 76).

Unframed glasses, if their edges are bevelled, sometimes show a narrow spectrum along their edge (Fig. 77), due to distant lamps.

For raindrops on eyeglasses, *see* § 118.

62. *Visual Acuity*

A normal eye has no difficulty at all in distinguishing Mizar and Alcor in the Great Bear, separated by approximately 12 minutes of arc (Figs. 61, 78A). The question now is how much farther this acuity will enable one to go. People with sharp eyes can distinguish points separated by half this amount, as in the double star α *Capricorni* (The Goat)—separation of components 6′, magnitudes 3·8 and 4·5.

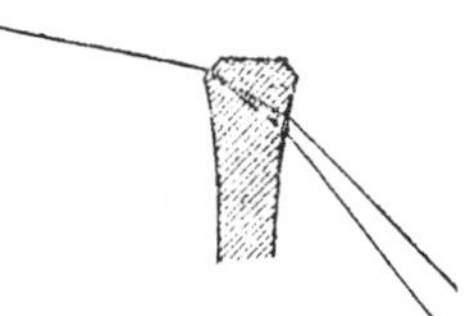

FIG. 77. How a spectrum is formed by eyeglasses.

Only a very few people can resolve separations of 4′ and 3′.

α *Librae* (The Scales)—separation of components 4′, magnitudes 2·8 and 5·3.

ε *Lyrae* (the Lyre)—separation of the components 3′, magnitudes 5·3 and 6·3.

Exceptionally good observers, of which there are very few, can distinguish an incredible amount of detail when the sky is bright and the atmosphere calm. One of them asserts that with the naked eye he can see α of the Scales as a twin star (distance nearly 4′). Saturn is, to him, distinctly oblong, and Venus crescent-shaped, at favourable moments, if he looks at it through a smoked glass or through a cloud of smoke that happens to be of the right transparency. He is

FIG. 78A. Some widely spaced double stars.

even able to see two of the satellites of Jupiter, though only in the dusk when the stars of the first and second magnitude are beginning to appear.

Dusk is the best time, too, for other observations: for example, the characteristics of the moon's surface are then much more clearly visible than at night, as one is less dazzled.

It is rather good fun to observe the narrow crescent of the moon as soon as possible after the new moon, and it has been done by some within fifteen hours only. It is of course very essential to know where to look. In our own country the sun must be at least 12° under the horizon.

63. *Sensitivity of the Direct and Peripheral Fields of Vision*

Which are the feeblest stars perceptible by you? Look at the square of the Great Bear and compare it with our Fig. 61. Most people can see the stars of the sixth magnitude, some those of the seventh magnitude. All such observations should take place far from city lights and with a clear sky.

We will now try to find out which stars remain visible if

you look at them directly and intently. A certain amount of will-power is required not to turn away your gaze, but to keep it directed exactly on the star. You will notice, to your surprise, that each weak star disappears as soon as you stare at it intently, but should your gaze move ever so slightly away from it, the star reappears ! To me personally, even the stars of the fourth magnitude become invisible, whereas those of the third magnitude remain visible (see here Figs.

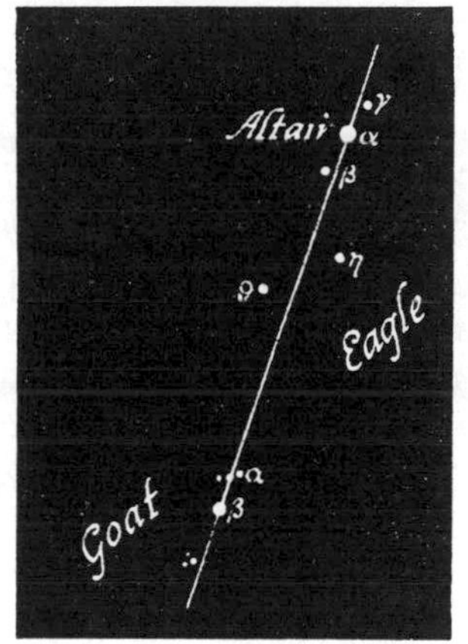

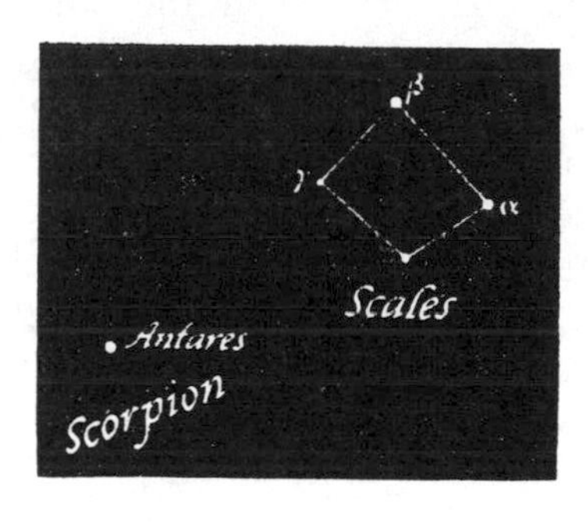

Fig. 78b. Some more double stars.

61, 62). There must, therefore, be a difference of as much as 3 magnitudes between the threshold value for the yellow spot and that for the surrounding retina, which corresponds to a factor 16 in the light-intensity ! This difference in sensitivity is to be ascribed to the fact that the central part of the yellow spot consists almost wholly of microscopic cells in the shape of small cones, and the peripheral retina of small rods which are much more sensitive. Even experienced observers will be amazed at the magnitude of this effect, so accustomed are we to letting our gaze wander unconsciously away from the star in order to see it better.

It is well worth while to see if one can follow a bright star or planet (e.g. Venus) at early dawn. As the sky grows lighter it becomes more and more difficult to distinguish the point of light, and a peculiar thing is that often one cannot see it, simply because one does not look in the right direction, though it is perfectly perceptible when one has once found it again. One has a similar experience when looking for a lark singing in the blue sky.

If one looks carefully one can often follow Venus until it is broad daylight, and see her the whole day. Sometimes the same thing can be done with Jupiter, but it is much more difficult, and it is exceptional to be able to observe him up to the time the sun reaches an altitude of 10°. Mars may be seen when the sun is low. These observations should be made especially when the planet is near the moon which is an excellent guide in finding the faint luminous points in the vast blue sky. Do not these observations contradict our deduction made from our experiments with stars, that the yellow spot has the lower sensitivity? By no means, for the rods only come into play in very feeble light, and are out of action during the day. By day the small groove of the yellow spot is most sensitive, while at night the outer parts are.

64. *Fechner's Experiment*

On a day when the clouds are light and hazy we will choose for our experiment a cloud that is only just perceptible against the background of the sky. Hold a piece of smoked glass or a uniformly fogged photographic film before your eyes, and you will see that that same little cloud is still only just distinguishable.

This led Fechner to conclude that the eye can distinguish two brightnesses if their *ratio* (not the *difference* between them) amounts to a definite and constant amount (the one about 5 per cent. greater than the other).

Repeat the experiment with *very* dark glass: the cloud is no longer visible and all the fine degrees of light have vanished. This shows that the fraction that was only just distinguishable is not absolutely constant.

A counterpart of Fechner's experiment is the daily disappearance of the stars. The *difference* in brightness of the star in regard to its surroundings is always the same, but the *ratio* in the daytime differs greatly from that at night. As a rule one can say that our visual impressions are mainly determined by the brightness ratios. This feature of our sense of vision is of the utmost importance for our daily life. Thanks to this the objects around us remain definite, recognisable things, even under changing conditions of illumination.

65. *Landscape by Moonlight*

If Fechner's law were rigorously valid and the eye could appreciate only ratios of intensity, a landscape by moonlight would convey an impression differing in no way from that of the same landscape in sunlight, for though all the light-intensities are thousands of times less, objects are illuminated in the same manner and by a light-source of practically the same shape and position.

It is clear from this that Fechner's law no longer holds when the brightnesses are very small. You should observe a landscape in the moonlight and note, especially, the differences compared to the illumination in the daytime. The main characteristic is that *all the parts not fully illuminated by the light of the moon, are almost uniformly dark*, whereas, in the daytime, various degrees of brightness are noticeable in these parts. This explains why if an underexposed photographic negative of a landscape in sunlight is printed too dark, the print looks like a landscape by moonlight. In a similar manner, painters suggest a nocturnal landscape by painting almost everything equally dark, which owing to the weakening of the contrasts gives us unconsciously the impression that the lighting must indeed be very feeble. *See* also § 77.

66. *Landscape in Brilliant Sunlight*

The brightnesses on a day in summer, at the seaside for instance, are so intense as almost to dazzle one. Here also the ratios seem smaller than in average illumination, for everything seems equally glaring in the blazing sunshine. This effect is frequently made use of by painters (cf. § 65).

67. *Threshold Value for the Observation of Brightness Ratios*

Windows reflect the sunlight and cast patches of light on the pavement in the road (§ 8). If the sun is also shining on the pavement, these patches are hardly visible, the ground being not quite even. But the light patch becomes noticeable at once whenever a window is moved, however little, or when our own shadow as we pass by glides like a film over the patch. (Is this not a remarkable psychological peculiarity? Our eye must certainly possess some special ability for noticing weak light phenomena which move.) A sheet of glass reflects 4 per cent. from each of its two surfaces, that is a total of 8 per cent.; if the incidence is oblique it is slightly more (§ 52). An increase of 10 per cent, in brightness is, therefore, evidently the threshold value of what can be distinguished by our eye in ordinary circumstances and without special measures.

We see a small pool of water in front of a sunlit wall, and expect to see a patch of reflected sunlight on the wall. But though, when the water is ruffled by the wind, lines of light run over the wall (§ 8) the patch of light itself is hardly visible, unless the wall has an exceptionally smooth surface. An increase in brightness of 3 per cent. is, therefore, observable only in very favourable circumstances (§ 87).

Stand some evening between two lamps, so close to one that your shadow cast by the other, just disappears. You can deduce the ratio of the lighting from your distance from both lamps, and therefore also the percentage difference in brightness necessary to make the shadow just discernible (§ 48).

68. *Veil Effect*

We go for a walk in the daytime. How is it that a transparent muslin curtain prevents us from seeing what is happening inside the rooms of houses? The veil-like curtain is strongly illuminated, and if the objects in the room have only a small percentage of that brightness they add to the uniform brightness of the veil a fraction too small to be perceived. This is, as you see, an application of Fechner's law (§ 64).

At night, when there is a light in the room, you can see through the curtain quite well. The side nearest to us is practically unilluminated, and only imposes a very feeble illumination on the objects of various brightness in the room.

For those who are in the room and look outside, the effect is in both cases the other way round. The same phenomenon occurs when an aeroplane, clearly visible in the moonlight, is no longer to be found when a searchlight is used! The air between our eye and the aeroplane is illuminated by a beam of dazzling light which prevents us from seeing the weak light-contrasts behind it.

69. *Visibility of Stars by Day*

We will begin by taking special note of the visibility of stars at night. The darker the sky, the more stars we can see; on the other hand, when the light of the moon spreads a uniform brightness over the heavens most of them disappear. This too is a real curtain effect (§ 68).

> Around the moon, so beautiful and full,
> The stars conceal again their radiance,
> While she spreads her silvery light
> Far and wide over the Earth.
>
> SAPPHO.[1]

A child once thought that a cloud before the moon would be sufficient to make the stars visible again. Why is this not the case? (Fig. 79).

Observations of those stars that are only just visible enable one to draw curves showing how the brightness in the sky near the moon is distributed.

[1] Note the excellent observation of detail in 'around the moon,' 'conceal again their radiance,' for when the moon rises, the stars that before were visible, disappear.

By day the sky is still more brightly illuminated, and the stars are then completely invisible. Moreover, our eye has already become adapted to the broad daylight, and is therefore thousands of times less sensitive.

A remarkable account, dating as far back as the time of Aristotle, tells us that, seen from the inside of deep wells, mine-shafts and wide chimneys, the air seems darker than we usually see it, and it should even be possible to observe some of the brighter stars. This phenomenon has since been mentioned by a number of writers, who relied, however, mostly on their memories or on the stories of others.

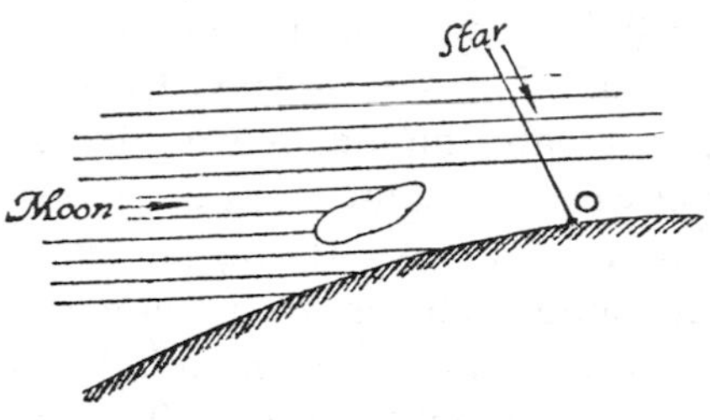

FIG. 79. A cloud in front of the moon is not sufficient to make the stars visible to the observer O.

Nowadays, there is not a single place where this phenomenon can normally be seen and studied, though it has been suggested that one might try with a cylinder 12 yards long, and 20 inches in diameter. The whole effect could only consist in the eye being less dazzled by light entering from its surroundings. This, however, makes little difference, seeing that the field of light seen directly by us, remains illuminated and is the deciding factor.

Still more improbable is the story that the stars can be seen by day, reflected in dark mountain lakes. The 'observers' of this phenomenon though they noticed how dark was the reflection of the sky, forgot that the light of the stars diminishes in exactly the same proportion owing to the reflection.

70. *Irradiation*

It seems as if the setting sun causes an indentation in the line of the horizon (Fig. 80).

When the first crescent of the moon appears and the remaining part of the moon's disc seems to glimmer feebly in the 'ash-grey light,' it strikes us that the outer edge of the crescent seems to be part of a larger circle than the outer edge of the ash-grey light (Fig. 80). According to Tycho Brahe's estimate, the ratios of the mutual diameters were as 6:5.

Also dark clothes make us look slimmer than white ones do.

Leonardo da Vinci, in his writings, says of this phenomenon: 'We can see this when we look at the sun through bare branches of trees. All the branches in front of the sun

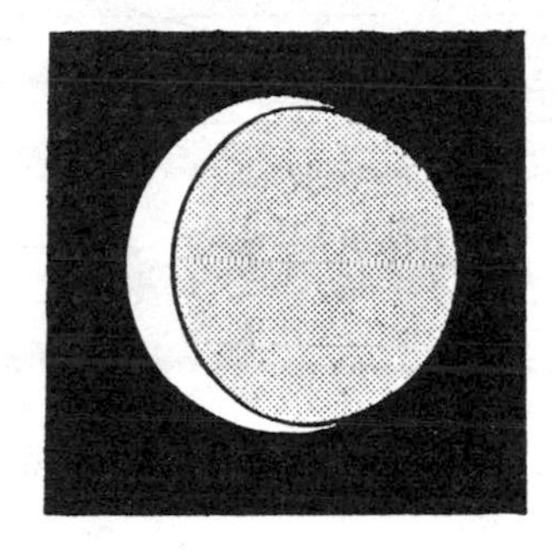

FIG. 80. Instances of irradiation: the sun when it sets, and the crescent of the moon.

are so slender that one can no longer see them, and the same with a spear held between the eye and the sun's disc. I once saw a woman dressed in black with a white shawl over her head. This shawl seemed twice as broad as the darkly clad shoulders. The crenels in the battlements of fortresses are of exactly the same width as the merlons and yet the former appear to be appreciably wider than the latter.'

One often sees two telegraph wires apparently intersecting at a very small angle when observed in a certain direction (Fig. 81, *a*). The remarkable thing about this is that with the sky as background, this point of intersection vanishes in the intense brightness surrounding them, contrasted with the double lines of dark wires to the right and left. Whenever the wind sways the wires, however little, the white gap moves to and fro along the wires (Fig. 81, *b*).

On the other hand the appearance is quite different when the background consists of parallel dark lines, such as steps or tiled roofs or brickwork; in this case the wire seems to be curiously swollen and broken wherever it crosses one of these dark lines. This effect occurs also when wires are seen against the sharp outline of a house (Fig. 81, *d*)—in short, when the straight edge of any solid object cuts acutely across a series of parallel lines.

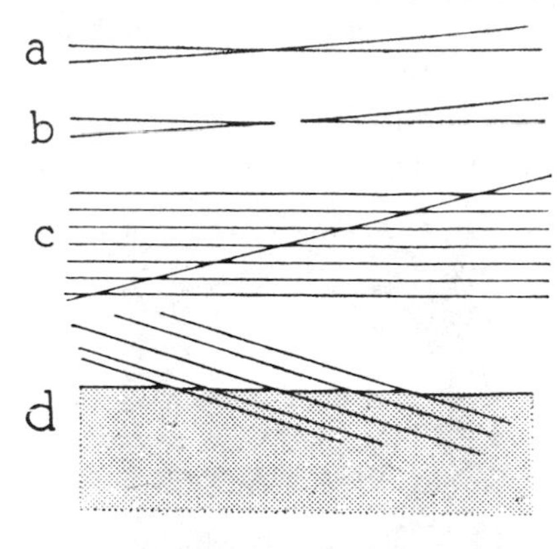

FIG. 81. Telegraph wires showing instances of irradiation.

The origin of all these deformations lies in the fact that the images in our eye are modified by refraction and imperfect reproduction. Mentally we place the borders between contiguous areas where the brightness changes most quickly, and if the image is made diffuse by diffraction this border differs from that given by ideal geometrical considerations. The border line therefore shifts systematically outwards in the case of bright fields on a dark field, and this shifting is known as 'irradiation,' a few instances of which have just been given.

71. *Dazzling*

Where the intensity of the light entering the eye is too great, 'dazzling' occurs. We understand by 'dazzling' two things: (*a*) the appearance of a strong source of light in the field of view, resulting in the other parts of the field of view being no longer clearly observable and (*b*) a feeling of giddiness or of pain.

An example of the first condition is given by the headlights of an approaching motorcar shining in our direction. We can no longer see the trees along the road and nearly collide with them. On closer inspection of the scene before us we discover that everything is covered by a haze of light many times stronger than the faint shapes of trees and other objects seen at night. This general haziness is caused by the scattering of the incident rays in the refractive media of the eye, which are always sufficiently granular and inhomogeneous to cause scattering. It even appears that dazzling light does not enter the eye through the pupil only, but also partly straight through the sclerotic. Moreover in the vicinity of the illuminated part the retina becomes much less sensitive; at an angle of 10° or more from the dazzling source, the latter effect becomes stronger than the scattering haze.

The second sensation caused by dazzling is felt quite clearly when we gaze at the sky in daytime. We should stand in the shadow of a house to avoid looking straight at the sun. The nearer our gaze approaches this celestial body, the more intolerable the fierce glare becomes, and if there should be white clouds the brilliance is hardly bearable. It is remarkable how much more sensitive one person is than another to the painful effects of dazzling.

VII

Colours

Everything that lives strives for colour.
GOETHE, *Theory of Colours*

72. *The Mixing of Colours*

When we look at the scenery from a window of a railway compartment we can also see the faintly reflected image of the scenery on the other side of the train. The two images overlap, so that one can study the colour-mixture formed. The reflection of the blue sky makes that of the green fields green-blue, and the resulting colour becomes much paler and less saturated, a phenomenon inherent in colour-mixing.

The panes of many shop windows nowadays are made without a frame, so that from the point O one can see through the pane the inside of the window-sill A and at the same time the reflection of the part outside, B (Fig. 82). If A and B are coloured differently, we get a splendid instance of mixed colours, and according as the position of our eye is higher or lower, the nearer the combined colour approaches that of A or that of B. This shows at the same time that a pane of glass reflects more light at large angles of incidence.

Nature mixes colours for us in yet another way. Flowers in a meadow seen at a distance merge into one single tint, so that dandelions on green grass may give a colour-mixture of yellow and green. The blossoms of apple-trees and pear-trees are a dirty white (yes, really, *dirty* white !) a colour-mixture arising from pink and white petals, green leaves, red anthers of the pear-trees and yellow anthers of the apple-trees, etc. The physical explanation of this mixing of colour is that our eye images every point of light more or less diffusely (cf. § 59) and that the patches of different colours overlap. Painters avail themselves of this physiological fact in their technique of pointillism.

73. *Play of Colour and Reflections*

Leonardo da Vinci, writing on painting, says: 'Therefore, painter, show in your portraits how the reflection of the garments' colours tints the neighbouring flesh. You desire to depict a white body, surrounded by air alone. White is no colour of itself; it changes and adopts part of the colours around it. If you see a woman clad in white, in an open landscape, she will be of such brilliance on that side towards the sun as to dazzle the eyes almost as much as the sun itself. That side of her, however, which is illuminated by the light from the sky will have a bluish hue. Should she stand near a meadow between the sunlit grass and the sun itself, the folds of her gown on which the light of the meadow falls will show the reflected rays of the green meadow.'

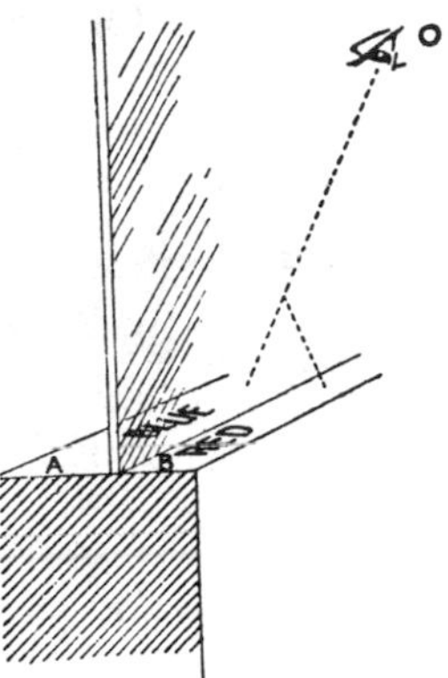

FIG. 82. Colour-mixing observed in shop windows.

74. *The Colours of Colloidal Metals—Violet Window-panes*

The windows of some old houses have beautiful violet-tinted panes. This violet tint is the result of many years of sunlight shining on the glass. Nowadays the same process of colouring can be carried out much more rapidly by exposing the glass to the fierce rays of a quartz-mercury lamp. The colour is to be ascribed to a minute quantity of manganese forming a colloidal solution in the glass; the tint depends not only on the optical properties of the metal, but also on the size of the particles. If you heat the glass the violet colour disappears.

Faraday tells us that in his time glass turned violet to a

very noticeable degree after the sun had shone on it for as short a time as six months.[1]

75. *The Colour of Discharge Tubes; Absorption of Light in Gases*

The many coloured advertising signs which at night transform our towns into fairy cities are glass tubes filled with highly rarefied gas, with an electric discharge passing through it. Orange light is given out by tubes filled with neon; blue and green, by tubes filled with mercury vapour, made of blue or green glass to weaken the different components of the mercury light, while yellow light is given out by helium tubes of yellow glass.

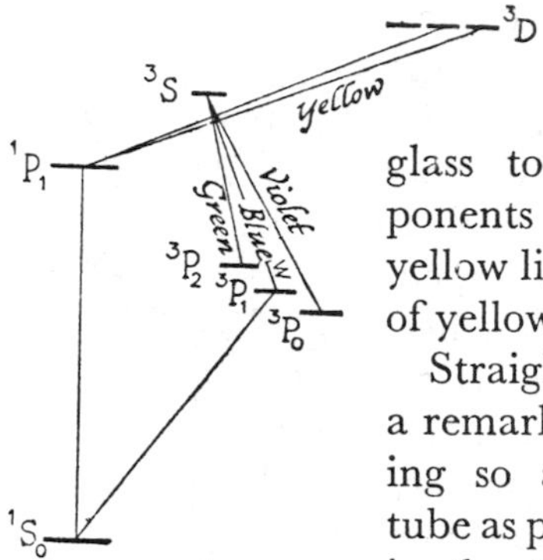

FIG. 83. Electron transitions in the mercury atom, to which the visible spectrum of mercury is mostly due.

Straight, blue discharge-tubes show a remarkable phenomenon. By standing so as to look as close along the tube as possible you will notice a change in the tint of the tube, which is now *blue-violet*, whereas seen in a perpendicular direction it is much more *blue-green*. The explanation is that the mercury light passing through the glass of the tube consists chiefly of three radiations, violet, blue and green, the first of which is faint. Their combined radiation through a thin layer of gas gives an impression of blue-green in our eye. But if we look through a very thick layer, for instance, by looking lengthways along a long tube, the light from the further end has to travel a long way through the vapour before reaching our eye, and in travelling such a long way the absorption of the green mercury-light is much greater than of the bluc, so that the ratio of the radiations becomes entirely different and the hue alters accordingly.

The green, blue and violet mercury lines together form a triplet arising from transitions between the ^3P and ^3S levels

[1] *Exp. Res. in Chem. Phys.*, p. 142.

(Fig. 83). The green and violet arise from transitions to the metastable levels $^3P_2$ and $^3P_0$, from which the electron cannot easily jump to lower energy levels; the number of atoms with electrons in such levels is therefore always abnormally large and the absorption very great.

For a similar reason the green tubes produce a more yellow impression seen lengthways. Here again two radiations are particularly strong, namely the green and the yellow mercury lines. Once again our observation confirms the fact that, of the two, more of the green light is absorbed.

76. *The Purkinje Effect. Cones and Rods*

Leonardo da Vinci noticed that 'Green and blue are invariably accentuated in the half-shadows, yellow and red and white in the light parts.'

Observe the contrast between the flaming red of geraniums in a border and their background of dark green leaves. In the twilight, and later in the evening, this contrast is clearly reversed, the flowers now appearing much darker than the leaves. You may wonder, perhaps, whether the brightness of red can be compared at all with the brightness of green, but the differences are so pronounced here as to leave no room for doubt.

If you can find a red and a blue in a picture gallery that appear to be equally bright by day, you will see that, in the twilight, the blue becomes by far the brighter of the two, so much so that it seems to radiate light.

These are examples of the *Purkinje effect.* This is due to the fact that in normal illumination our eyes observe with the cells in the retina called *cones*, but with the cells known as *rods* in very weak illumination. The former are most sensitive to yellow, the latter to green-blue, and this explains the reversal in brightness-ratio of various coloured objects when the illumination varies in intensity.

The rods only provide us with the impression of light, not of colour. Illumination by the moon is so weak that, practically speaking, the rods, only, are at work, and colours in a landscape are no longer perceptible: we have become colour-blind. This colour-blindness is still more complete on dark nights (cf. § 63).

77. *The Colour of Very Bright Sources of Light tends to White*

In our towns, we can often see during the evening how various sources of light are reflected in canals and spread out into pillars of light (§ 14). It is surprising how easily differences of colour between them can now be observed, e.g. between those from incandescent gas-light and electric lamps, while the sources of light themselves look nearly equally white. Similarly the differences between the colours become more distinct when the lamps are seen through fog or blurred windows. And by a curious property of our eye their colours tend to white, when the light is concentrated into one point of considerable brightness.

78. *The Psychological Effect of a Landscape seen through Coloured Glasses*

Goethe, in his *Farbenlehre*, says: With yellow 'the eye rejoices, the heart expands, the spirit is cheered and we immediately feel warmed.' Many people feel an inclination to laugh when looking through yellow glass. Blue 'shows everything in a sad light.' Red 'shows a bright landscape in an awful light. This is the colour that will be cast over heaven and earth on the Day of Judgement.' Green looks very unnatural, very likely because green sky so seldom occurs. Vaughan Cornish tried to distinguish between the colours in a landscape giving a sensation of 'warmth' and those giving a sensation of 'cold.' He finds that red, orange, yellow and yellow-green belong to the former category and blue-green, blue and violet to the latter.

For the psychological effect of colours and tints in a landscape see Vaughan Cornish, *Scenery and the Sense of Sight* (Cambridge, 1935).

79. *Observation of Colours with the Head Downwards*

There is an old prescription among painters for seeing more life and greater richness in the colours of a landscape, and that is to stand with your back to the landscape, your legs wide apart and bend forward so far as to be able to see between them. The intensified feeling for colour is supposed to be connected with the greater quantity of blood running to the head.

Vaughan Cornish suggests that lying on one's side would produce the same effect. He ascribes this to the fact that the well-known over-estimation of vertical distances is neutralised (§ 110), so that the tints apparently show steeper gradients. The question is whether this applies also to the much stronger effect while bending.

VIII

After-images and Contrast Phenomena

80. *Duration of Light Impressions*

While we are sitting in a train, another train races past in the opposite direction. For a few moments we can see the country right through the windows of the other train, distinctly, almost without flickering, only not quite so bright.

Or again, we can see right through the windows of a train while we are standing on the platform, or see the country reflected in the windows. In both cases, if only we look steadily in front of us, the images will appear without any flickering.

In order to ascertain the rate at which light and dark must alternate to eliminate any flickering, let us go and walk along the side of a high and long fence regulating our step in such a way as to obtain an impression of uniform lighting, taking care meanwhile, to stare through the fence in the same direction all the time.

The speed at which flickering just vanishes depends on the ratio of brightness between 'light' and 'dark,' and also on the ratios between duration of lighting and duration of screening. In reality the light-impression does not vanish suddenly but decreases gradually. The continual process of waxing and waning of the light-impressions in a cinema must, therefore, be very complicated.

A classical instance is that of falling snowflakes.

Leonardo da Vinci noticed this: 'The snowflakes near to us seem to fall faster, those some distance away slower; and the former appear to hang together like white cords, the latter, however, do not seem to hang together.'

Raindrops, which fall so much faster than snow, always appear lengthened into long, thin lines.

81. *The Railings (or Palisade) Phenomenon*[1]

A surprising pattern is shown by the spokes of a fast rotating wheel, seen through a railed fence. Strange to say, this pattern is symmetrical so that it is impossible to gather from it the direction of rotation (Fig. 84). Though the wheel possesses a quick forward and rotatory motion, the pattern remains practically at rest. When a train slows down at a station and one happens to see the large wheels of the engine through the railings of a fence, the phenomenon is seen to perfection. It is most striking when the illumination of the rim is strong, that of the spokes rather poor, and when the openings between the bars are narrow. One does *not* see the pattern if one looks through the railings at a wheel that is revolving *but not rolling*; the combination of both the rotatory and translatory motion is essential.

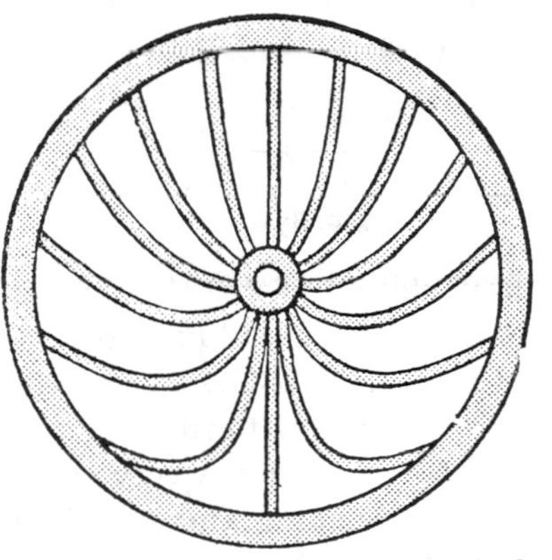

FIG. 84. The railings or palisade phenomenon—a rolling wheel seen through a long railing.

In order to explain the phenomenon, we start from the fact that *the observer follows the wheel with his eyes, to which he therefore refers everything he sees.* This is the condition which has to be fulfilled and which is brought about by the above-mentioned mode of lighting, etc. Imagine, therefore, the wheel to revolve round a fixed axis O, but the openings on the railings to move uniformly past it (Fig. 85A).

[1] P. M. Roget, *Philos. Trans.*, **115**, 131, 1825.

Suppose that in their initial positions a certain definite opening cuts a definite spoke at A; then part of the spoke at A will be seen through the opening. A few moments later the spoke will lie along OB, while the opening will also

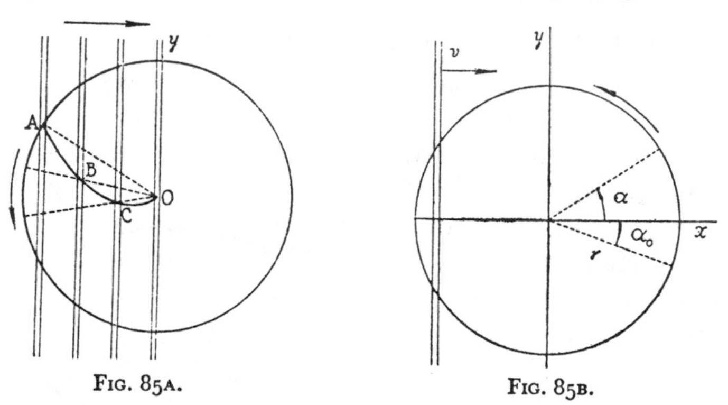

FIG. 85A. FIG. 85B.

have shifted so that they intersect at B. Still later, the point of intersection will have moved to C. In this way the whole of the curve ABCO will be described point by point. Each curve of the pattern is therefore the locus of the points at which we see, for a very short time, the intersection of one definite opening and one definite spoke. It is owing to the persistence of the visual impression that the whole of the curve appears to be seen simultaneously, provided the wheel is revolving sufficiently fast.

Each following spoke, after becoming visible in its turn through the same opening, will describe a curve of the same family, but with a different parameter; this means that a complete pattern will arise. If the next opening reaches the position of the one preceding it in exactly the same time as that in which the next spoke reaches the position of the spoke preceding it, the same set of curves will evidently be described again, and the whole pattern is then at rest. But if the distances between the openings are slightly different, each spoke will arrive at the opening just a bit too early (or too late). In this case each curve is transformed into another one, belonging to the same family, but characterised by a different parameter. We shall then see a pattern, slowly changing, in the same or opposite direction as the rotation, but this change is not a rotation of the pattern as a whole, for this remains symmetrical about the vertical. Finally, there is the possibility of the distances between openings being much too large or too small, for instance, twice as small. Then we shall see twice as many curves as there are spokes and the pattern will be once more at rest, if the spacing is regular.

It will be clear from the above argument that, generally speaking, the slowly changing pattern will occur most frequently. In reality the railing is often so short that the whole phenomenon is over in a second or less, so that there is hardly time to realise this changing of the pattern at all. Personally, I have seen it a number of times.

The equation of the set of curves is easily deduced. Choose the axes of co-ordinates as in Fig. 85B and let v be the velocity of the openings in the railings. Let α_0 be the initial inclination of the radius vector (i.e. a spoke) to the x axis, and α the inclination after a time t. Then the co-ordinates of the point of intersection of spoke and opening at the instant t are

$$x=vt \qquad y=x \tan \alpha$$

Also from the connection between rotatory and translatory motion we have, if r is the length of a spoke

$$\frac{vt}{r}=\alpha-\alpha_0, \text{ or } x=r\,(\alpha-\alpha_0)$$

Eliminating α gives for the equation to the family of curves

$$y=x \tan \left(\frac{x}{r}+\alpha_0\right)$$

As appears from this expression, y remains the same, when α_0 and x change their sign simultaneously, that is the pattern is symmetrical about the y axis.

More complicated patterns arise when one of the large wheels of a cart is seen through the other wheel. As soon as the direction of vision is a little to the right or to the left, so that the one wheel no longer covers the other one entirely, the most curious curves can be observed. They were noticed by Faraday, and they reminded him of magnetic lines of force. They are the loci of the points where two spokes cross.

82. *Flickering Sources of Light*

Among the illuminated advertising signs flaming so fantastically at night in our large towns the orange neon tubes are the most conspicuous. They are fed by alternating current of a frequency of 50 cycles per second. That means that the light-intensity alternates 100 times a second, since two maxima of light correspond to a cycle. The flickering is so rapid that, as a rule, we do not notice it at all.

But if you move some shining object to and fro in the light of neon tubes, you will see its light-track as a rippled luminous plane. The faster you move the object the further apart the ripples will be. The number of ripples enables one to compute the frequency of the alternating current. When, for example, one describes a circle four times a second, with a shining pair of scissors and the light-track shows 12 maxima, the frequency of the current-impulses is $12 \times 4 = 48$, and of the alternating current itself 24 cycles per second.

This experiment can also be carried out by reflecting the source of light in a rapidly oscillating mirror, or piece of glass (of your watch, for example), or one can rapidly move one of the lenses of one's spectacles in a small circle in front of one's eye (cf. § 40). Finally, the flickering can be seen with the naked eye, by fixing one's gaze first on a point in the neighbourhood of the neon tube and then changing its direction quite suddenly; the image of the source of light moves in this case over the retina and each maximum is perceived separately. This sudden changing of the direction of the gaze, without ceasing to concentrate your attention on the source of light, is surprisingly difficult. Sometimes one succeeds; sometimes one does not.

Investigate also electric lamps fed by alternating current. When you swing a silver pencil in the light, the ripples will be clearly visible, proving that the light from, and the temperature of, the filament increase slightly at every current-impulse, and decrease in between (Fig. 86). When the lamp is fed by direct current, no ripples are to be seen at all.

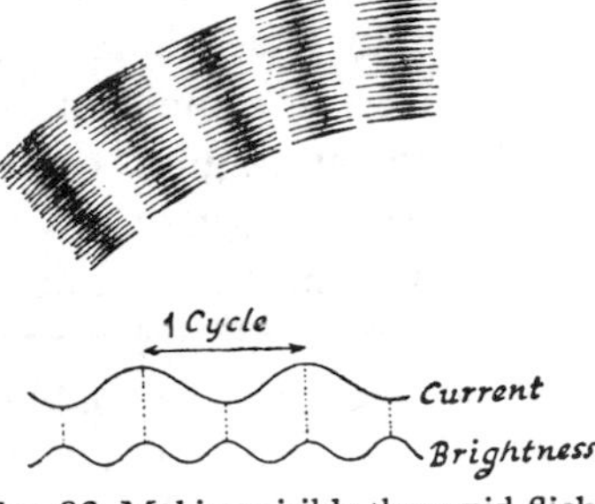

FIG. 86. Making visible the rapid flickering in the light of our electric lamps.

Sometimes when looking out of the window of a railway compartment at night at sodium lamps, used sometimes for illuminating important streets, a very distinct rippling can be seen under the following conditions. There should be about 6 feet between you and the window, which should be wet or blurred and the wet film smeared slightly in a vertical direction. As soon as the light of the distant lamp reaches certain parts of the pane, the rippling becomes visible. The explanation is that the wet film is not of the same thickness all over; owing to the wiping, a series of thin prisms has been formed with vertical edges and refracting angles, which change from point to point. This causes irregular and sometimes sudden displacements of the images of the lamp and, since these lamps are fed by alternating current, ripples can be seen exactly as in the case of the rapidly moving eyeglass.

83. *The Merging-frequency for the Central and for the Peripheral Field of View*

The following remarkable experiment can be performed in places where the frequency of the alternating current supply is low (20–25 cycles per second). Look first at a lamp and you will see a steady light, while the wall illuminated by the lamp flickers. Then look at the wall; its illumination becomes steady, but the lamp in its turn flickers.[1]

It is clear that the perceptive power of our direct and of our peripheral field of view must be different. It might be possible that the fluctuations in the intensity of the lamp are very slight and that the differential threshold for intensities is lower for the peripheral field of view. In order to ascertain this, we describe in the light of the same lamp a circle with some shining object. The fluctuations in brightness at regular distances in its light-track are now easily visible, even when we watch the track intently (cf. § 82). This means that our direct field of view is sufficiently sensitive to small differences of intensity, but that it is not capable of keeping pace with the fluctuations.

Laboratory experiments have also proved the existence of this peculiarity of our eye.

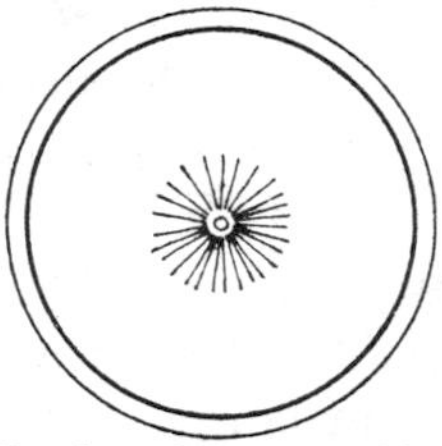

Fig. 87. What a rapidly revolving bicycle wheel looks like.

84. *The Bicycle Wheel apparently at Rest*

The wheels of a passing bicycle look more or less as shown in Fig. 87. Our eye can only follow the motion of those parts of the spokes which are near the centre, because there they move slower.

However, settle down comfortably at the side of a road where many bicycles are sure to pass. Look intently at a fixed point in the road. The moment the front wheel of a bicycle enters your field of view you suddenly see a number of spokes quite clearly, even when the bicycle is moving quickly. It is a very striking phenomenon. The chief thing is to look steadily in one direction and not to watch the approaching bicycle.

The explanation is that the point of the circumference where the wheel touches the ground is momentarily at rest, because there the ground grips the wheel (Fig. 88). The ends of the spokes near this point will then be nearly at rest, too, whereas points further from the ground will

[1] Woog, *C.R.*, **168**, 1222; **169**, 93, 1919.

move rapidly along curved tracks, owing to their combined rotation and translation. If, therefore, we can manage to look steadily at a definite point on the ground, the *lower*

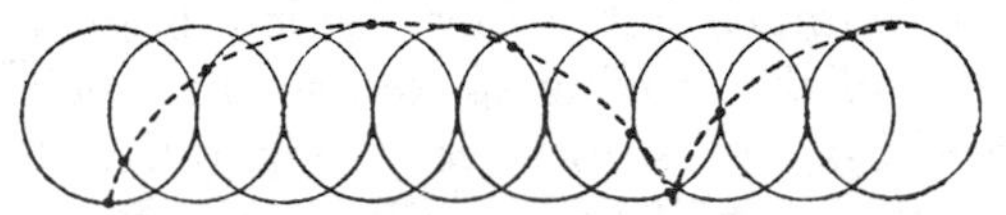

FIG. 88. Trajectory of a point on the circumference of a rolling wheel. As one sees, the point remains stationary for a moment during each revolution, namely when it touches the ground.

parts of the wheel should appear to us to be more or less at a standstill, and this is indeed what we actually observe.

It is my belief that we can see the spokes clearest of all when they appear in our peripheral field of view. It is, therefore, quite possible that the capacity for perceiving rapid fluctuations of light plays a part in this matter also.

85. *The Motorcar Wheel apparently at Rest*

When a motorcar comes along, even at a moderate speed, the spokes of its wheels cannot be distinguished. At each fixed point of our retina the flickering of light and dark is so rapid that the impressions merge into each other; the muscles of our eye cannot make the direction of our gaze describe a cone at the rate required to follow separate spokes.

Now and again, however, it happens that the spokes become visible for an infinitesimal instant, as in a snapshot. As a rule only a few spokes can be seen, but at times I seem to see the whole wheel clearly. In this case the explanation of the bicycle wheel at rest would not be satisfactory. This is such a striking phenomenon that it has been held that the wheel is actually at rest at certain moments, which is obviously a sheer impossibility !

One realises very soon, however, that the momentary visibility of the spokes of the motorcar wheel occurs practically whenever we set our foot firmly on the ground; you can also see it when you tap your spectacles (supposing that you are short-sighted) or jerk your head. It is, maybe, that in these cases the eye, or the direction of your gaze, is set into very rapid, damped vibrations, which happen to follow exactly some of the moving spokes, so that their images on the retina will be at rest for a very short time. Is it perhaps the axis of the eyeball that carries out a slight to-and-fro movement; or does the eyeball as a whole shake in its socket (translatory motion)? May one assume that the eye, on receiving little shocks such as these, is capable of performing very rapid random rotations round its axis?

FIG. 89.

A direct proof of the theory that the eye vibrates is the following fact. If one walks at night with a vigorous, swinging stride, and fixes one's gaze steadily on a distant lamp, one notices then that the light describes at every stride a little curve, more or less like that of Fig. 89.

At times the phenomenon is also visible when, *standing still*, one looks at a passing motorcar. The explanation of this must be found in sudden, slight, unconscious movements of the eye. The fact that the eye frequently moves with such little jerks can be proved by looking for an instant and very cautiously into the setting sun. The after-images will then turn out to consist of a number of black spots, not of a black, continuous band (cf. § 88).

86. *The Screw of an Aeroplane apparently at Rest*

A traveller by air noticed that he could distinguish the propeller of his aeroplane, despite its rapid rotation, when he looked sideways at an angle of 45°, thereby observing with his peripheral field of view. And yet the screw made twenty-eight revolutions per second, which means that it flickered fifty-six times a second. The 'seeing' of the propeller is nothing but the perception of this very high flickering frequency. The fact that this is more easily done in the peripheral than in the central field of view is a noteworthy confirmation of our remarks in § 83.

87. *Observations on Rotating Bicycle Wheel*

As a rule, the spokes of a rotating bicycle wheel are not visible separately; they are blurred into a thin veil, darkest close to the centre and lighter towards the rim. The shadow of the wheel on an even road shows a similar distribution of light. How dark is this shadow? The spokes are each 0·08 inches thick, and their distance apart at the rim is, on the average, 2 inches. The time during which a point on the road is illuminated depends on the ratio of the unobstructed area of the wheel to the whole of it, so from the foregoing figures we may write:

$$\frac{\text{Time during which road is illuminated}}{\text{Total time light is falling on the wheel}} = \frac{2}{2 + 0{\cdot}08} = \frac{100}{104}$$

By Talbot's law this makes the same impression upon our eye, as if the shadow cast by the revolving wheel had a constant brightness, equal to $\frac{100}{104}$ of the brightness of the unshadowed road. But the sun does not shine at right angles to the wheel, so that in the shadow the spokes are closer together, although their thickness remains the same. It will be clear then that, even near the rim, the shadow will be as much as 4 per cent. to 8 per cent. less bright than the surrounding ground, and that nearer to the centre this reduction is probably as great as 10 per cent. to 20 per cent. And yet it is difficult to perceive any difference of brightness at all, the dark shadow of the tyre forming much too marked a separation between the two fields under comparison. The gradual decrease in brightness towards the centre is hardly noticeable, because we are always inclined to see a clearly circumscribed, coherent figure as a whole, and this psychological tendency overlooks the actual difference in brightness.

On closer inspection, however, we discover as a rule one or more light rings in the shadow of the wheel (Fig. 90). They are often open curves of limited length. Get off your bicycle and investigate carefully where the light arc is formed. It corresponds to where two spokes cross; indeed, at each of these points one spoke has, so to speak, disappeared, so that the average shadow must be less. But what a slight difference it is! And yet how plainly our eye perceives it, now that the two brightnesses to be compared adjoin without a line of separation between. It is difficult to describe the intertwining of the spokes. Mostly, four at a time form a group, which is repeated round the wheel. The crossing of two spokes describes a definite curve, which becomes visible as a small, bright arc. After the wheel has rotated over four times the distance between two spokes, the small arc will be formed again. If, moreover, in each group two crossings should occur, one of which happens to follow the track of the other, the small arc would appear particularly bright. In the former case, it would be 1 per cent. brighter than the shadow round it, in the latter case 2 per cent. Since, however, the spokes in the shadow are as a rule projected closer together and the small bright arc is, moreover, very often at some distance from the rim, these percentages will probably be 3 and 6. These amounts are, therefore, the smallest differences of brightness still perceptible when two fields are directly adjacent. Though the unevenness of the road, which here acts as a projection-screen, is a serious drawback, the result agrees well with our former estimates (§ 67).

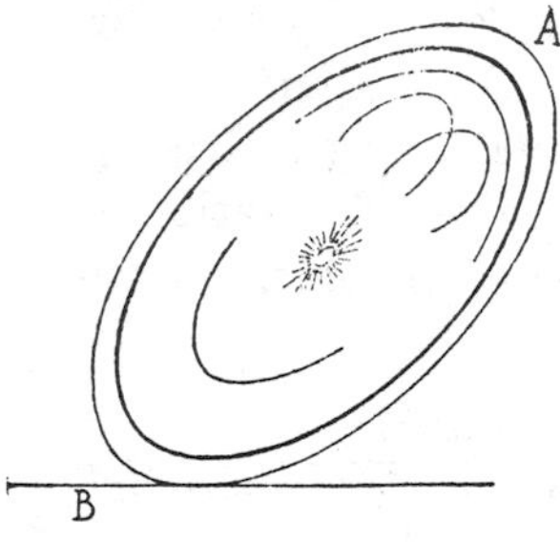

FIG. 90. Light or shadow curves in a rotating bicycle wheel.

Try to account for the fact that the arcs and rings of light are usually at their brightest near the end A of the elongated wheel shadow, and investigate why the pattern at A is not the same as at B.

When, instead of looking at the shadow of your wheel, you look directly at the wheel of someone cycling at your side you will see these same arcs and rings still better, because now they stand out quite sharply without any blurring (cf. § 2). Against a bright background the spokes appear dark so that the rings are brighter; when, however, the wheel is illuminated by the sun against a dark background the rings are darker.

This is by no means the last of the remarkable effects exhibited by rapidly rotating bicycle wheels. It may occur when you look at the *shadow* of the wheel, that you see the sharp lines of the spokes flash out as quick as lightning, if your eyes happen to make a rapid circular movement, so that, involuntarily, you follow the shadow exactly at its own speed (cf. § 85). If you wear spectacles small sudden displacements of the glasses will suffice to make you see the separate spokes move in the queerest jerky manner. But the most remarkable shadow of all is to be seen when you cycle on an uneven cobbled road. Notwithstanding the roughnesses of the background, you perceive clearly a set of radial but *curved* lines, nearly always in the same part of the shadow. They appear also when you yourself cycle on an even road and the shadow falls on the uneven cobbles of the footpath. The unevennesses of this kind of projection-screen correspond obviously to the part played by the slight taps on our spectacles. But how does the curvature of the lines arise? And why do they usually appear in the same portion A of the shadow?

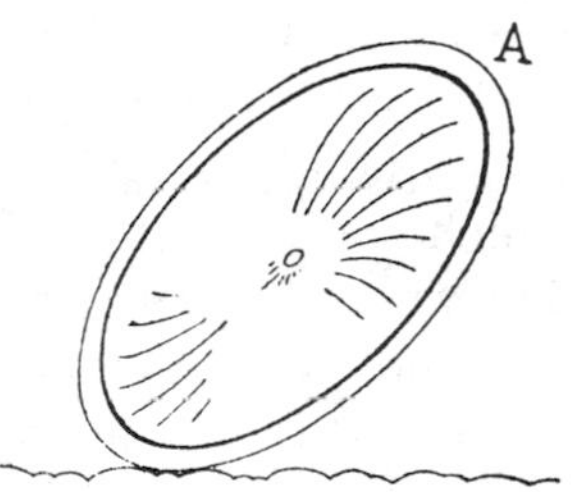

FIG. 91. Curved lines in the shadow of a bicycle wheel passing over a cobbled road.

Quite apart from the curves already mentioned, there is still one more peculiar light figure, which however can only be seen when the sun shines on a bicycle with *glittering* brand-new spokes.

88. *After-images*

Be very cautious during these observations! Do not

overstrain your eyes! Do not carry out more than two tests in succession!

Look carefully at the setting sun and then close your eyes.[1] The after-image consists of several small round discs, a proof that the eye must have moved with little jerks during the short time your gaze lasted. The discs will strike you as being remarkably small, for, owing to the fierce brightness of the sun, you are accustomed to 'seeing' it looking larger than it really is; its true dimensions are shown in the after-images.

Open your eyes again and you will see after-images everywhere you look. The farther away the objects are on which you project them, the larger the after-images appear. The angular diameter, of course, is always the same, but if you know that a certain object is a long way off and yet see it subtend the same angle as another one closer to you, you conclude unconsciously, on the grounds of daily experience, that in reality the object in the distance must be the larger of the two.

An after-image on a dark background is light (a positive after-image), as can be seen very well by closing the eyes and, because the eyelids are transparent, covering them with the hand. *Vice versâ*, an after-image on a light background becomes dark (a negative after-image). The fierce light has apparently stimulated our retina locally and the impression remains, but at the same time that part of the retina has become less sensitive to new light-impressions.

Sources of light feebler than the sun give likewise feebler after-images. The stimulus acting on the retina then becomes very much weaker after a few seconds or even fractions of a second, and only the fatigue remains, so that only the reversed after-image on the light background can be observed.

The change from white to black corresponds, in the case of coloured light-sources, to the change of the after-image into its complementary colour, so that red changes to green-

[1] Goethe, *Theory of Colours* (1840). Titchener, *Experimental Psychology* (New York), I, 1, 29; I, 2, 47.

blue, orange to blue, yellow to violet, green to purple and *vice versâ*.

Twilight is the best time in which to see after-images; all Goethe's typical examples of after-images were observed in the evening. The eye is well rested then, and the contrast between the light in the west and the dark in the east is at its sharpest.

Goethe in his *Farbenlehre* writes: 'One night as I entered the room of an inn, a fine girl came towards me. Her face was of a dazzling whiteness, her hair black and she was wearing a scarlet bodice. I looked at her intently in the dim twilight while she was standing some distance away from me. When, after a moment, she left me I saw on the white wall opposite me a black face, surrounded by a bright glow, and the dress of the very well-defined figure was a beautiful sea-green.'[1]

It is related that to people who had been gazing for half an hour at the orange-yellow flames in a fire, the rising moon seemed blue.[2]

The after-image of a flash of lightning seen during a thunderstorm at night can sometimes be observed as a fine black snake-like line on the background of an illuminated white wall, or of the feebly diffused light of the sky.[3]

When one stares into the distance and scans the horizon, while standing on the beach at nightfall, there comes a moment when the difference between the light sky and the dark sea is no longer really visible. This is clearly due to the fact that the longer a light excites the eye, the weaker its stimulating effect becomes; the retina grows fatigued. That this is indeed true is shown the moment we transfer our gaze a little higher; the negative after-image of the sea takes the form of a light streak against the sky. If we let our gaze descend a little, we can see the dark after-image of the sky against the sea.[4]

[1] Goethe, *Theory of Colours*.
[2] Cassell's *Book of Sports and Pastimes* (London, 1903), p. 405.
[3] *Nat.*. **60** 341, 1905. [4] Helmholtz, *Physiologische Optik*, 3rd Ed., II, 202.

89. *The Phenomenon of Elisabeth Linnaeus*

Elisabeth Linnaeus, the daughter of the famous botanist, noticed one evening that light was emitted by the orange-coloured flowers of the nasturtium (*Tropaeolum majus*). It was thought to be an electric phenomenon. This observation was confirmed by Darwin with a kind of South African lily; also by Haggrén, Dowden and earlier explorers, their observations always taking place during the dusk of dawn or twilight. Canon Russell repeated the observation with the marigold (*Calendula officinalis*) and the burning bush (*Dictamnus fraxinella*), remarking at the same time that some people see it better than others. And yet it seems that this phenomenon, concerning which a whole series of treatises was published at the time, is to be ascribed simply to after-images ! After-images were seen by Goethe when he stared intently at brilliantly coloured flowers and then at the sandy road. Peonies, oriental poppies, marigolds and yellow crocuses gave lovely green, blue and violet after-images,[1] especially at dusk, and the flame-like flash became visible only on glancing sideways for a moment, all of these being details to be expected from after-images.

When anyone thinks he can see this phenomenon very clearly, he should hold brilliantly coloured paper flowers close to the real ones, and see whether the former show the same phenomenon.

90. *Changes of Colour in After-images*

The rapidity with which after-images fade differs for different colours, particularly when the light-impression has been very strong. This explains why the after-images of the sun and intensely white objects are coloured. As a rule, the after-image on a dark ground is green-blue at first, and afterwards purple coloured.

'Towards evening I entered a smithy, just when the glowing lump of iron was placed under the hammer. After watching intently for some time I turned round and

[1] Goethe, *Theory of Colours.*

happened to look inside an open coal-shed. An enormous purple-coloured image floated before my eyes, and when I looked away from the dark opening at the light woodwork the phenomenon appeared to be half-green, half-purple, according to whether the background was dark or light.'[1]

When looking at snow in the sun, or reading a book while the sun is shining on it, every bright object near us looks purple; afterwards in the shadow every dark object looks a lovely green. Here also the after-image on a bright ground has the complementary colour of that on a dark ground. Some observers speak of 'blood-red' instead of purple.

This may partly be due to another factor; light from the sun usually shines not only in our eye, but also on our eye, and that part shining on our eye partly penetrates the eyelids and the eyewall, thereby becoming blood-red. Our field of vision is entirely filled with this general red illumination, and we see it clearly whenever the surrounding objects are dark; black letters, for instance, look red. If, now, we go into the shade, or go indoors, our retina remains still tired as regards red light, and all the bright parts look green.

For an observer who does not take any special precautions, the coloured after-images of white light (that has passed through the pupil) and the fatigue as regards red light (that passed through the cornea) produce a combined effect.

Black letters have been seen to look red[2] in the evening light, probably because the low sun was shining in the reader's eyes.

91. *Contrast-border between different adjacent Brightnesses*

The outlines of a dark row of houses seen against the lighter sky appear to be edged with light, particularly in the evening. This can be explained by assuming that small involuntary movements are made by the eye and that the bright after-images of the houses cover the surrounding sky and make it lighter. Only a small part of the effect, however, is explained in this way, the decrease of sensitivity in

[1] Goethe, *Theory of Colours*. [2] *ibid.*

the retina surrounding an illuminated part being of much more importance (§ 72).

'I was once sitting in a meadow, talking to a man standing a little way off, his figure outlined against the grey sky. After looking at him for some time intently and steadily, I turned away and saw his head surrounded by a dazzling glow of light.'[1]

Pater Beccaria, while carrying out some experiments with a kite, perceived a small cloud of light surrounding it and also the string attached to it. Every time the kite moved a little faster, the small cloud seemed to lag behind and float for a moment to and fro.[2]

A very striking example of optical contrast can be seen on undulating moors, whose successive ridges become lighter and lighter owing to aerial perspective, and finally become lost to sight in the hazy distance (Plate VIII, *b*). Each ridge looks darker along the top than along its base, the effect being so convincing that everyone is bound to notice it. And yet it is only an illusion, arising from the fact that each ridge is bounded at the top by a lighter strip and along the base by a darker one. To prove this, a piece of paper should be used (dotted line in Plate VIII, *b*) to screen off the upper part of the landscape; this will be quite sufficient to make the contrast effect disappear.

92. *Contrast-edges along the Boundaries of Shadows*[3]

Everyone knows that a piece of cardboard held in the sun will cast a shadow on a screen, and that between the light and the shadow there exists a half-shadow, caused by the finite dimensions of the sun's disc (§ 2). But does everyone know that this half-shadow has *a bright edge* near the transition from light to half-shadow?

Experiment while the sun is low, and therefore not too bright, holding the screen about 4 yards behind the piece of cardboard, and moving it slightly to and fro to smooth out

[1] Goethe, *Theory of Colours.* [2] *ibid.*
[3] K. Groes-Pettersen, *Ash. Nachr.*, **196**, 293, 1913.

(*a*) Contrast-border along roofs in the evening

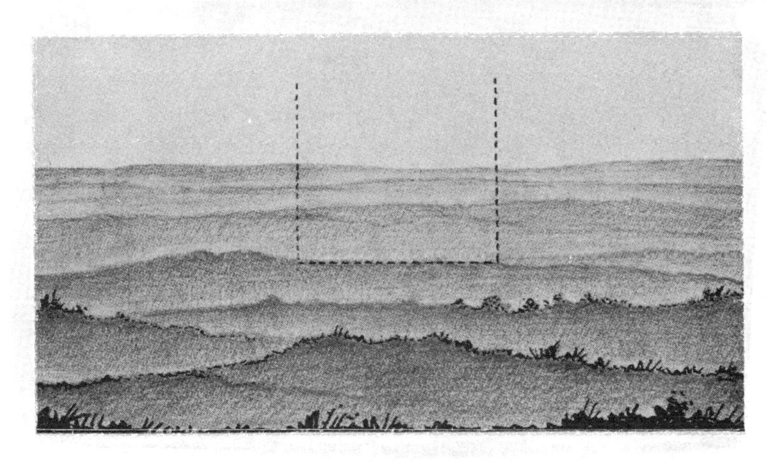

(*b*) Contrast phenomena observed on undulating ground. The illusion can be made to disappear by screening off part of the landscape as shown by the dotted lines

local unevennesses. The effect is very clear. The distribution of light observed is shown by the solid line in Fig. 92.

Can you understand this? The distribution of light to be expected can be deduced from the following considerations.

From the successive points 1, 2, 3 of the illuminated screen the sun's disc is seen to be covered less and less by the cardboard. Their brightness is in proportion to the increasingly uncovered part, and should therefore follow the dotted curve of Fig. 92. There cannot, therefore, be a brighter edge at all, and the whole thing must be due to optical illusion.

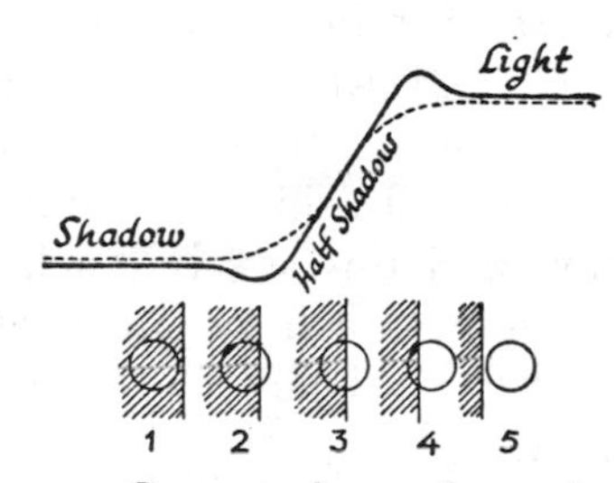

FIG. 92. Contrast-edges along the boundary line of a shadow.
------ Actual distribution of brightness.
——— Apparent distribution of brightness.

And, indeed, all the circumstances tend to favour this. Mach showed that contrast-bands invariably become visible when the brightness does not change at a uniform rate—that is, where the brightness graph is curved. The contrast-band always appears to be an exaggeration of the curvature. That this is so can be easily understood either by assuming that small movements are continually being made by the eye, or from the weakened sensitivity of the retina in the neighbourhood of its illuminated parts.

The instances mentioned in § 91, also agree completely with the theory of Mach; we need only regard an angle in the brightness-curve as an increase of the curvature.

And, finally, from time to time a very special opportunity occurs to test this theory—namely, during a partial eclipse of the sun. Repeating the experiment, one will now obtain various unusual distributions of light along the edge of the half-shadow, according as the sun is screened off by the moon and as the position of the shadow-casting cardboard is altered. Each of these distributions shows its (apparent) contrast-bands and in each of these cases Mach's law is satisfied. It is no wonder that the shadows look so unusual as to arouse the interest of even the most casual observers (cf. § 3).

93. *Black Snow*

Watch the snow-flakes drifting down out of a grey sky. Seen against the sky, the flakes look decidedly dark. One must bear in mind that the only quality in which black, grey and white differ is their brightness, for which the surrounding background provides the standard of comparison. In this case we refer all brightnesses to that of the sky, and this sky is much brighter than one would have thought, much brighter, anyhow, than falling snowflakes seen from below. The phenomenon was mentioned by Aristotle.

94. *White Snow and Grey Sky*[1]

When the sky is uniformly grey, it looks much duller than the snow-clad earth. And yet this effect is evidently misleading, for it is this same sky that illuminates the earth, and the illuminated object can never have greater surface-brightness than the source of light. When tested by the photometer, the greater brightness of the sky is confirmed beyond any doubt. If you take a little mirror and place it in such a way that the image of the sky is seen next to that of the snow, you will notice that the snow is indeed grey in comparison with the white sky. Be sure to carry out this experiment; it is as convincing as it is surprising!

And *yet* the illusion of the contrast is not destroyed, though we know that in reality it is the other way round. The contrast between the snow and the very much darker woods, or shrubs, or houses, is the decisive factor here.

Similarly, on a dull day, a white wall can appear brighter than the sky. Photographs and pictures not in accordance with this illusion strike us as being very unnatural.

[1] *J.O.S.A.*, **11**, 133, 1925.

95. *Colour Contrast*

In a variety of cases in which one definite colour in the surroundings predominates, the complementary colour, in its turn, will appear to be more accentuated. Sometimes this can be explained in the same way as the contrast-border—namely, by the involuntary motions continually performed by the eye. But the fact that those parts of the retina which have been stimulated by the predominating colour make the adjoining parts less sensitive to that colour, is much more important in this connection. It is the same as if our eye were now more sensitive to the complementary colour, which therefore gives us an impression of greater freshness and saturation. When considered in this way, the colour contrast is one more example of the general law that colour and brightness are only judged in connection with the complex of all the images delineated on our retina.

An observer has noticed how, in a courtyard paved with grey limestone with grass growing in between, the grass became an infinitely beautiful green when the evening clouds cast a reddish, scarcely noticeable glow on the pavement.[1]

When, as we walk through the fields under a moderately clear sky, the colour green predominates on all sides, the trunks of trees, mole-hills and paths look reddish to us.

A grey house seen through green blinds looks reddish. Again, when the waves of the sea show a beautiful green, the parts in shadow look purple (cf. §§ 212, 216).[2]

If your surroundings are lighted by paraffin lamps or candles, of which the light is reddish, the light of arc lamps or of the moon will appear greenish-blue. When the sources of light are not too fierce, this contrast is particularly striking, as, for example, when one sees simultaneously reflections of the moon and of gas flames in water.

The spots where sunbeams, after penetrating the green foliage of the wood, strike the ground, seem to us light

[1] Goethe, *Theory of Colours*. [2] *ibid.*

pink compared to the general green of the surroundings.[1]

Leonardo da Vinci noticed how 'black clothes make a face appear whiter than it is, white clothes darker, yellow clothes bring out the colours in it, and red clothes make it paler.'

The colour-contrast is most pronounced when the brightnesses of areas do not differ widely. What the result is when brightnesses *do* so differ can be seen splendidly in the evening twilight, when the rows of houses stand out darkly against the blazing orange of the western sky. From a distance, one can only see their uniformly dark silhouettes, all details and differences of brightness having disappeared. Branches and the foliage of trees are outlined in the same way, like dark velvet, their own colours gone (cf. § 220); and this is not because the illumination in itself is too slight, for at the same time the colour of every detail on the ground, for instance, can be distinguished clearly.

After we have walked for a few hours in the snow, during which time white and grey were about the only colours to be seen, other colours give us the impression of being particularly saturated and warm. Our eyes were 'rested for colours.'

'For the rest, these phenomena occur to the observant everywhere, even to the point of annoyance,' says Goethe, in the *Theory of Colours*.

96. *Coloured Shadows*

When a pencil, held perpendicularly on a sheet of paper, is illuminated on one side by candle-light, on the other by moonlight, the two shadows exhibit a striking difference of colour; the former is bluish, the latter yellowish.[2]

It is true that here there is a *physical difference* of colour, for where the first shadow falls the paper is only illuminated by the moon, and where the second shadow falls, only by the candle; and moonlight is whiter than candle-light. But, in any case, it is *not* blue. The real difference of colour

[1] Helmholtz, 'On the relation of Optics to Painting.' *Popular Scientific Lectures*, 2nd Series (London, 1873).

[2] Goethe, *Theory of Colours*.

between the two shadows is evidently accentuated and modified by *physiological* contrast.

Similarly, at night we can observe the difference in colour between our two shadows, the one cast by the moon and the other by a street-lamp.

How relative the 'orange' of electric lamps is in comparison with the light of modern sodium lamps can be observed very well in places where both illuminations are mixed. The shadow cast by the sodium lamp is beautifully blue; that cast by the electric lamp is orange! As soon as we are illuminated by sodium lamps only, our shadow appears black; walking on, we see it suddenly turn into blue when we approach an ordinary electric bulb; and *vice versâ*, we see our black shadow cast by the electric light change suddenly into orange, when we come close to a sodium lamp. The eye evidently adapts itself to its surroundings, and has a tendency to take the predominating colour there for 'white'; every other colour is then judged relative to this white.

Goethe remarks that the shadows of canary-coloured objects are violet. Physically speaking, this is most certainly not true, but, owing to the phenomenon of physiological contrast, it may appear to be so—for instance, when the illuminated side of such objects is turned towards the observer, so that, to him, its shadow is in juxtaposition to a fierce yellow.

One might ask why shadows cast by the sun at midday show practically no colours, since the blue of the sky differs so widely from the colour of the sunlight. The answer is that the difference between the brightnesses of shadow and light is too great. When, however, the screen on which the shadow is cast is inclined so much that the sun's rays strike it almost grazingly, the colour-contrast will become much more pronounced.

A classical case is that of shadows on snow, the purity of their colours being particularly evident. They are blue because they receive only the light of the blue sky; their blueness equals even the blue of the sky itself. And considering that we see them next to the snow in the yellowish light of the sun, they ought to look still more blue. But their hue is less pronounced than one might expect, owing to the great difference in brightness. But watch the shadows when the sun sets over a landscape, especially the last minutes before it disappears. As the sun turns orange, then red, then purple, the shadows become blue, green and green-yellow. These tints are so pronounced because, this time, the difference in brightness between the shadow and the surrounding snow is much less than in daytime. For the rays of the sun strike the snow at very small angles and the diffuse light from the sky becomes relatively more important. Moreover, the sun's colours become more and more saturated.

'During a journey in the Harz, in winter, I descended the Brocken at nightfall; there was snow on the white fields above and below me and the heath was covered with snow; all the sparse trees and projecting cliffs, all the groups of trees and masses of rock were completely covered with hoarfrost; the sun was just setting over the lakes of the Oder. By day, when the snow was tinged with yellow, the shadows were only a faint violet, but now, when the illuminated parts reflected an intenser yellow, they were decidedly bright blue. When, however, the sun was at last on the point of setting and its light, toned down by the atmosphere, coloured everything around me a most glorious purple, the tint of the shadows was seen to change into a green which for purity could compete with the green of the sea, for beauty with that of the emerald. The phenomenon became increasingly vivid, the surroundings changed into a fairy-world, for everything was covered with these two bright and beautifully matched colours, until finally, the sun once set, the gorgeous scenery changed to a grey twilight and afterwards to a clear night with moon and stars.'[1]

The phenomenon of coloured shadows on snow is partly,

[1] Goethe, *Theory of Colours.*

and in a curious way, of a psychological nature.[1] In daytime, when the sky is blue, the shadows show a much more saturated blue, if one is not aware that it is snow. A shaded patch of snow in the distance can give the impression of 'white snow in the shade,' as well as of 'a blue lake.' In the same way, snow-shadows appear to be much bluer on the ground-glass plate of a camera than in the landscape so that one does not recognise them at once! An observer who, from a dense, dark fir wood, saw the hoar-frost on some distant shrubs, was evidently *unbiassed*, for the hoar-frost did indeed appear blue to him, the circumstances being identical to those when looking through a tube with an opening at the end (§ 174).

Psychologists know very well that colours can be reduced to their true hue by looking at them through a small aperture. They then give the same impression as if they were lying in the plane of the diaphragm. The moment, however, we imagine an object in its own surroundings and illuminated in the ordinary way, we compensate their influence automatically, so that one and the same object appears to us remarkably constant even under changing conditions.

A very curious description of this same phenomenon as observed by children, i.e. by unprejudiced observers, is given by a Russian writer. I do not doubt for a moment but that this description is taken from reality, though some of the details must have been omitted by the author, who wrote from memory: at least part of the sky must have been blue, while the snow was falling and the sun was hidden.

' "Galja, look! . . . Why is there a blue snow falling? Look! . . . It is blue, blue! . . ."

'The children became excited and began to shout to each other in sheer delight:

' "Blue! Blue! . . . Blue snow! . . ."

' "What is blue? Where?"

[1] I. G. Priest, *J.O.S.A.*, **13**, 308, 1926.

'I looked round at the snow-covered fields, the snow-clad mountains, and felt excited too. It was extraordinary; the snow came whirling and floating down from all sides—in the distance and quite close in blue waves. And the children shouted in happy excitement:

' "Has the sky come down in pieces? Yes, Galja?"

' "Blue, blue!"

'And once more I was struck by the keen and poetical power of perception of the little mites. Here was I, walking along with them without noticing that floating blue. I had lived through many winters, had many times revelled in the delight of falling snow and not once had I realised this immeasurable azure flight of the snow circling above the earth.'[1]

97. *Coloured Shadows arising from Coloured Reflections*

When coloured objects are illuminated by the sun they often cast so much light around them that this gives rise to shadows, which then show the complementary colour. A little pocket-book is the ideal instrument for tracing these light effects. Open it so as to form a right angle; one side of this dihedral angle screens off the sunlight, while the other side catches the coloured reflection. The pencil belonging to the pocket-book is held in front of the paper; its shadow assumes a complementary colour, and is, therefore, an extremely sensitive indicator as to whether the incident light is coloured. Green paint on a wall or a green shrub casts pink shadows. A yellow wall casts blue shadows (these were once traced as far as 400 yards), and the ochre-coloured side of a mountain did the same.[2]

98. *Unexplained Contrast-phenomena*

An observer tells us[3] that on a clear night he saw from his ship the moon, which was 20° above the horizon, reflected by the waves in the shape of a light triangle, extending from

[1] From the Dutch translation: Fj. Gladkow, *Nieuwe Grond* (Amsterdam, 1933), p. 161.

[2] *C.R.*, **48**, 1105, 1859.

[3] Cl. Martins, *C.R.*, **43**, 763, 1856.

the ship to the horizon. The remarkable thing, however, was that he saw a similar triangle, upside down and dark, descending from the moon to the horizon. The effect was, beyond doubt, a physiological and not a true one for various reasons; for it appeared also when mountains on the coast reached almost as high as the moon; it disappeared when the lower triangle of light and the moon were screened off. And when, after turning round he suddenly looked again, the illusion only reappeared after a few seconds. Moreover, in the dark region, stars of the second magnitude were no longer visible, 'in reality, therefore, this region of the sky was more strongly illuminated' (?).

Who will observe this again and try to give us a satisfactory explanation?

Plassmann says that when one looks repeatedly at the setting sun, it often appears as a red disc, with a central white patch.[1]

Sharpe remarks that two days after new moon a faint light edge is visible at the side of the faintly luminous disc opposite the slender crescent.[2]

These phenomena are, of course, due to contrast, but their complete explanation is not known.

[1] *Met. Zs.*, **48**, 421, 1931.
[2] *Phil. Mag.*, **4**, 427. Also *J.B.A.A.*, **28**, **29**, **45**.

IX

Judging Shape and Motion

99. *Optical Illusions regarding Position and Direction*

Let us suppose that we can distinguish in our field of vision two groups of objects. Within each group the objects are either parallel or at right angles to each other, but the two groups are inclined relative to each other. Then one group appears to be 'dominant,' and we tend to regard it as the true standard for determining horizontal and vertical directions.

If a train happens to stop or to move slowly on a bend in the line, so that the compartment is inclined sideways, all posts, houses, and towers appear to me to be tilted the opposite way. I am evidently aware of the inclined position of my compartment, but only *to a certain extent.*

In the corridors of a ship, heeling under a wind blowing from one side, a man meeting me seems to me to be inclined relative to the vertical.

A cyclist will have similar experiences in judging slight inclines in the road.[1] That part of the road where he is cycling will appear to him invariably too horizontal; cycling down a steep hill, a sheet of water at the side of the road will not give him the impression of being horizontal, but of sloping towards him. On a gentle downward slope, the road seems to rise further on, whereas in reality it is flat; a distant rise in the road appears too steep, a downward slope, on the other hand, appears too slight. What my eyes show me more in particular is the way the slope in front of me *changes*, and the visual impression is often at variance with what I gather from the resistance felt while pedalling.

[1] Bragg, *The Universe of Light* (London, 1933), p. 66.

We can observe a remarkable illusion in a train with the brakes on. Fix your attention on chimneys, houses, the frames of the windows or any other vertical objects; the moment the train slows down appreciably you will get the impression that all these vertical lines are tilted forwards and most distinctly of all just when the train comes to an abrupt standstill; immediately afterwards they stand upright again. Under these conditions even a horizontal meadow appeared to incline and then to become horizontal again. The explanation is that, while the brakes are applied, we feel ourselves sway slightly forwards as if the direction of gravity were altered. Relative to our muscular sense of this new 'vertical,' the real objects are inclined forwards (Fig. 93).

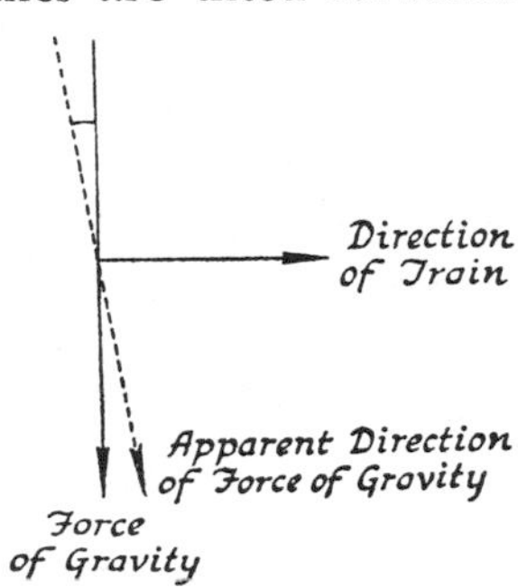

FIG. 93. Apparent change in the direction of gravity as a train slows down.

100. *How Movements are Seen*

As a rule, people think that movements are made apparent when a change in the position of an object relative to a fixed point is observed. But this is not invariably true; a velocity can be observed as a single impression, just as well as a length or a duration of time. When you watch the moving clouds, you get at once an impression of their direction and of their velocity.

It has been found that speeds as small as 1 or 2 minutes of arc per second are perceptible, but only when there are fixed points of reference in the field of vision (though we may be unaware that we are referring to them). Without such points the observation of speed is about ten times as uncertain. The fixed system for comparison in this case is your eye, the muscles of which make you feel that it is at rest, and relative to this frame of comparison you realise by your sense of vision that the images move over your retina.

Study the passing clouds and try, during the first quiet moments of watching, to determine at once the direction in which they move. Vary the conditions: high clouds, low clouds, gentle wind, strong wind, moon, no moon. A velocity of 2′ per second means that it takes the edge of a cloud 15 seconds to pass completely over the disc of the moon.

If one watches a net with wide meshes hanging out to dry, one can plainly see each gust of wind that passes over it, but when one fixes one's gaze on one of the meshes, hardly any movement is perceptible at all. It appears that the eye is very sensitive to a complex of small, mutually connected movements.

101. *Moving Stars*[1]

In the year 1850 or thereabouts, much interest was aroused by a mysterious phenomenon; when one looked intently at a star, it sometimes seemed to swing to and fro and to change its position. The phenomenon was said to be observable only during twilight, and then only when the stars in question were less than 10° above the horizon. A brightly twinkling star was first seen to move with little jerks, parallel to the horizon, then to come to a standstill for five or six seconds and to move back again in the same way, etc. Many observers saw it so plainly that they took it to be an objective phenomenon, and tried to explain it as a consequence of the presence of hot air striae.

But any real physical phenomenon is entirely out of the question here. A real motion of $\frac{1}{2}$° per second, seen by the naked eye, would easily be magnified to 100°, or more, by a moderately powerful telescope; that means that the stars would swing to and fro and shoot across the field of vision like meteors. And every astronomer knows that this is sheer nonsense. Even when atmospherical unrest is at its worst the displacements due to scintillation remain below the limit of perceptibility of the naked eye. *Psychologically*

[1] *Pogg. Ann.*, **92**, 655, 1857. For more recent literature concerning 'autokinetic visual impressions' *see Hdb. d. Phys.*, Vol. 20, *Physiologische Optik*, p. 174.

speaking, however, the phenomenon has not lost any of its importance. It may be due to the fact of there being no object for comparison, relative to which the star's position can be easily observed. We are not aware that our eye continually performs little involuntary movements, so that we naturally ascribe displacements of the image over our retina to corresponding displacements of the source of light.

Somebody once asked me why a very distant aeroplane appears invariably to move with little jerks when followed intently with the eye. Here the same psychological cause obviously comes into play, as in the case of the 'moving' stars, and 'very distant' seems to point to the fact that this phenomenon, too, occurs most of all near the horizon.

And how can we account for the fact that, suddenly and simultaneously, three people saw the moon dance up and down for about thirty minutes ?[1]

102. *Illlusions concerning Rest and Motion*

A very familiar illusion arises when, from the stationary train in which you are sitting, you see the train next to yours begin to move. You think for a moment that it is your own train that is slowly leaving the station. Or again, after looking for some moments past a high tower at the passing clouds, it seems as if the clouds are standing still and the tower is moving. In the same way some people can see the moon racing through motionless masses of clouds. Take care when crossing a narrow plank over a brook not to look at the running water underneath, for fear of dizziness; here your judgment of rest and motion is upset because an unusually large part of your field of vision is in motion. On one's first sea voyage, one sees the things hanging in one's cabin swing to and fro, while the cabin itself is at rest.

In all these cases the illusion is closely connected with the one in § 99. A closer psychological investigation has shown that we have a tendency to consider those things in motion which we know from experience to be usually the

[1] *Nat.*, **38**, 102, 1888.

moving elements in the landscape. But there is, besides, another very important and more general law: the notion of rest is, for us, automatically connected with the wider frame, the enclosing elements in our field of vision, whereas motion is connected automatically with the elements enclosed. In several cases mentioned above this second law clashes with the first one, and, as our illusions prove, completely overrules our common everyday experience.

I am sitting near the window of a railway compartment gazing dreamily at the ground as it races past, when the train stops, and while I am perfectly sure that it has come to a standstill, yet, on looking out, I receive the irresistible impression that it is gliding slowly backwards. Not, however, in such a way as to make the whole field of view shift at the same rate! Close to me the motion appears to be quicker; further away, slower; and somewhat to the right and to the left of the point at which I am gazing the motion is also slower. The entire landscape seems to rotate slowly round the point where I am sitting, but, in the manner of some elastic substance, it stretches and shrinks while it rotates. This rotation takes place in a direction contrary to the one when the train was in motion (§ 107). It would be amusing to go and sit quickly at the opposite window the moment the train stops; the rotation ought then to go on in the same direction.

It may be possible that unconsciously our muscles have grown accustomed to following the objects that come racing past, and that, when the train stops, these involuntary movements of the eye do not stop at once, so that for some time we add, so to speak, a constant 'compensating velocity' to the actual velocities. But it is impossible ever to explain by one single movement of the eye the manner in which the velocity changes towards the boundaries of the field of view. Experiments have been carried out in which an observer watched for some time small objects continually moving away from a central point in every direction; when the movement ceased, the points of light seemed to move back again towards the centre from all sides. This cannot possibly be explained by one single movement of the eye. It is therefore more likely that our 'mind' which has been taught to reduce the velocity by a definite amount in every part of the field of vision, continues to do so after the movement has ceased.[1]

The above phenomenon occurs also when we look steadily at a little spot on the window of the compartment, thereby eliminating the movement of the eye, provided the train does not go so fast as to make the outside objects resemble mere streaks.

On the other hand, however, an old observation of Brewster[2] points very positively to involuntary movements of the eye. Looking out of a train window, little pebbles nearby seemed drawn out into short streaks, but looking quickly at the ground somewhat further away, the pebbles appeared for a very short moment at rest, as if illuminated by an electric spark. This definitely proves, in my opinion, that our eye does indeed follow the moving objects, though not at exactly the same rate.

Brewster made still another observation (*loc. cit.*) while looking through a narrow slit in a piece of paper at the pebbles flying past; he noticed that when he suddenly turned away his eyes, while still looking through the slit, so that the image of the pebbles fell in his indirect field of vision, everything became clearly visible for a moment. What is the explanation?

When passing by a playground on my right, with a very long fence, I keep my head turned to the right while looking at the children. After one or two minutes I look straight ahead again, and see the stones in the road and other objects in front of me move from right to left. When I try to repeat the experiment, looking steadily at the railing this time, instead of at the children, the phenomenon is much less striking. When carrying out observations of this kind, one generally notices that one need not follow the rapidly

[1] Von Kries in Helmholtz. [2] *Proc. Brit. Ass.*, p. 47, 1848.

moving objects themselves with one's eye, but that it is better to gaze at some neutral background while images with strong contrasts of light and dark move across the retina.

I watch the falling snow-flakes and fix my gaze on one of them as it falls, then I look up quickly and select another flake, and so on for several minutes. If I now look at the snow-covered ground, I see it literally rise, while I feel as if I myself were sinking.

Look for a few minutes at the surface of a fast-flowing river or at floating blocks of ice as they drift, all the while fixing your gaze on the top of a mooring post, for example, or on some detail on an island. When you now look at the firm ground again, you see an 'anti-current movement.' Similarly, after admiring a waterfall for a time, the banks seem to move upwards. Purkinje, after looking out of his window for some time at a procession of men on horseback, had the impression that the row of houses across the road was moving in the opposite direction. When you walk along a narrow path through corn and look at the distant moon, the conditions are again particularly favourable for the occurrence of this illusion.

These conditions are briefly: (*a*) the movement should last at least one minute; (*b*) it should not be too fast; and (*c*) the eye should look steadily at an object either moving or stationary, and always in such a manner that the images moving over the retina show contrasts and well-defined details.

103. *Oscillating Double Stars*

This phenomenon was observed by the famous Herschel. Look through ordinary opera-glasses at the last star but one of the Great Bear. You clearly see a faint star close to the bright one (Figs. 61, 78). Carry out the experiment preferably when the faint star is more or less vertically under the bright one (though you may also succeed when it is in other positions). Move your opera-glasses gently a little to the left, then to the right, and back again to the left, and so on, just fast enough for the images of the stars to remain visible as little dots of light. It will then seem as if the faint star lags slightly behind the bright one each time, as if it were attached to it by a piece of string, and performs an oscillating motion (Fig. 94).

FIG. 94. Apparent swinging motion of double stars observed by means of opera-glasses moving to and fro.

The explanation lies in the fact that it takes some time for the light to stimulate our retina, and the brighter the star is, the shorter the time required; by the time we locate the fainter star, the brighter one has already moved a little farther.

This same phenomenon has been used in recent years by Pulfrich for the construction of a new type of photometer.

104. *Optical Illusions concerning the Direction of Rotation*

When the sails of a windmill turn in the twilight, and we look at their silhouette (Fig. 95, *a*) from a direction inclined relative to the plane of the sails, we can imagine their rotation to be clockwise just as well as counter-clockwise (Fig. 95, *b*). The change from one rotation to the other requires a momentary concentration of attention, but it is usually sufficient simply to go on looking at it quietly, when the rotation will appear to change of its own accord. Most meteorological stations are equipped with Robinson's anemometer, i.e. a little windmill, with a vertical axis of rotation. When I look quietly from a distance at its rotating vanes, without any conscious effort of my will they appear to reverse the direction of their rotation every twenty-five or thirty seconds. A wind-vane swinging to and fro can cause us to doubt in the same way, especially if it is not too high up (Fig. 95, *c*).

In all these cases our judgment of the direction of rotation depends on which parts of the track appear closer to us and which parts further away. Those parts which happen to claim our attention most seem, generally speaking, nearer to us. The reversal of the apparent direction of the rotation must, therefore, be ascribed to a sudden change in our attention.

105. *Stereoscopic Phenomena*

Looking through the pane of a railway carriage window of inferior quality, one can observe an amusing phenomenon. Wait till the train stops and look attentively at the pebbles on the ground. Keep your eyes close to the glass, and your head

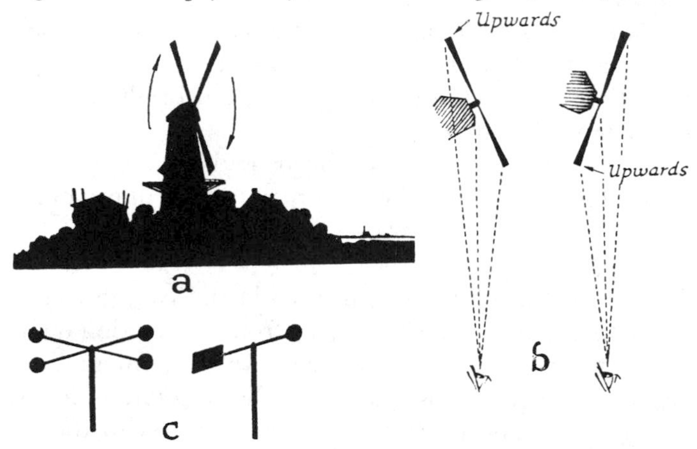

FIG. 95. The silhouette of a windmill in the evening: (*a*) What the observer sees. (*b*) How he can interpret his observation. (*c*) Other deceptive silhouettes.

steady and shake off the preconceived idea that the ground must look flat. You will suddenly perceive that it seems to undulate, even to undulate very strongly. If you move your head slowly parallel to the pane, the undulations seem to shift over the ground in the opposite direction; if you move away from the window, they seem to remain about equally high, but to become wider.

The explanation is that the window-pane is not perfectly flat, but of a very slightly varying thickness. As a rule these undulations run parallel to some definite direction, as a result of the pane having been made by rolling out the red-hot molten glass under steel rollers. Such an undulation is equivalent to a prism with a small refracting angle, and causes the rays of light to deviate slightly. In Fig. 96 the eyes L and R are assumed to look at the point A of the ground, so that the uneven thickness is not apparent. When they look at B, however, the ray BR is no longer straight, but broken along BCR. The result is that the eyes are directed as if they were looking at B′, which lies nearer to them than B. In another part of the pane the deflection of the rays will be different, so that then the object apparently recedes. This enables one to understand that a slight unevenness of the pane can cause the illusion of a strong undulation of the objects outside, though sometimes the way in which the effects on the separate eyes combine is fairly complicated.

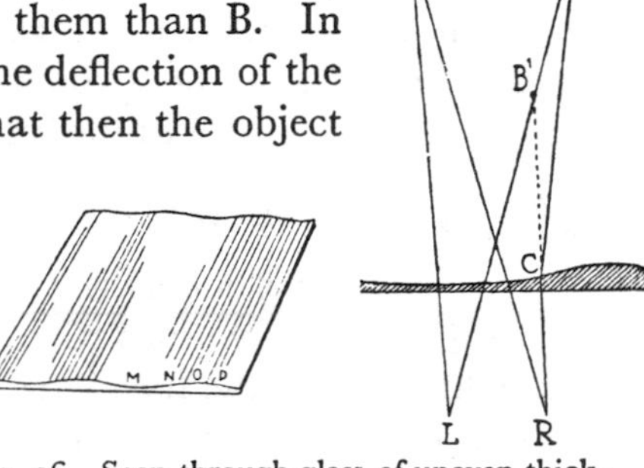

FIG. 96. Seen through glass of uneven thickness, the ground seems to undulate.

If, for example, the left eye looks through a flat part and the right eye through an uneven part of the pane, the details of the way in which the stereoscopic effects arise can easily be traced. Shut your left eye and rock your head slightly to and fro; the pattern on the ground will rock in the same direction for those parts where the pane is concave (M, Fig. 96), and in the opposite direction where it is convex (O). (Why?) If you now open both eyes, the parts M and O correspond to places on the ground that we see at normal distances. Looking with our right eye through N, we see a crest, through P a trough. Try to check all this by your own observations and to account for all the details.

A phenomenon closely related to this is to be observed when one stands very near a slightly rippled surface of water. The eyes try to fix the reflected image of, say, a branch of a tree; but as the two eyes do not look at the same point of the undulating surface, two images are seen at a continually changing angular distance from each other and it is impossible to adjust the axis of the eyes on it properly. This causes a very peculiar sensation, difficult to describe. As soon as we close one eye the surface of the water is scarcely perceptible and we can imagine that we see the tree itself moving in the wind, instead of its reflection. When we look with both eyes we become suddenly aware of the rippled surface itself, but this surface *glistens*; this is a characteristic phenomenon when the two eyes receive widely different images, the one light and the other dark.

106. *The Man in the Moon*[1]

The man in the moon is an excellent warning to us to carry out our observations with due objectivity! The dark and light spots on the moon are really flat plains and mountains, and their distribution is obviously very haphazard. Unconsciously, we try to distinguish in this fantastic distribution of light forms more or less familiar; we fix our attention on certain peculiarities so that these become clearer and more striking, whereas shapes to which no attention is paid become less distinct. Thus in the full moon at least three aspects of a human face can be seen: side view, three-quarter face and full face. Also, a woman's figure, an old woman with a bundle of twigs, a hare, a lobster, etc.

Illusions of this kind have played tricks on the best observers, the famous case of the canals on Mars being one instance out of many. It is as well to bear this in mind in connection with many a fantastic description of mirages or *fata morgana.*

107. *The Revolving Landscape and the Accompanying Moon*

Let us fix our attention on two trees or two houses at unequal distances from us; as soon as we move we notice that *the one farther away from us moves with us, and the one nearer to us remains behind.* This is a simple example of parallax, a geometrical phenomenon, having no special physical background.

One of the first things that struck me when I was sitting

[1] Harley, *Moon-Lore* (London, 1885). Titchener, *Experimental Psychology.*

in a railway train as a child was the way the landscape seemed to revolve. Suppose I look to the right out of the train, every object near to it races away to the right, whereas every object in the distance moves with me to the left. The whole scene seems to turn round an imaginary point, the point at which I happen to be gazing. Whether I direct my gaze into the distance or to a point close by, the points beyond the one I happen to be looking at seem to move with me, while those closer to me remain behind. Try this experiment yourself! It is clear that these visual impressions are a consequence of parallax; but the new element in addition to this, is that we refer everything to that point on which our gaze is fixed. This is a psychological peculiarity of our visual observations. Whether we walk, cycle, or travel by train, we see the faithful moon 'keeping us company' on the distant horizon. The sun and the stars too, only we don't notice these so much. This proves that our attention is fixed on the landscape, and therefore, owing to parallax, the celestial bodies beyond appear to move with with us, relative to the landscape.

108. *Searchlight Phenomenon.*[1] *Cloud-bands*

A searchlight casts a slender beam of light horizontally over a wide open space. Although I know that the beam runs in a perfectly straight line, I cannot get rid of the illusion that it is curved, highest of all in the middle and sloping down to the ground on both sides. The only way to convince myself that the beam really is quite straight from one end to the other is to hold a stick in front of my eyes.

What is the cause of this illusion? This tendency of mine to see the course of light as a curve is due to the fact that on one side I see it slope downwards towards the left, and on the other side towards the right. As if the straight lines of an ordinary horizontal telegraph wire did not behave in the same way! However, looking at the beams of light at night I have no point of reference in surrounding objects to enable

[1] Bernstein, *Zs. f. Psychol. und Physiol. der Sinnesorgane*, **34**, 132.

me to estimate distances and nothing is known to me, *a priori*, of the shape of the beam.

A similar phenomenon can be seen at night along a row of high street-lamps, especially when there are no parallel rows of houses, or when these are hidden by trees. The row of lights then looks curved just like the beam of a searchlight.

Immediately related to this is the observation that the line connecting the horns of the moon, between its first quarter and full moon, for instance, does not appear to be at all perpendicular to the direction from sun to moon; we apparently think of this direction as being a curved line. Fix this direction by stretching a piece of string taut in front of your eye; however unlikely it may have seemed at first, you will now perceive that the condition of perpendicularity is satisfied.

The rows of clouds which seem to radiate from a point on the horizon and meet again on the other side of the heavenly vault, run in reality straight, horizontal and parallel to each other. See also § 191.

If one stands near a lighthouse at night, with one's back towards it, one can see a most impressive sight. As the great beams sweep over the landscape, they seem to converge to an imaginary 'anti-light-source point' a little below the horizon, and to rotate about this point.[1] Observing one of these beams of light, I conclude that it lies in a definite plane determined by its true position in space and the point occupied by my eye. When the beam revolves, the position of this plane in space changes continually, but continues to pass through the line joining the lighthouse, my eye and the 'anti-light-source point.' Instead, therefore, of seeing the beams as horizontal lines radiating from a point behind me, I can imagine them to be 'rays with their lower part cut off revolving about an anti-light-source point, which lies below the horizon.' The fact that I involuntarily adhere

[1] G. Colange and J. le Grand, *C.R.*, **204**, 1882, 1937, are erroneously of the opinion that the phenomenon is only visible under the very exceptional circumstances prevailing at the powerful lighthouse at Belle-Isle. It could, however, be observed very well near the small lighthouse at de Koog in the Netherlands.

to the second point of view is psychologically remarkable, and arises from my tendency to consider converging lines as belonging together and to prolong them up to their vanishing point.

109. *Apparent Flattening of the Vault of Heaven*[1]

When we survey the sky from the open fields, the space above us does not generally give us the impression of being infinite, nor of being a hollow hemisphere spanning the earth. It resembles, rather, a vault whose altitude above our heads

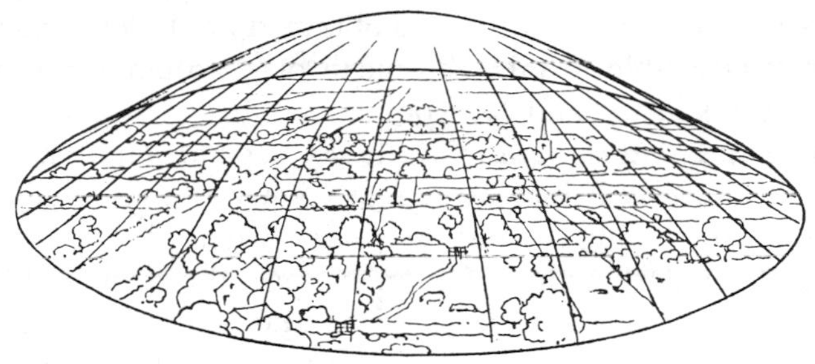

FIG. 97. The sky seems to arch over the earth like a kind of dome.

is less than the distance from ourselves to the horizon (Fig. 97). It is an impression and not more than that, but for most of us a very convincing one, so that its explanation must be psychological and not physical.

It is naturally impossible actually to measure this flattening in any way; we can, however, form estimates:

(*a*) We begin by asking what the ratio $\frac{\text{eye to horizon}}{\text{eye to zenith}}$ may appear to be; the answers lie mostly between 2 and 4, according to the observers, and to the circumstances under which they observed.

(*b*) We estimate, as well as we can, the direction in which the centre *of the arc* joining zenith to horizon appears to lie. When this estimated centre is checked, it turns out

[1] For the very extensive literature on this subject and the following one *see* A. Müller, *Die Referenzflächen der Sonne und Gestirne*; E. Reimann, *Zs. f. Psych. u. Physiol. der Sinnesorgane*, 1920; R. von Sterneck, *Der Sehraum auf Grund der Erfahrung* (Leipzig, 1907).

not to lie at a height of 45°, but much lower, mostly at heights from 20° to 30°; values as low as 12° or as high as 45° have been given, though rarely.

It is important that unprejudiced observers should be found, to whom it should be made clear that they must divide, *not the angle but the arc* into two parts. It is also very important to estimate the position of the zenith correctly; the best way to do this is to turn first towards one point of the compass, and then towards the one opposite, and then to see if the two estimates thus obtained are in agreement.

It is advisable to take the mean of 5 values for each of the figures in (*a*) and (*b*) above.

The apparent flattening of the sky depends on a variety of circumstances. It increases greatly when the sky is clouded, and in the twilight, and decreases when the stars shine brightly on dark nights. On an average, the estimated lower 'half' of the angle between horizon and zenith is 22° by day and 30° by night. Note that observations in this connection made at sea are of special value, the view there being open on all sides, and there being no distractions to interfere with the estimate.

Seen through a piece of red glass (large enough for its edges not to be disturbing), the sky seems flatter; seen through blue glass, it seems higher and more hemispherical.[1]

More detailed estimates can furnish still more accurate information as to the arched shape which we unconsciously ascribe to the sky. To many observers it appears to have a shape like a kind of helmet.

110. *Over-estimation of Heights* (Fig. 98)

The apparent flattening of the vault of the sky seems to be connected with the fact that we generally over-estimate heights above the horizon. It is evident that we invariably confuse, unconsciously, the measurement of the *arc* and the measurement of the *angle*; point M, chosen in such a way

[1] Dember en Uibe, *Ann. Phys.*, **1**, 313, 1920.

that HM=MZ, is much less than 45° above the horizon, though to us it appears to lie midway between zenith and horizon.

The sun in the winter seems to be fairly high in the sky at noon, and yet its height in our latitudes is only 15° above the horizon. In summer it appears to reach nearly to the zenith, whereas in reality its height hardly exceeds 60°.

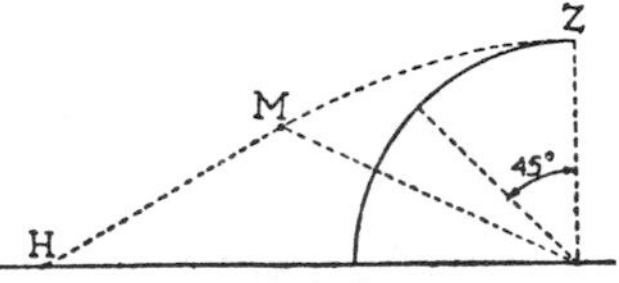

FIG. 98. The division of the apparent arc from zenith to horizon into two parts.

In a similar way we over-estimate the height of hills and the steepness of a rising slope in front of us. Observers have even described the 22° halo round the sun or moon (§ 134) as being higher than it is broad.

These illusions can be avoided to a great extent by looking at the landscape with half-closed eyes, when light and dark parts are seen only as large masses.

111. *The Apparent Increase in Size of the Sun and Moon on the Horizon*

This is one of the strongest and most generally known optical illusions. The rising moon can be very large, but when it is high in the sky it is quite small! And the sun, 'the huge, tomato-red sun, grows larger and larger!'

But is it, after all, an illusion? Let us project an image of the sun and then measure it. Take a spectacle-lens with a focal length of about two yards,[1] fix it in a slit in a cork and put it on the window-sill in the rays of the setting sun (Fig. 99A). The window should be open or its glass will impair the sharpness of the image. We hold a sheet of paper at about 2 yards behind the spectacle-lens to catch the rays of light, and on it a fine sharp picture of the sun appears! Should it not be perfectly round, the spectacle-lens cannot have been standing quite perpendicularly to the incident sunbeams; turn it to and fro and incline it more and then less. When you have once ascertained exactly where the

[1] Opticians call it '+0·50.' Ask for a round, unfinished glass with raw edges.

paper should be held so as to obtain as sharp a sun-picture as possible, indicate the diameter with two pencil-marks and then measure it with a ruler, to within 0·5 mm. Measure,

FIG. 99A. Projection of an image of the sun with a spectacle-lens of long focal length.

FIG. 99B.

preferably, the *horizontal* diameter, because the vertical one is slightly reduced by atmospheric refraction. Repeat these measurements a few times and then take the average.

Now carry out the same experiment when the sun is high. The arrangement is a little more difficult. Nail the cork, with the lens still on it, high up on a post; by choosing the right side of the post, and by turning the cork round the nail, one can adjust the spectacle-lens quite perpendicular to the rays of light (Fig. 99B). Measure the image of the sun; it is (within the errors of observation) just as large *when the sun is high as when it is low.* Even the most accurate measurements with the most powerful telescopes do not show the slightest trace of difference.

It is clear, therefore, that the increase in the size of the sun and moon near the horizon is a psychological phenomenon. But even that is governed by fixed laws and can be estimated numerically. Take a white cardboard disc of about 12 inches in diameter and stand so far away from it that the disc appears the same size as the moon. Of course, this cannot be done by direct *comparison* or you would find, just as in the case of an actual measurement, that the size is invariably the same. You must, therefore, first look at the moon and impress strongly on your brain how large it looks, then turn round and compare that impression with the apparent size of the disc. A still better way is to fasten a series of white discs on a black background and then go and stand always at the same distance from it. Carry out these estimates when the moon is high as well as when it is low.

Similar comparisons can be made with the sun. Use a dark glass—a heavily-fogged photographic plate, for instance—to avoid being dazzled, and look afterwards at the discs with your naked eye. The observations are difficult because the psychological phenomenon is influenced by a variety of subtle factors, variations in the attention you give to them, etc. Notice how much better you succeed after a little practice!

The numbers obtained in this way show us that the sun and the moon look as much as 2·5 to 3·5 times as large near the horizon as they do high in the sky! The difference between the physical and the psychological phenomenon is therefore very striking indeed. The effect is still greater in the twilight and when the sky is clouded.

The apparent increase in the size of the setting sun is much more striking where the country is flat than when the sun sets behind high mountains; at sea, however, the increase is only slight.[1]

112. *The Connection between the Apparent Increase in Size of the Heavenly Bodies near the Horizon and the Shape of the Heavenly Vault* (Fig. 100)

Attempts have been made to reduce the phenomenon just described to the apparent flattening of the vault of the sky. The idea is, that we imagine the sun and moon to be as far away from us as the sky around them; therefore the sun, when low, would seem to be several times farther away than when it is high; the fact that, nevertheless, its angular diameter is the same causes us unconsciously to ascribe to it a

[1] Vaughan Cornish, *Scenery and the Sense of Sight* (Cambridge, 1935), Chap. II, which contains an interesting theory about the phenomenon.

size several times larger. From Fig. 100 we see that, since α is the same for both positions of the sun,

$$\frac{s_1}{s_2} = \frac{r_1}{r_2}$$

In order to check this connection, the apparent sizes of the sun and the moon at various heights above the horizon were estimated (cf. § 111). These experiments are difficult. Results obtained in the daytime when the sky was blue, and on cloudless starry nights, prove that the size of the sun and moon seem indeed to vary more or less in proportion to the distance of the vault of the sky. The proximity of clouds (*not* the terrestrial objects outlined against the horizon) makes the low-lying sun seem larger; the reason for this is that a clouded sky looks much more flattened than a cloudless sky, and, therefore, at the horizon, much farther away from us, and unwittingly we push the sun, as it were, so far back as to prevent our thinking it to be in front of the clouds. In the same way the moon, when low, is thought of as being larger by day if in proximity to masses of clouds. It is very remarkable that when the sky is clear the moon looks much larger in the twilight than by day or at night: this agrees with the greater flattening of the vault of the sky in the twilight. If the night is misty, so that the moon illuminates the neighbouring parts of the sky strongly, it seems to us as if the only slightly flattened night sky is once more replaced by the flat twilight shape and the moon again seems larger. Anyone inclined to the opinion that the apparently larger size of the moon when near the horizon or surrounded by mist is connected with its decrease in intensity of the light, can be set right by the two following observations: (*a*) the *crescent* of the moon does *not* look larger in the mist, which is easily understood, seeing that the crescent illuminates the

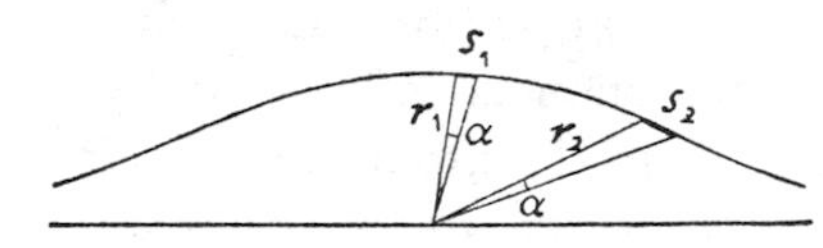

FIG. 100. Where the vault of the sky seems farther away, the sun's disc looks larger.

surrounding sky only slightly; (*b*) during a lunar eclipse the moon, high in the sky, does not look larger. From all that has just been said, it is clear that the sky in the background is the main essential and that this determines our estimate of the size of the sun and moon. And yet we must admit that there are also objections to thus closely connecting these two phenomena: many persons see the sun and moon on the horizon 'closer than ever,' or are absolutely incapable of saying anything as regards the apparent distance, although they may have a strong impression of the increase in size. Objections of this kind need not, in my opinion, be decisive, for it is quite probable that by putting the question as to the distance so directly, we arouse other psychological impulses than those by which their unconscious judgment is chiefly influenced.

113. *The Concave Earth*

This is the counterpart of the visual impression made by the vault of the sky. When the air is clear the earth surveyed from a balloon appears to curve upwards, so that we seem to be floating above a huge concave plate. The horizontal plane through our eye appears to us invariably as a flat plane; other horizontal planes in the distance, above or below it, seem to curve towards this fixed plane.

When the balloon is sailing a few miles above banks of clouds these too seem curved, their convex side being turned towards the earth and the concave side turned upwards. Should we happen to be between two layers of clouds, one above us and one below, we feel as if we are floating between two enormous watch-glasses. Similar observations can be made from an aeroplane.

114. *The Theory of Under-estimation*

Sterneck very cleverly succeeded in finding a mathematical formula for that apparently vague psychological phenomenon of 'the celestial vault.' Though it may be true that he does not give a definite explanation of it, he at least connects it with a large group of observations familiar to us in our daily experience.

The farther away the objects are, the more difficult it becomes to distinguish their distances. Street-lamps farther away than 160 to 170 yards seem at night to be all at the same distance. None of the mountains on the horizon, and none of the celestial bodies seem farther away than any of the others. The average, uninstructed observer under-estimates all long distances, e.g. a fire at night, the lights of a harbour seen from the open sea.

For objects close to us, this under-estimation is slight; it increases according to the increase in distance of the objects; finally, the apparent distance approaches a limit. Rectangular fields seen from a train resemble trapezia (with side a at the top) because the angle subtended by side a is in accordance with its true distance, but is too small for its apparent distance. As the train approaches a tunnel and you look out of the window at the brick wall of the entrance to the tunnel, the bricks will be seen to swell and become larger. The explanation is that, if the true distance becomes twice as small, the bricks subtend twice as large an angle, but the apparent distance appears to be only one and a half times smaller (for instance), and therefore it seems as if the bricks themselves had increased in size.

Von Sterneck attempted to connect the apparent distance d' and the true distance d by the following simple expression:

$$d' = \frac{cd}{c+d},$$

c being a constant for each particular case, the greatest distance which can be estimated under the given circumstances of illumination, etc.; c varies from about 200 yards to ten miles. We see that by this formula d' is practically equal to d so long as d is small compared with c; if d becomes of the same order as c an increase in under-estimation arises; if d is larger, the apparent distance approaches a limit. The formula therefore provides a good qualitative description of experience, and more detailed observations showed also a surprisingly good quantitative agreement.

The theory of under-estimation explains how an observer O, standing at the bottom of a mountain, over-estimates the steepness of the slope, estimating the distance OB as if it were OB′—that is, seeing AB as AB′; and it demands, as a logical consequence of this, an under-estimation of the downward slope by an observer standing at the top (Fig. 101). We shall now see how it attempts to explain the apparent shape of the celestial vault and with this also the apparent increase in size of the celestial bodies near the horizon.

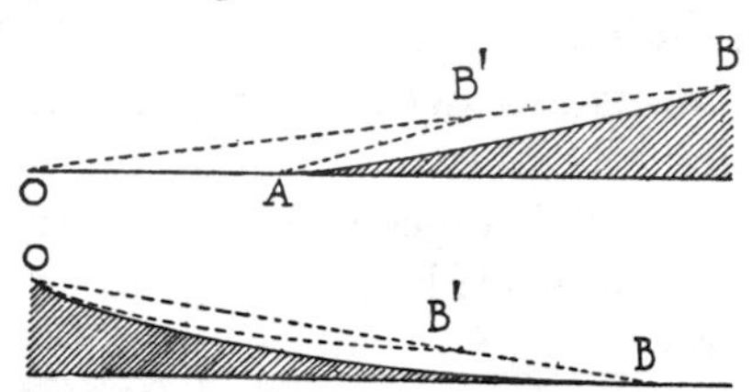

FIG. 101. Observer O over-estimates the upward slope and under-estimates the downward slope.

Let us imagine a cloud-bank at a height of one and a half miles above our head. This bank ought to resemble an extremely flat plate, for, owing to the curvature of the earth, our eye is at a distance of about 110 miles from the layer of clouds on the horizon and only 1·5 miles from the layer of clouds at the zenith. The clouded sky, however, does not look at all like this! The short distance is under-estimated slightly, the long distance very much. Assume that we estimate the ratio $\frac{\text{eye to horizon}}{\text{eye to zenith}}$ to be about 5; this would mean that under these circumstances $c = 6{\cdot}6$ miles: the formula for the under-estimation theory furnishes the correct values (try this yourself!). It follows from this that we ought to see the clouded sky as a kind of vault, a hyperboloid of revolution, which does indeed agree with our general impression of it. Note therefore that we actually do not see the celestial vault flattened, but, on the contrary, relatively *higher* than it is!

But what about the blue sky, and the starry sky? Von Sterneck simply takes a new value for the constant c each time and his formula describes the observation in each definite case with surprising accuracy. However, it is difficult to understand how in these cases we can speak of a certain 'distance' being under-estimated. And this leads us to the more general questions: how do we get any impression at all of distance in the case of such indefinite objects as clouds? And blue sky? And a cloudless sky at night? The under-estimation theory may be true as far as terrestrial objects are concerned whose dimensions and distances are known to us by every sort of experience; it is, however, very doubtful whether it can be applied to the sky above. Besides, no light has as yet been thrown on the origin of the under-estimation.

115. *Gauss' Theory of Visual Direction*

In connection with the above paragraph there are a number of observations which show that the shape of the celestial vault and the apparent increase in size of the celestial bodies near the horizon depend on the direction of our gaze in relation to our body. Therefore Gauss assumed that the experience of many generations has made us better adapted to the observation of those things that are in front of us than of those above us, and that this influences our estimation of distances and dimensions.

When the full moon is shining high up in the sky, we sit down in an easy chair, or on the ground with our backs against something sloping. If we lean far back, but hold our head in its usual position with regard to the rest of our body, and observe the moon, it seems to be appreciably larger. If we get up suddenly, so that our gaze has to be directed upwards, the moon seems to be smaller again. *Vice versâ*, the full moon near the horizon looks much smaller when we lean forwards.

Both phenomena can be seen alternately when the sun is 30° or 40° high and its light is tempered by mist. Lean backwards and forwards and the disc appears alternately larger and smaller. Lie with your back flat on the ground;

the sky appears compressed on that side towards which your head is lying *now*, whereas on the side opposite it seems perfectly spherical (Fig. 102). This shows us clearly that (relative to our body) the gaze directed downwards and the gaze directed forwards are, in the present connection, practically equivalent, whereas the upward gaze makes the objects seem compressed.

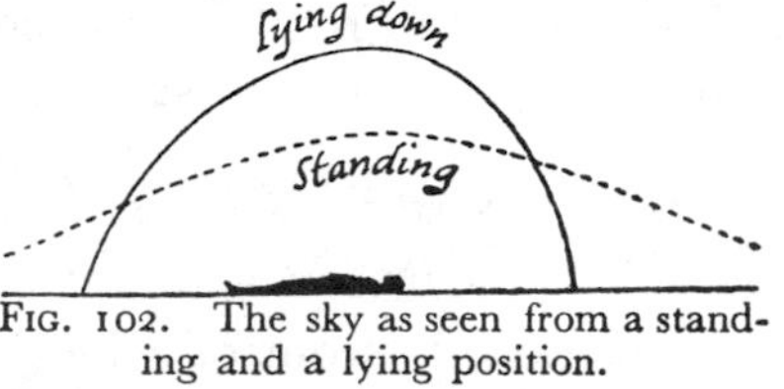

FIG. 102. The sky as seen from a standing and a lying position.

Hang by your knees from the horizontal bars and look round you while your head is hanging down. The sky will look hemispherical.

All these observations confirm each other. In addition to this, constellations seen through a telescope, and therefore free from any external influences in the landscape, likewise seem larger when they are lying low on the horizon; the only thing that can have any influence here is the direction of the gaze.

Now, do not try to check this further by judging the apparent size of sun and moon in a mirror so that in this way you see, for instance, the moon high up while your gaze is directed horizontally. If the observer is conscious of the presence of a mirror in any way, the illusion is partly lost. For this reason this kind of experiment is very difficult to perform.

Various other theories concerning the visual impressions just mentioned are easily refuted. For example, it has been maintained that a 'physical theory' as to the shape of the celestial vault can be given, and this theory amounts to the principle, an incomprehensible one, that the brighter the sky is the farther away it seems to be, the distance varying as the square root of the brightness. The sky when blue is darker at the zenith than along the horizon and this would make it look lower. This theory is, however, sufficiently refuted by the fact that the sky when uniformly clouded proves to be brighter at the zenith than on the horizon, and yet appears flattened to us. And, besides, when the sky is clouded, the part of the clouds in front of the sun which looks brighter than the rest always seems nearer to us than the surrounding parts of the sky.

116. *How Terrestrial Objects Influence our Estimation of the Distance to the Celestial Vault*

If you stand in front of a long row of houses and look at those immediately facing you, the sky there will seem much nearer to you than it does above those at the end of the row.

Apparently we estimate the distance of the sky to be about 50 to 60 yards! But the fact that we see objects which we *know* to be very distant is sufficient to make their background, the sky, seem much farther away. To a certain extent, each terrestrial object has its own background in the sky. This shows how purely psychological all these phenomena must be, and how impossible it is to speak of an ideal 'surface of reference' which for us would be *the* celestial vault!

Look down a long railway track or a wide road bordered by trees which make us realise the great distance: the sky seen in that direction seems much farther off than at other points of the compass. If, however, you screen off the landscape as far as the horizon with a sheet of paper, the sky there at once seems nearer.

As a counterpart of this, we can direct our gaze in a *vertical* direction in the same manner; the sky will then seem higher. This phenomenon is particularly striking when seen from the foot of a high tower or, still better, near the tall, slender aerial towers of a large wireless station. The sky above seems to arch until it forms a kind of cupola; between three towers the whole sky seems to be pressed upwards.[1] Different observers draw, independently of each other, the apparent shape in the same manner (Fig. 103).

If when looking towards one of these towers you divide the arc from the horizon to zenith into two parts (§ 109), the lower part will appear to be much greater than if you made

[1] H. Stücklen, *Diss.* (Göttingen, 1919).

your estimate with your back to the tower and at some distance from it. The angles subtended by the lower part will now seem to be greater than 45°, even as great as 56°, which means that the vault of the sky looks higher than a hemisphere!

However convincing these observations may be, remember that in themselves they can never explain the shape of the celestial vault nor the apparent increase in size near the horizon. Even when observed through a very dark piece of glass, the sun when high will always appear small, and large when it is low, even though no more of the landscape is to be seen.

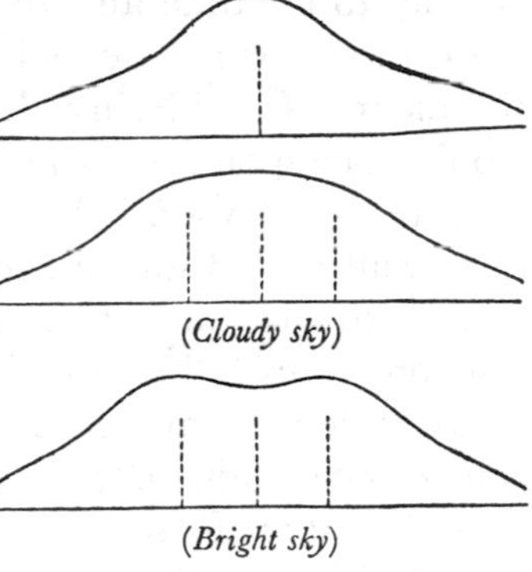

FIG. 103. The apparent shape of the sky above aerial towers.

117. *The Apparent Size of the Sun and Moon given in Inches. The Method of After-images*[1]

We know that we cannot estimate the size of the sun and moon in linear measure; we can only designate the angle subtended by them at the eye. And yet it is remarkable that a great number of people maintain that the celestial bodies are to them as large as soup-plates, while there is a minority that suggests dimensions of the order of a coin. Should you feel inclined to smile at this, remember that even a scientifically minded man *feels* how perfectly impossible it would be to state that the diameter of the moon looks like 1 mm. or 10 yards, whereas he *knows* quite well that 1 mm. at a distance of 4 inches, or 10 yards at a distance of 1,000 yards, would cover the moon exactly. The psychological factors playing a part here are as yet very little understood.

Everyone knows that after-images of the sun can be obtained by glancing swiftly at it and then blinking (§ 88). This after-image is projected on every object we gaze at

[1] G. ten Doesschate, *Nederl. Tijdschr. voor Geneesk*, **74**, 748, 1930.

afterwards. On a wall near to us it looks very small and insignificant, on objects farther away it seems larger. (But note: we do not estimate the angle it subtends but the size of 'the thing itself.') This effect is perfectly comprehensible for if an object at a distance is to subtend the same angle as an object near to us, it must be larger in linear measure. When does the after-image look as 'large' as the sun itself? According to the opinion of different observers this occurs whenever the wall is from 55 to 65 yards off and this applies equally to daytime and to night-time: this therefore shows the distance which we feel to be between us and the sun or the moon. Considering that the angle subtended is $\frac{1}{108}$ this would correspond to a diameter of 18 to 22 inches.

In the same way, it has been shown that the after-image on a wall at a distance greater than 65 yards still looks as large as that on the sky immediately above it (that is, near the horizon), while the after-image projected high in the sky looks decidedly smaller than on a wall 65 yards off. This shows once again that to us the distance to the sky above looks less than to the horizon and that 65 yards is about the 'limiting distance' in the theory of under-estimation (cf. § 114).

X

Rainbows, Haloes and Coronae

RAINBOWS

The following simple observation is an introduction to the study of the rainbow. What we see occurring in a single drop of water is also visible in millions of raindrops and gives rise to a glowing arc of colour.

118. *Interference Phenomena in Raindrops*[1]

Many people obliged to wear glasses out-of-doors complain that raindrops distort the images and make them unrecognisable. It may perhaps console them if we call their attention to the splendid interference phenomena visible in these same raindrops. All they need to do is to look at a source of light in the distance—a street-lamp, for example. A raindrop that happens to be exactly in front of the pupil becomes strangely distorted, a spot of light with extraordinary projections and indentations, with a border of very beautiful diffraction fringes in which colours, too, are distinguishable (Fig. 104, *a*).

One remarkable thing about it is that the spot of light remains in the same place even when the eyeglass is moved slightly to and fro. Another is that the general shape and protruding curves of the spot of light seem, at first, to bear no relation whatever to the shape of the raindrop. The explanation is simple. Regard the eye as a small telescope forming an image of a source of light in the distance, and the drop of water as a group of small prisms held in front of the objective. It is then clear that each small prism refracts a group of rays laterally, independently of its position on the

[1] Larmor, *Proc. Cambr. Philos. Soc.*, 7, 131, 1891.

objective (provided it is still within the opening of the objective); the shape of the light-patch will depend, however, on the value of the refracting angle and on the orientation of each small prism. A drop of water extended vertically does indeed give a horizontal streak of light.

But now the diffraction fringes! These would not exist if the drop of water formed an accurate lens and so imaged the source of light exactly at a point, for then all parts of the light wave-front, since they left the source simultaneously, would arrive at the image together with no change of relative phase. But since the surface of the water is curved irregularly the refracted rays do not meet in a focus, but are enveloped by a *caustic* (Fig. 104, *b*). In such a case one

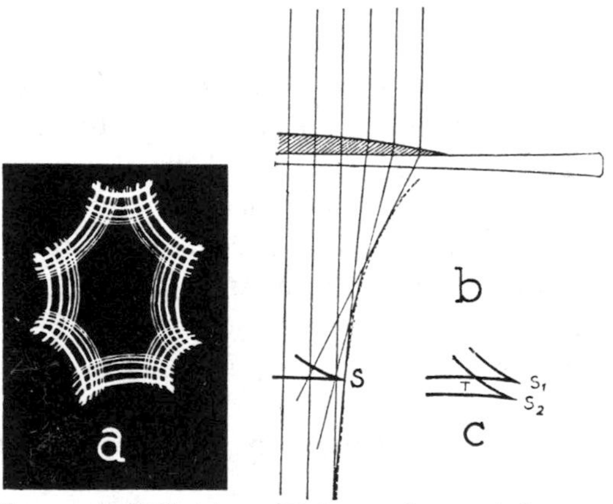

FIG. 104. Diffraction of light rays by a raindrop on an eyeglass. (*a*) Interference pattern. (*b*) Path of the light rays; the dotted line is the enveloping caustic; the thicker line is the wave surface, having its cusp at S. (*c*) Two consecutive wave-fronts, both passing through T.

always finds that through a point in the neighbourhood of the caustic there pass two different rays which have traversed light-paths of different lengths; interference therefore occurs. On drawing the wave surface one finds a point of reversal giving a cusp; at any moment, therefore, there will always pass through a point T two wave-fronts with a definite difference of phase (Fig. 104, *c*).

The distances of the dark fringes measured from a fixed point are given by the formula $\sqrt[3]{(2m+1)^2}$ where $m=1, 3, 5 \ldots$ These distances are therefore in the same ratio as 2·1; 3·7; 5·0; 6·1, etc.

119. *How a Rainbow is Formed*

> My heart leaps up when I behold
> A rainbow in the sky.
>
> WORDSWORTH.

It is a summer afternoon and oppressively warm. Dark clouds on the western horizon: a storm is brewing! A black arch of clouds is rising rapidly, behind which the sky in the distance seems about to clear; the front edge has a light border of cirrus clouds with fine transverse stripes. It spans the entire sky and passes, awe-inspiring, over our heads with a few peals of thunder. All at once the rain comes pouring down; it has grown cooler. The sun, already low, shines again. And in the storm, drifting eastwards, appears the wide arch of a many-hued rainbow.

Whenever it may occur, a rainbow is always formed by light playing on drops of water. They are generally rain-drops, occasionally the fine tiny drops of mist. In the finest, smallest drops of all, those which constitute the clouds, it can never be seen. So that if ever you hear anyone say that he has seen a rainbow in the falling snow or when the sky was quite clear, be sure that the snow was half-thawed or that one of those showers of drizzle was falling which occur sometimes without there being any clouds. Try to make more of these interesting observations yourself! The drops in which a rainbow arises are usually not much farther than a half to one and half miles away from us (Plate IX, *a*; p. 202). On one occasion, I saw the rainbow standing out clearly against the dark background of a wood at a distance of 20 yards from my eye: the rainbow itself was therefore still nearer. A case is known in which a rainbow was seen in front of a wood 3 yards away![1]

[1] *Nat.*, 87, 314, 1913.

According to an old English superstition, a crock of gold is to be found at the foot of every rainbow. There are, even nowadays, people who imagine that they can really reach that foot, that you can cycle up to it, and that a peculiar twinkling light is to be seen there. It must be clear that a rainbow is not at one fixed point like a real thing; it is nothing but light coming from a certain direction.

Try to photograph the rainbow on an ortho- or panchromatic film with yellow light filter; speed $\frac{1}{10}$ second, with stop F/16.

120. *Description of a Rainbow*

'Rubens' rainbow . . . was dull blue, darker than the sky, in a scene lighted from the side of the rainbow. Rubens is not to be blamed for ignorance of optics, but for never having so much as looked at a rainbow carefully.'

RUSKIN, *The Eagle's Nest*

The rainbow is a part of a circle; the first thing that occurs to us is to make a rough estimate as to where its centre lies, that is to say the direction in which we see that central point. We notice immediately that this central point lies below the horizon and we easily find that it is the point to which the prolongation of the straight line from the sun to the observer's eye (after penetrating the earth) is directed: that is *the anti-solar point*. This line is the axis to which the circle of the rainbow is attached like a wheel (Fig. 105). The rays from the rainbow to the eye form a conical surface; each of them makes an angle of 42° (=half the angle at the vertex of the cone) with the axis.

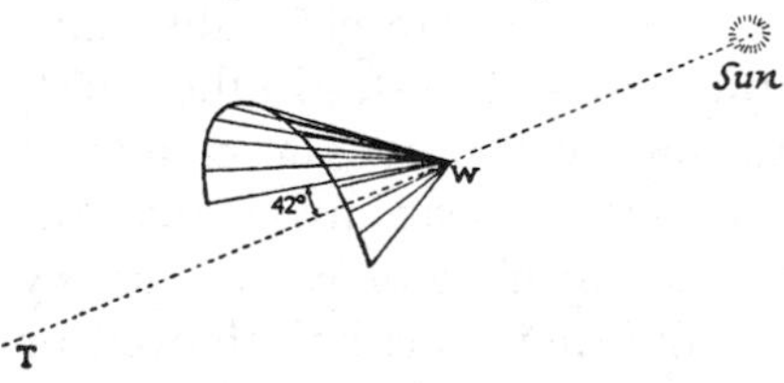

FIG. 105. The direction, in relation to the sun, in which we see the rainbow.

The lower the sun descends, the more the anti-solar point, and therefore the whole rainbow, ascends, more and more of the circumference appearing above the horizon until it becomes a semicircle as the sun sets. On the other hand, it disappears completely below the horizon when the sun is higher than 42°; this is why no one in our part of the world has ever seen a rainbow in the summer at about midday.

Measure half the angle at the vertex by pinning a postcard on to a tree and turning it in such a way that one of its edges points exactly to the top of the rainbow; the shadow of the pin then gives the direction of the line between sun and observer, so that the angular distance of the rainbow to the anti-solar point can be read at once (Fig. 106).

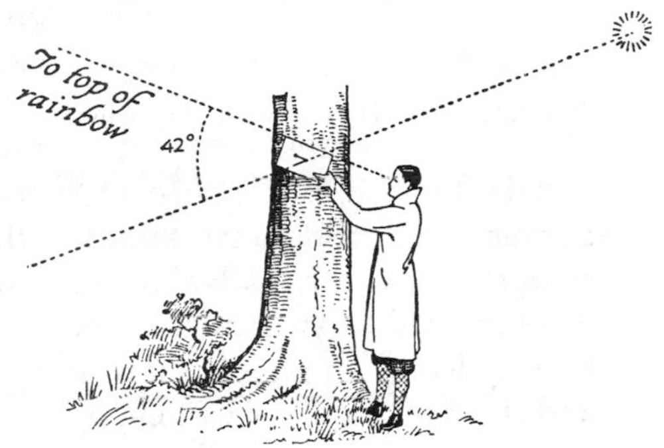

FIG. 106. Measuring the angular distance from the rainbow to the anti-solar point.

One can also apply one of the methods in § 235 for the determination of the angular height of the top above the horizon (Fig. 107) and the angle $2a$ between the two ends of the arc, making a note of the time of observation. Later on one computes the height of the sun then, which furnishes at the same time the angle H of the anti-solar point T below the horizon. From these measurements one finds three values for the angular radius to be determined, of which one can take the average value, namely:

$$r = H + h$$

$$\cos r = \cos a \cos H$$

$$\tan r = \frac{1 - \cos a \, . \cos h}{\cos a \, . \sin h}$$

Properly speaking, the rainbow should not appear as an arc, but as a closed circle; we cannot follow it farther than the horizon because we can see no raindrops floating below us. In *Physica* it was stated that one ought to be able to see the complete circle from an aeroplane, with the shadow of the plane in the centre. This grand sight has indeed been observed.

A secondary rainbow round the primary rainbow is regarded by many people as exceptional. As a matter of fact, however, it is nearly always visible, though naturally much weaker than the primary rainbow. It is concentric with it, and therefore also has the anti-solar point as its centre, but its rays form an angle of 51° with the axis of sun and eye.

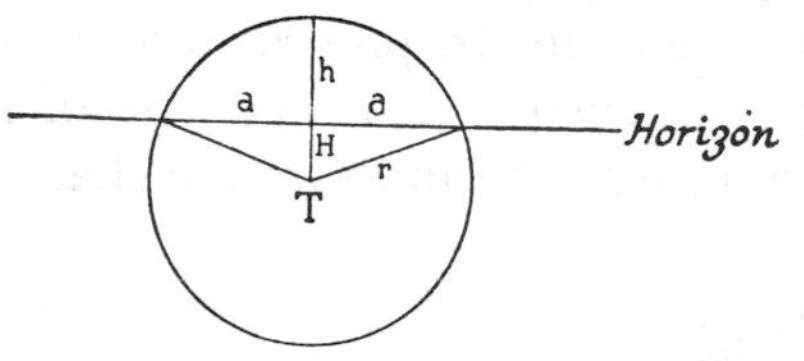

FIG. 107. a, h, H, r are all *arcs* which are measured in degrees.

The 'seven colours of the rainbow' exist only in the imagination; it is a figure of speech which is long-lived, because we so seldom see things as they are! In reality, the colours merge gradually into one another, though the eye involuntarily sorts them into groups. It is a striking fact that different rainbows show great differences; even one and the same rainbow can change while it is being observed, and the top become different from the lower parts. In the first place, great differences are found when one simply measures the total breadth of the colour band in angular measure (*see* Appendix, § 235). Further, although the order of colours is always red, orange, yellow, green, blue, violet, the relative breadths of the different colours and their brightnesses vary in every possible manner. My impression is that different observers do not always describe the same rainbow in the same way; therefore, in order to be sure of the difference between the rainbows, either the observations of a single person should be compared or one should ascertain beforehand whether two observers generally agree in the impressions they form.

This unprejudiced description of the colours of the rainbow brings to our notice the remarkable fact that often, beyond the violet on the inside of the bow, there are several *supernumerary bows as well*; as a rule they are to be seen best of all where the rainbow is at its brightest, near the highest point. Their colours are usually alternately pink and green. As a matter of fact, their name has been wrongly chosen, for although they are weaker they form just as much part of the rainbow as the 'normal' colours do. These supernumerary bows often change rather quickly in intensity and breadth, this being an indication of alterations in the size of the drops (§ 123).

The order of the colours in the secondary bow is the reverse of that in the primary rainbow: the red of one bow faces the red of the other. The secondary rainbow is very seldom so bright that its supernumerary bows become visible; they lie beyond the violet and therefore outside the outer edge of the secondary rainbow.[1]

121. *The Rainbow close to our Eye*

The way in which a rainbow arises in a mass of drops of water is directly visible to us when we see the sun shining on the fine spray floating above fountains and waterfalls. Along the side of a steamer, where the waves break against the bow and splash into foam, rainbows are sometimes seen to accompany the ship for quite a long time, sometimes strong, sometimes weak, according as the cloud of fine drops is dense or thin; especially when the ship's course is in the direction of the sun is there a good chance of your being able to see this phenomenon.

Here are some simple methods for reproducing in our garden a shower of rain in which the sun will form a rainbow: (*a*) garden-hose; (*b*) Tyndall's apparatus,[2] in which a jet of water under pressure splashes on to a round metal disc and scatters into small drops; or (*c*) Antolik's vaporiser (Fig. 108); all that is required is to blow forcibly with the mouth at *a*. The size of the drops can be regulated by moving the small tube *bcd* a few millimetres up or down the wider tube *ef*, which is done by shifting the perforated cork disc *g*; an important factor, too, is the size of the aperture at the end *u*. Water can be added through the wider tube *a* without having to open the apparatus. My own experiments with this little apparatus have been very satisfactory.

[1] Observed by Brewster in 1828. [2] *Phil. Mag.*, **17**, 61, 1883.

The tiny drops sprayed by the vaporisers used for our hot-house plants are so fine that a real rainbow cannot be seen in them, only a white mist-bow with blue and yellow edges (cf. § 128). Only here and there a few patches of large drops occur accidentally and the ordinary rainbow is momentarily visible.

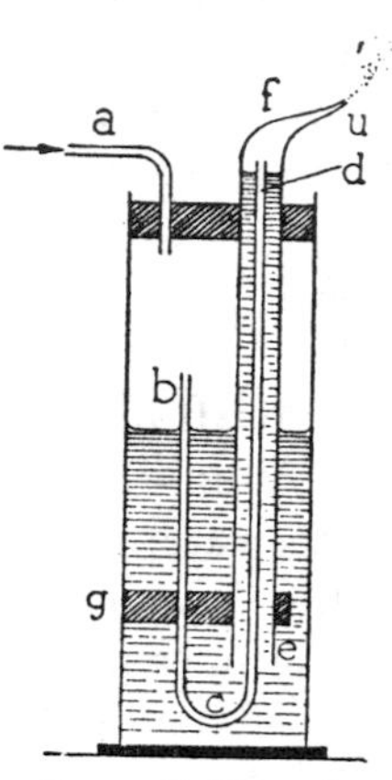

Fig. 108. Vaporiser for reproducing the rainbow.

Always look for rainbows in a direction 42° away from the anti-solar point, and preferably against a dark background.

Experiments of this kind are excellent materials for observation. Rainbows can often be seen as closed circles, when there is a sufficient number of drops of water *below* the line of our horizon too. If we move, the rainbow moves with us; it is not an object, it is not visible at a definite *place*, but in a definite *direction*; one might say that it behaves as something infinitely far off, which moves with us in the same way as the moon does. If one stands very near the cloud of drops, as, for instance while spraying with the hose, *two* rainbows can be seen to cross. How is this? Shut your eyes alternately; it will appear that *each eye* sees its *own* rainbow (which follows also from the fact that the rainbow moves along with us). The secondary rainbow and the supernumerary bows can often be seen splendidly. If the direction of the jet is altered or the rainbow is seen in other places in the spray, the relative importance of colours in the bow will change: the reason being that the average size of the drops is different.

122. *Descartes' Theory of the Rainbow*

In order to investigate the path of light in a drop of water, we fill a flask with water and hold it in the sun (Fig. 109, *a*). On a screen AB, provided with a round aperture (a little larger than the flask), a faint rainbow R will then appear. It has the shape of a closed circle, its angular distance is about 42°, and the colour red is on the outside, just as in a real rainbow.

This experiment can be carried out with equal success with an ordinary tumbler, which must, however, be more or less cylindrical. The time should be either morning or evening while the sun is low. The image on the screen will not be circular, but will show parallel stripes.

In front of the flask at S hold a little screen attached to a thread and you will see a shadow on the lower part of the rainbow (Fig. 109, *b*). If you press your moistened finger on the flask somewhere near V you will see a darker spot at the place corresponding to this on the lower part of the rainbow. It is evident, therefore, that the rainbow is formed by rays incident at the distance SC from the central line and reflected at V in the drop of water behind. If you hold a ring a few millimetres thick and in diameter 0·86 times the diameter of the flask, carefully centred in the incident beam, the rainbow will disappear altogether (Fig. 109, *c*).

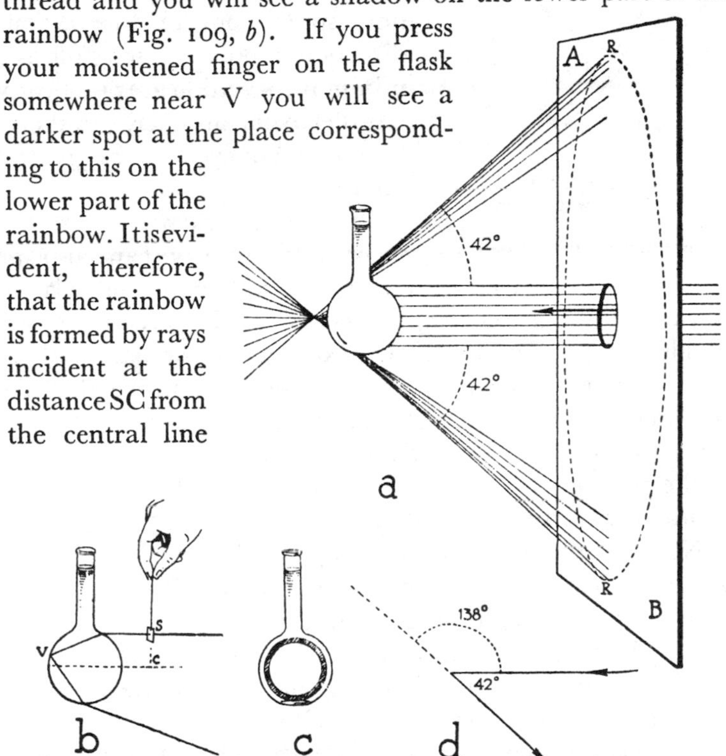

FIG. 109. Reproduction of a rainbow by means of a flask filled with water.

Fig. 110 shows the exact course of the rays, calculated from the ordinary laws of reflection and refraction. One can see how the light-rays incident upon the drop of water emerge in different directions according to the point where they strike it; one of these rays is deviated less than the others, namely through 138°, i.e. it makes an angle of $180° - 138° = 42°$ with the axis. The emergent rays are now spread out in the various directions; only those suffering minimum deviation are practically parallel to one another, and therefore reach our eye with the greatest 'density.'

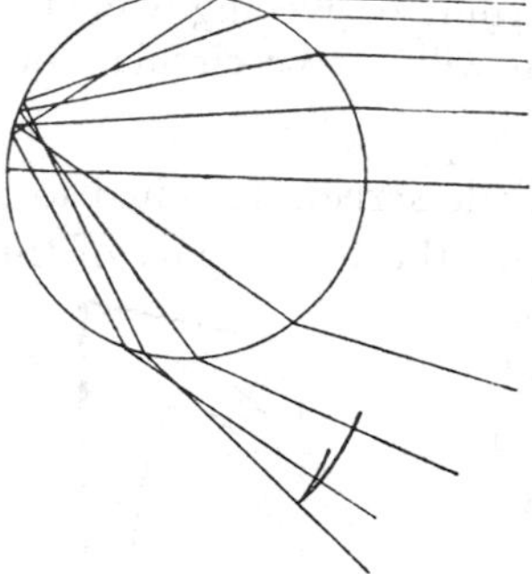

FIG. 110. The path of light-rays in a drop of water and how a rainbow is caused. The thick line indicates the wave-front.

In a well-darkened room the secondary rainbow can also be seen on the screen at an angle of 51° with the axis, or deviated through $180° + 51° = 231°$ from the direction of the incident rays (Fig. 111). By experiments similar to those performed with the primary bow, one can prove that the secondary bow arises from *twice* reflected rays. The order of the colours is the reverse of that in the primary rainbow, exactly as in reality.

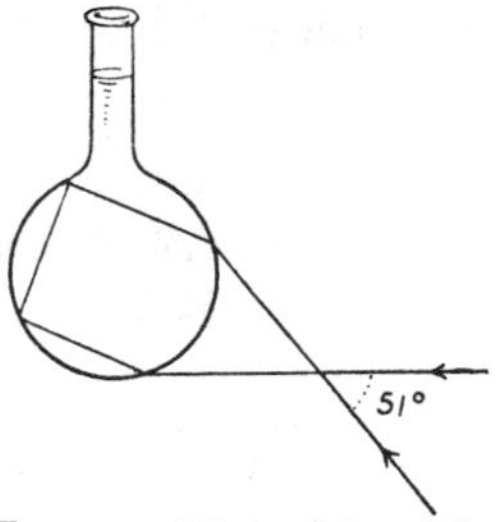

FIG. 111. The origin of the secondary rainbow.

Now imagine each of the drops in a cloud reflecting, in the way described, a great deal of light into a cone of 42° and rather less into a cone of 51°. All drops seen at an angular distance of 42° from the direction of the incident rays from the sun are in such a position that they send the light of their primary rainbow towards our eye, while from those seen at an angular distance of 51° from the incident sunlight we receive the twice reflected rays. In this way, therefore, the primary and the secondary rainbow are formed (Fig. 112).

123. *The Refraction Theory of the Rainbow*

In Descartes' theory only the rays suffering minimum deviation were considered, as if they were the only ones. In reality, however, there are a number of rays of greater deviation, enveloped completely by a curved *caustic*. These are exactly the conditions under which *interference* occurs as shown near the caustics of raindrops on an eyeglass (§ 118).

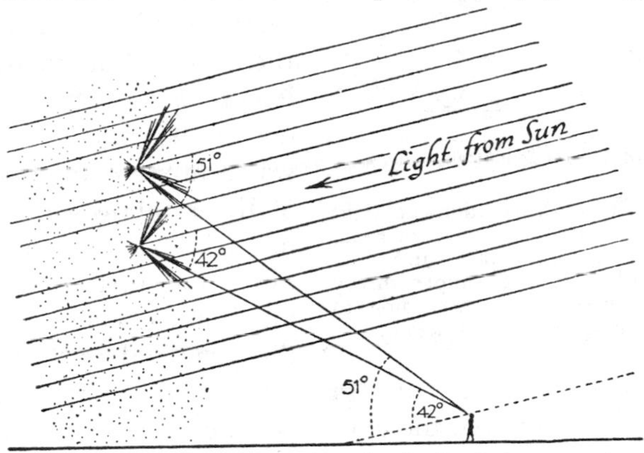

FIG. 112. Sunlight falling on a cloud of raindrops produces a primary and a secondary rainbow.

And, especially when dealing with tiny drops, the considerations of *light-rays* do not suffice, but the *wave-front* where it shows a *cusp* in the neighbourhood of such a caustic must be investigated (Fig. 110).

According to Huygens' principle, the points of the wave-front are regarded as centres of radiation and the problem is then to investigate how the vibrations reaching our eye from every part of the wave-front interfere mutually. This investigation, carried out by Airy, completed and applied by Stokes, Möbius and Pernter, leads to the famous rainbow integral,

$$A = c\int_0^\infty \cos\frac{\pi}{2}(u^3 - zu)\,du,$$

A being the amplitude of the light-vibration entering our eye, as a function of the angle z with the direction of the rays of minimum deviation. The integral is computed by developing it in a series and the light intensity which we see in a direction z is then simply given by A^2.

Fig. 113 shows for one colour how the distribution of light found with large drops (*a*) is altered by diffraction when the drops are small (*b*). The phenomenon is still, in the main, determined by the rays of minimum deviation ($z=0$) but a number of lesser maxima have appeared besides. Now one must draw such curves separately for a number of colours, shifting them according to the wave-length; for any given angle of deviation z, one obtains in this way a mixture and the colours of the rainbow can therefore never be truly saturated hues. Since the first and strongest maximum of each colour plays the principal part and these principal maxima shift gradually with increasing wave-length, we observe the colours of the rainbow practically in the same way as we would expect from the elementary theory. The modifications due to diffraction consist in the colours being slightly different according to the size of the drops, and that inside the bow, supernumerary bows appear. Finally, one must bear in mind that the sun is not a mere point and that therefore, the sunbeams are not strictly parallel (§ 1), so that since they spread over an angle of fully half a degree, the colours of the rainbow become slightly obliterated. The diffraction theory enables one, on seeing a rainbow, to state at once the approximate size of the drops to which it is due. The chief features are the following:

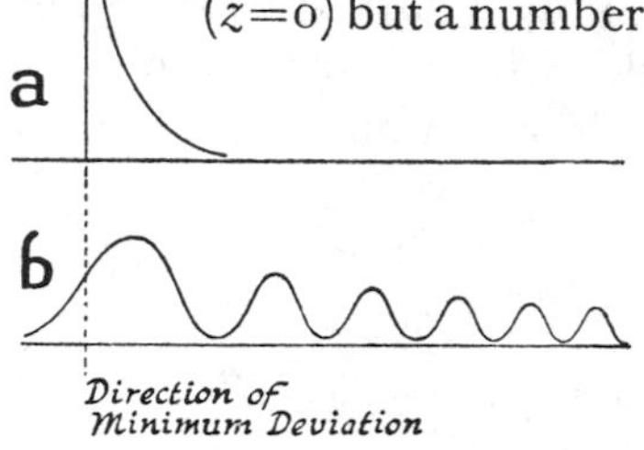

FIG. 113. Distribution of light in the pencil of rays emerging from a drop of water. (*a*) According to the simple theory of Descartes. (*b*) According to the theory of diffraction.

Diameter.

1–2 mm. Very bright violet and vivid green; the bow contains pure red, but scarcely any blue. Spurious bows are numerous (five, for example), violet-pink alternating with green, merging without interruption into the primary bow.

Diameter.	
0·50 mm.	The red is considerably weaker. Fewer supernumerary bows, violet-pink and green again alternating.
0·20–0·30 mm.	No more red; for the rest, the bow is broad and well-developed. The supernumerary bows become more and more yellow. If a gap occurs between the supernumerary bows, the diameter of the drops is 0·20 mm. If there is a gap formed between the primary bow and the first supernumerary bow the diameter of the drops is less than 0·20 mm.
0·08–0·10 mm.	The bow is broader and paler, and only the violet is vivid. The first supernumerary bow is well separated from the primary bow by a fairly wide gap and clearly shows white tints.
0·06 mm.	The primary rainbow contains a distinct white stripe.
<0·05 mm.	Mist-bow (cf. § 128).

124. *The Sky Round the Rainbow*[1]

An attentive observer will notice that the sky between the two rainbows is darker than the sky outside them. There is, of course, a background of clouds varying in brightness, but the effect is, nevertheless, generally distinctly visible (Plate IX, *a*; p. 202).

The explanation is that, besides the rays of minimum deviation, every drop reflects rays in other directions which deviate more from the incident direction. These are shown faintly in Fig. 114. Note that in the secondary bow the rays are deviated on the side opposite to that on which they are deviated in the primary bow. The observer will therefore first see faint diffuse light from that part of the sky inside the primary rainbow, owing to the once reflected rays which are deviated more than 138° and therefore make

[1] *Nat.*, **109**, 309, 1922.

an angle of less than 42° with the axis; and then faint diffuse light from that part of the sky outside the secondary rainbow, owing to the twice reflected rays which are deviated more than 231° and therefore make an angle of more than 51° with the axis. Sometimes radial streaks of light, devoid of

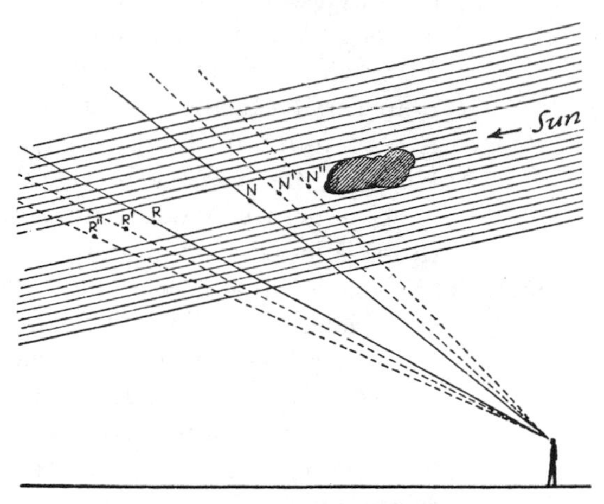

FIG. 114. Fragments of clouds between the sun and the shower of rain cause radial stripes in the sky.

colour, are visible in the diffuse glow between primary and secondary bows.[1] These are analogous to the crepuscular rays and to rays on moving water (§§ 191, 217). This phenomenon is easily explained by imagining that somewhere between the sun and the raindrops a small cloud is floating (Fig. 114). The drops lying inside the shadow behind the cloud cannot, in that case, radiate any light towards the observer. The rainbow which he sees is built up of light from all the drops in his line of vision, and lacks therefore the contribution from the drops R; and similarly the secondary bow lacks the light from the drops N, while in the diffuse light the contributions from drops like R′, R″ . . . and N′, N″ . . . are wanting. In the plane containing his eye, the sun and the cloud all phenomena are for that reason weaker: a ray-like shadow arises which, prolonged,

[1] S. Thompson, *Nat.*, **18**, 441, 1878.

passes through the point exactly opposite the sun, i.e. through the centre of the rainbow.

125. *Polarisation of Light in the Rainbow*[1]

It is very interesting to try to see a rainbow reflected in a piece of glass—not in a silver-backed mirror, which would be unsuitable for this purpose, but in an ordinary piece of glass, either blackened or backed with a piece of black paper.

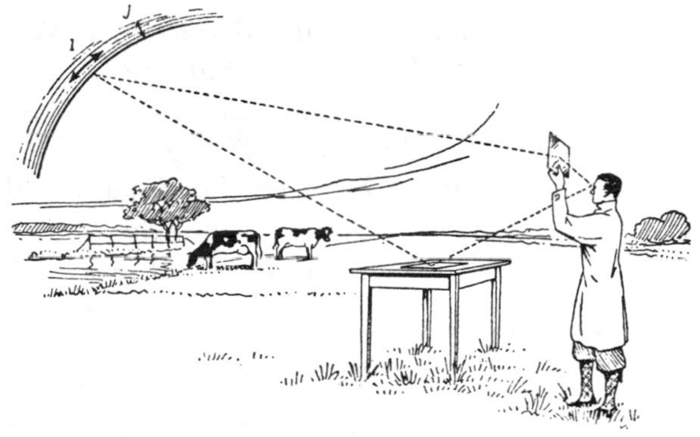

FIG. 115. How to observe the polarisation of light in a rainbow.

It should be held close to the eye and so that one looks in it in an oblique direction, say at 60° to the normal. One can hold the glass either horizontally or vertically, as in Fig. 115. If we look at the top of the rainbow, we shall see that with the glass horizontal the reflection of the bow is very distinct and bright, while with it vertical the reflection is so weak as to be almost imperceptible. This shows that the light of the rainbow has different properties in different directions perpendicular to its direction of propagation, that is, it is 'polarised.'

There is a still easier way of making this observation, namely by studying the rainbow through a nicol prism, a small apparatus which enables one to distinguish at once whether or not the light is polarised. The nicol is rotated about its axis: in one position the rainbow is very bright and in the other very weak. We can imagine the composite

[1] F. Rinne, *Naturwiss*, **14**, 1283, 1936.

light as made up of light vibrating in a certain direction i and light vibrating in a direction j at right angles to i. We find the ratio of the intensities $i:j$ to be 21 : 1, i.e. the polarisation is very complete. It is not so pronounced in the secondary rainbow, though still very distinct, the ratio being 8 : 1. Both these results agree with theory.

126. *The Effect of Lightning on a Rainbow*

A striking observation was carried out by J. W. Laine. Each time it thundered he noticed that the boundaries of the colours in the rainbow became obliterated. The change was particularly noticeable in the spurious (or supernumerary) bows: the space between the violet and the first spurious bow disappeared entirely and the yellow grew brighter. It was as if the whole rainbow vibrated. According to the table in § 123, these alterations indicate an increase in the size of the drops.

This optical effect did not occur simultaneously with the lightning, but several seconds later, together with the sound of the thunder. One might imagine that owing to the vibration of the air, the drops tended to merge into each other; but this tendency is so slight that a really perceptible effect seems improbable. It is also possible that the electric discharge brings about a change in the surface-tension of the drops, so that they merge together more easily; but then it would be a mere coincidence if the space of time required for this change should equal that between the lightning and the thunder.

127. *The Red Rainbow*

During the last five or ten minutes before sunset all the colours of the rainbow are seen to fade, excepting the red: and, finally, only an all-red bow is left. Sometimes it is amazingly bright and remains visible even for ten minutes after the sun has set; the lower part of the bow is naturally screened off by that time, so that it appears to begin at some height above the horizon. Nature is here showing us the spectrum of the sunlight, and demonstrating how its composition changes during sunset. This change is due to the scattering of the shorter light-waves (§ 171).

128. *The Fog-bow or White Rainbow*[1]

When the drops are very small, the appearance of the rainbow is quite different. This can be seen very well if one stands on a hill with one's back to the sun and with fog in front and below. The bow has then the appearance of a white band, as much as twice the width of an ordinary rainbow, orange on the outside, bluish on the inside. On the inside, with a space between them, one or even two supernumerary bows are to be seen, the order of the colours being, strange to say, the reverse of those in the normal primary bow (first red and then green).

These features are in surprising agreement with the theoretical calculations for drops with a radius of 0·025 mm. and less (cf. § 123). With these very small drops, the radius of the rainbow is no longer 42°, but begins to diminish, and since 'small' here means 'approaching the wave-length of light' the effect is more prominent with red rays than with blue. Hence the red of the supernumerary bow will be much smaller in diameter than the blue and so will lie on the inside.

Those who are so fortunate as to see this beautiful phenomenon should carry out a few measurements to determine the diameter 2 θ of the bows (in degrees) (cf. § 235). The dark ring between the primary and the first supernumerary bow can be measured the most accurately; from the value obtained the radius of the drops (in millimetres) can be computed with the aid of the formula:

$$a = \frac{0{\cdot}31}{(41^\circ\ 44' - \theta)^{3/2}}$$

(Alternatively, one can take the average value between the blue and the orange edges of the primary bow, but then the numerator must be altered to 0·18.)

[1] *Phil. Mag.*, **29**, 456, 1890.

Strangely enough, the fog-bow has been seen even when the temperature was very low (0° F !), which proves that the drops of water in the atmosphere can be very strongly supercooled.[1] It has also been seen when the fog was so thin that the observer who saw it declared that there was no fog !

A fog-bow nearly always appears when the dazzling beam of a searchlight behind us penetrates the mist in front of us. Even ordinary street-lamps *frequently* give rise to it, though feebly and only against a dark background. Tyndall observed it once, using a candle as source of light. If a mist is seen backed by dark ground, the fog-bow can at times be seen as a complete circle, the few yards between our eye and the ground at our feet evidently being enough to produce this phenomenon.[2] On very rare occasions a double fog-bow has been observed.[3] Compare also §§ 165, 121.

129. *The Dew-bow or Horizontal Rainbow*

The heather, on autumn mornings, is covered with millions of cobwebs, which, otherwise unnoticeable, are now

FIG. 116. The dew-bow.

sprinkled with dew-drops and lit up by the sunbeams. In that play of light we may see a rainbow standing out on the ground in front of us, not as a circle, but in the shape of a wide-open *hyperbola* (Fig. 116).

The explanation is simple: the light reaches our eye from all directions, forming an angle of 42° with the axis through

[1] Ch. F. Brooks, *M.W.R.*, **53**, 49, 1925. G. C. Simpson (**38**, 291, 1912), mentions the appearance of a fog-bow at a temperature of −29° C.

[2] *Phil. Mag.*, **17**, 148, 1883. [3] *Onweders*, etc., **52**, 54, 1931.

the sun and our eye. The cone so formed intersects the surface of the ground in a hyperbolic curve as long as the sun is low. It would in the course of the day become an ellipse, although this has very seldom been seen. You might ask someone to help you by marking out and measuring the curve on the ground and then verify, with the help of the sun's altitude (deduced from the time of observation), that the curve is indeed a hyperbola corresponding to a cone with a vertical angle of 42°.[1] Observe how the coloured band increases in width, the farther it is from our eye. One single instance is known of the fog-bow and the supernumerary bows being observed in the dew.[2]

The dew-bow has also been observed in the following circumstances: (*a*) On a pond covered with duckweed; on a lawn. (*b*) On a pond with an oily surface on which dew-drops can lie *without mixing with the water*; a surface of this kind may be formed, for instance, by the sooty smoke from factories. In one case the size of the drops varied from 0·1 mm. to 0·5 mm., 20 drops per cm.^{2} showing a very distinct dew-bow.[3] (*c*) On a lake or on the sea, early in the morning, when the air has grown cooler but the water is still warm, so that a thin mist hangs above the surface of the water. The whole of the dew-bow is not always visible then, but only its two extremities. (*d*) On a frozen surface which can, apparently, become covered by dew-drops of suitable shape. How is this possible?[4]

This observation also has a remarkable psychological aspect. Why does the rainbow appear to us to be *circular* and the dew-bow *hyperbolic*, while in both cases the light-rays reach our eye from the same direction? 'It is a question of combining observation and expectation. When we see a dew-bow we are influenced by the thought that the light-phenomenon is spread out in a horizontal plane, and we unconsciously ask ourselves: what must be the true shape of

[1] A. E. Heath, *Nat.*, **97**, 6, 1916.
[2] W. J. Humphreys, *Journ. Frankl. Instit.*, **20**, 661, 1929.
[3] *Nat.*, **43**, 416, 1891.
[4] Clerk-Maxwell, *Papers*, II, 160.

the curve of light on the grass for us to see the phenomenon as we do see it? The answer is of course an ellipse or an hyperbola. But if, on the other hand, we ask: *how do we see the dew-bow?*—then our answer depends on the observation combined with our interpretation of it. If we saw only the light-phenomenon and knew nothing about its origin, only a circular shape would occur to us' (Stokes). A stereoscopic estimate of the distance of individual drops and clusters of drops would certainly help us to locate the dew-bow in a horizontal plane (cf. § 152).

For the reflected dew-bow, *see* § 131.

130. *Reflected Rainbows*

If we see a rainbow in the direction of a point A in a cloud, then when we observe the reflection of the landscape in

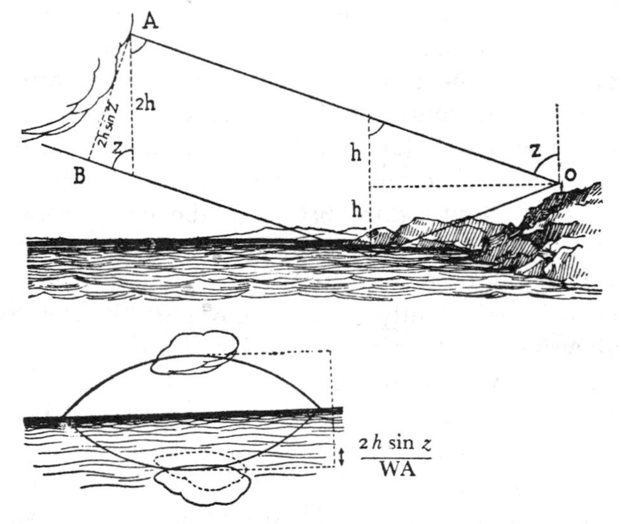

Fig. 117. The reflected rainbow.

smooth water we will see the rainbow in the direction of point B, so that it appears lower on the reflected cloud than on the cloud directly viewed (see Fig. 117).

This is because as already stated, the rainbow does not exist as a real object in the plane of the clouds but is, as it were, at infinity. Properly speaking, therefore, it is the cloud that is displaced, whereas the reflection of the rainbow is perfectly symmetrical relative to the horizon. The shift of the cloud will be more easily observable if we are at a certain height h above the water. We are then even able to compute the distance OA of the cloud from an estimate of its displacement in angular measure, for the angular displacement = $\frac{2\,h \sin z}{\mathrm{OA}}$.

An entirely different effect occurs, however, when the sunbeams are reflected before they produce the rainbow. A shifted arc WS will then appear round the centre T′, which is the reflection of the anti-solar-point T (Fig. 118). This arc extends over more than a semicircle. The distance between the tops of the two arcs is equal to that between T and T′, that is twice α, the height of the sun above the horizon. In many cases the shifted arc is only partly visible—for example, only the top or the two ends. When you observe an extraordinary rainbow you should therefore think first of the possibility of such a reflection. Consider next, where big pools are to be found in the neighbourhood and try to account for the incompleteness of the arc by the location of these pools. The two bows produced by reflection complete each other so as to make a closed circle (Fig. 118).

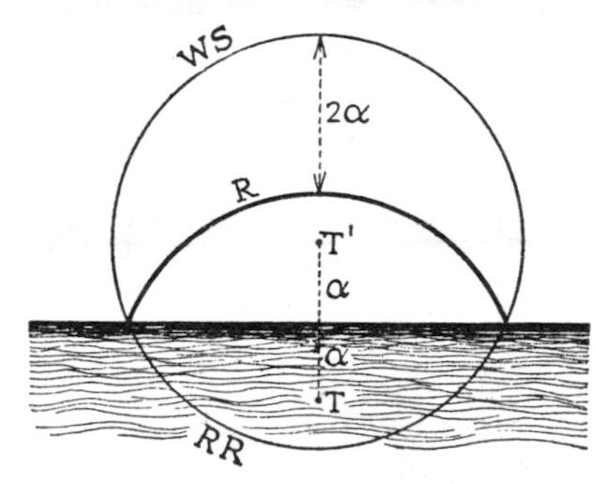

FIG 118. R = Rainbow. RR = Reflected rainbow. WS = Rainbow formed by the reflection of the sun.

131. *The Reflected Dew-bow*[1]

The dew-bow can also be reflected in water and the hyperbola with its beautiful colours, formed by the drops floating

[1] W. J. Humphreys, *Journ. Frankl. Instit.*, **207**, 661, 1929.

on the surface, is then double. The fact that the weaker of the bows is caused by reflection becomes very clear if one happens to observe a dew-bow on a frozen surface, because then the second bow disappears.

The angular distance of the bows is again twice that of the sun's height, but because the drops are this time on the surface of the water itself, it is not possible to ascertain straight away whether the reflection has taken place before or after the rays of light have passed through the drops of water. Both cases would give a hyperbola (cf. Fig. 119; in both figures the reflected ray rises at an angle of $42° - \alpha$).

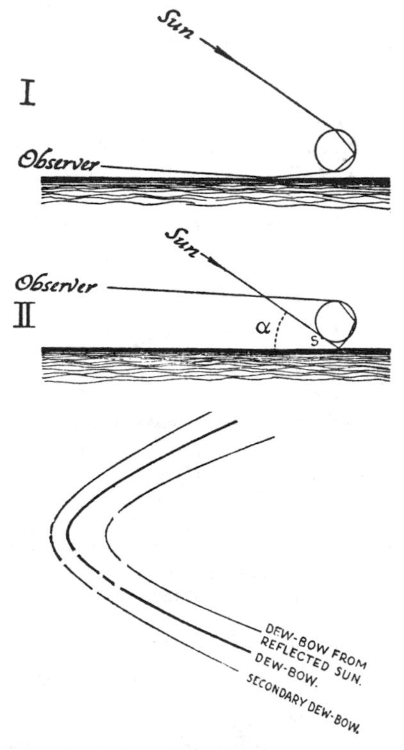

FIG. 119. Reflected dew-bows. I, The dew-bow is reflected. II, The reflected sun forms a dew-bow.

However, when the sun is fairly high (21° to 42°) two criteria exist:

(*a*) The part of the reflected bow near the top is absent. The explanation is that when the rays follow path II the incident bundle of rays is partly screened off by the drop itself at S, before it is reflected, and then penetrates the drop. If the path is as in I this characteristic peculiarity does not occur.

(*b*) If two neighbouring points of the two bows are observed with a nicol, it is found that the directions of the light-vibrations differ very greatly and are as a rule not horizontal.

It can be demonstrated that this will only be so if the reflection takes place before the refraction.

The question remains: Why is it more usual for the rays to be reflected first? The answer is simply that the emerging rays in the case of path I run grazingly over the water and are intercepted by the neighbouring drops.

When the sun is low, the rays of light will first penetrate the drop and then be reflected; the top of the reflected bow is again screened off, but the amount of polarisation is different. As yet no careful study has been made of this case.

132. *Abnormal Rainbow Phenomena*[1]

Here follow a few figures of exceptional shapes of rainbows, due partly to reflection in surfaces of water, but for which, in my opinion, there is no satisfactory explanation.

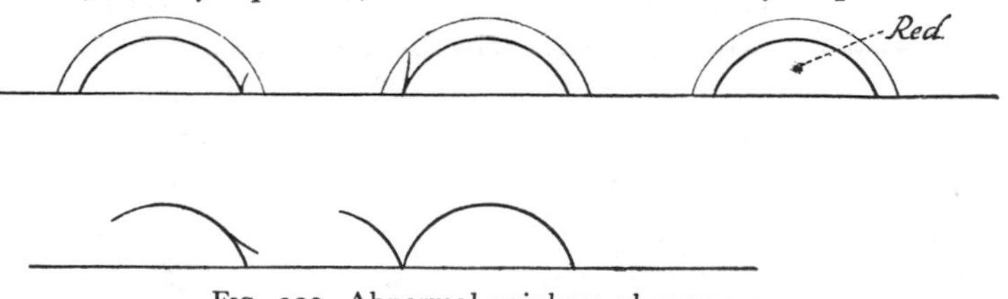

FIG. 120. Abnormal rainbow phenomena.

One more reason to keep one's eyes open for similar phenomena! Note especially the mutual position of the red and the violet sides of the abnormal bows.

133. *The Moon-rainbow*

Rainbows are formed by the moon as well as by the sun, though these moon-rainbows are naturally very weak. That is why they can be seen practically only when the moon is *full* and why they are seldom coloured, just as feebly illuminated objects usually appear colourless at night (§ 77).

Do not confuse them with haloes! The rainbow is only visible at the side *opposite the moon*!

The radius of the moon-rainbow can be determined very

[1] *Onweders*, etc., **21**, 54, 1900; **24**, 160, 1903; **29**, 110, 1908. *Hemel en Dampkring*, **27**, 359, 1929.

accurately if there happens to be a bright star somewhere near (cf. § 135).

HALOES

134. *General Description of Halo Phenomena*[1]

After a few days of fine bright spring weather the barometer falls and a south wind begins to blow. High clouds, fragile and feathery, rise out of the west, the sky gradually becomes milky white, made opalescent by veils of cirro-stratus. The sun seems to shine through ground glass, its outline no longer sharp but merging into its surroundings. There is a peculiar, uncertain light over the landscape; I 'feel' that there must be a halo round the sun!

And as a rule I am right.

A bright ring with a radius of rather more than 22° can be seen surrounding the sun; the best way to see it is to stand in the shade of a house or to hold your hand against the sun to prevent yourself from being dazzled (§ 160). It is a grand sight! To anyone seeing it for the first time the ring seems enormous—and yet it is 'the small halo'; the other halo phenomena develop on a still larger scale. Put out your arm and spread your fingers wide apart; you will see that the distance between the top of the thumb and the little finger is almost as wide as the radius of the halo round the sun (cf. § 235).

You can see a similar ring round the moon, too. I do not mean a corona a few degrees in diameter with red inside and blue outside, but the same large ring as that just described as a halo round the sun. Only once was an observer so fortunate as to see a ring round the setting sun and one round the rising full moon at the same time!

Rings of this kind are observable more often than one would think. It is fairly certain that a practised observer in our part of the world who is on the watch all day long will be able to see on the average one halo every four days, and

[1] For everything concerning haloes one should refer to the book by R. Meyer, *Die Haloerscheinungen* (Hamburg, 1929).

in April and May he may even see one every two days; the most observant see haloes on 200 days a year! Does it not seem incredible, therefore, that there are still so many who have never noticed a halo round the sun?

Besides the small halo, there are still numbers of other light-bows and concentrations of light in spots, each with a name of its own, which combine to form the halo phenomenon; the most important are shown in Fig. 121 as if they were outlined on an imaginary celestial globe. We shall now discuss these briefly in turn. It must be borne in mind, however, that as a rule only a few of them are to be observed at the same time. Most of those that have been seen were due to the sun; those belonging to the moon are much fainter, their colours being practically imperceptible (cf. §§ 77, 133). They are generally formed in veils of cirro-stratus or in cirrus clouds and seldom in cirro-cumuli or alto-cumuli; they can be seen in thunder-cirri, though not often. All clouds giving rise to haloes are composed of small ice-crystals and the regularity of the shapes of these crystals is responsible for the beautiful symmetry of the light-phenomena. The reason why so many ice-clouds show no halo phenomena at all is that little snow-stars and globular clusters of crystals are of the wrong shape to refract the light in the same way as a prism does, and that with too small crystals the halo phenomena are obliterated by diffraction.

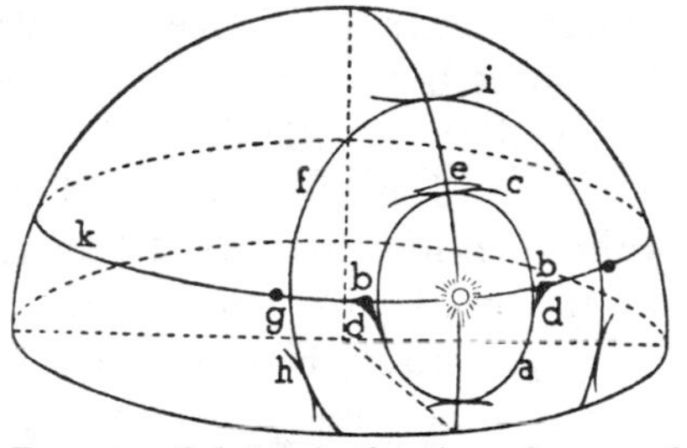

FIG. 121. Schematic drawing of some of the most important halo phenomena.

The photography of haloes is of importance for scientific purposes: it is of use in accurate measurements of angles and in determining light-intensities. For these purposes, however, the photographic plate should be exactly perpendicular to the axis of the camera, and the distance between the plate and the objective accurately known. An objective with a large opening should be used, a yellow filter, and panchromatic plates with an anti-halation layer. Lantern-plates may perhaps answer the purpose, too. Time of exposure with strong yellow filter and *f*/12 is about 0·01 second.

135. *The Small Ring or the* 22° *Halo* (Fig. 121, *a*; Plate IX, *b*)

This is the most common of all halo phenomena. The ring is complete, except when the cirro-stratus clouds are unevenly distributed over the sky; it is usually brightest at the top or bottom or to the left or right, rather than in the intermediate positions. The inner edge is fairly sharply defined and red in colour, then follow yellow, green, and white, running to blue. The radius of the small ring can be measured with one of the simple devices mentioned in § 235, preferably from the centre of the sun to the red inner edge; the best measurements give 21°50′.

The radius of the halo surrounding the moon can be determined very accurately on some nights, if one happens to observe that the position of some definite star coincides with the inner edge, for example, or with the maximum of brightness of the halo. One has then merely to note the name of the star (if necessary identifying it by means of a star-map) and the time. Afterwards any astronomer will be able to calculate at once how far apart the two celestial bodies are at that moment (cf. Fig. 125).

Observe that the sky inside the small halo often seems darker than outside it; whenever this is not so it is because the halo is superposed on a diffuse light, decreasing gradually in brightness from the sun outwards. This phenomenon reminds us very much of that observed in the rainbow (where the sky is darker between the two bows), and it is caused in a similar manner.

The small halo arises from refraction of the sunlight in a cloud of small ice-crystals, whose shape is known to be often that of a hexagonal prism. In every direction we look, innumerable minute prisms of this kind float about in every possible orientation (Fig. 122). Such a six-sided prism refracts the light as if it had a refracting angle of 60°: according to its orientation with respect to the incident rays, it will deflect them to a greater or lesser extent, but if the path

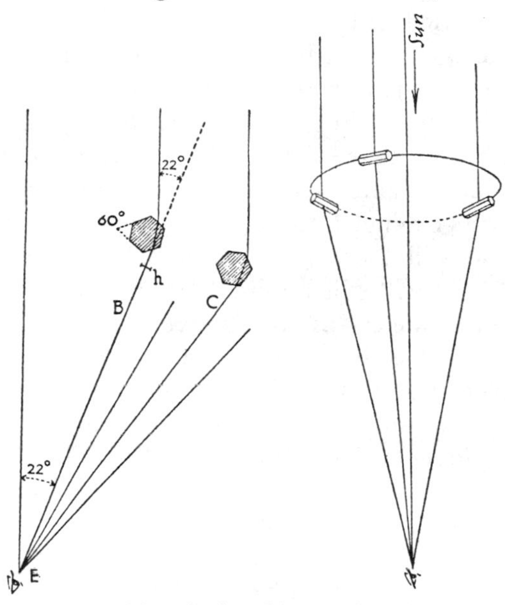

FIG. 122. How the 'small,' or '22° halo' arises.

of the rays through the crystal is symmetrical the deviation will have a minimum value D given by the well-known formula—[1]

$$n = \frac{\sin \frac{1}{2}(A + D)}{\sin \frac{1}{2}A}$$

where n is the refractive index of the material of the prism and A its refracting angle. For $A = 60°$ and a refractive index $n = 1 \cdot 31$, we obtain $D = 22°$: exactly the radius of the small halo !

Indeed, it is easily seen (as in the rainbow) that the rays OB which suffer minimum deviation will contribute by far the most to the brightness, since in that position the direction of the refracted ray changes only very slowly as the prism

[1] The formula is given in any textbook of physics, under 'Minimum Deviation of a Prism.'

turns. There are, therefore, comparatively far more ice-crystals sending the light to our eye in directions that lie close to this direction, than close to other directions. Our calculations were carried out for yellow rays; for red ones the minimum deviation is rather less; for blue ones somewhat greater. For that reason the inner edge of the halo is red, the outer edge blue. Since, however, the rays EC with a greater deviation than the minimum also contribute a certain amount of light, the green and blue 'rays of minimum deviation' will be mingled to a certain extent with yellow and red light and will therefore show a pale colour. A small quantity of light will still be visible everywhere outside the ring, but not inside it—as we observed already· the sharp inner edge as well as the hazy outer edge are thus explained. But whenever the crystals are not distributed haphazard in every possible direction, but assume definite *positions of preference*, a differentiation in the luminosity outside the small halo takes place and certain light-spots or arcs appear, which we will examine presently.

Let us first consider, however, whether or not the *diffraction phenomena* play a part here as they do in the rainbow.[1] Theoretically they should; the ice-crystal transmits only a narrow pencil of light of width h (Fig. 122), and therefore diffracts the light-waves in the same manner as a slit of width h would do. Very small ice-crystals would cause a white halo with a red edge just as small drops of water give rise to a fog-bow (§ 128). Moreover, one can expect to see supernumerary bows appear beside the small ring (§ 123), and indeed these have been seen at times; but calculation shows that they must be weaker than in the rainbow and are formed on the outside as well as on the inside of the ring. Those on the inside are the easier ones to see as they are outlined against a dark background. Observations so far indicate a variation in the colour and width of the small ring but more observations are needed. Often the best way to

[1] Visser, *Proc. Acad. Amsterdam*; Summary in *Hemel en Dampkring*, **15**, 17, 1917, and **16**, 35, 1918.

judge the colours is to look at them through smoked glass; estimate the width of each colour separately and of the whole of them together. Name them according to your own independent opinion! Will any two observers always call the colours of one and the same halo by the same names? Red and orange are often confused; also blue and violet; note how seldom yellow occurs in halo phenomena!

According to the simple theory of refraction, there should be practically no blue in the small ring and absolutely no violet, and this should apply to the upper tangential arc and the mock suns as well (§ 136). But observation sometimes shows us blue quite strong, especially in the upper tangential arc and the mock suns, the tints of which are always vivid. The theory of diffraction explains how blue and violet can appear, if only the size of the crystals is right, and also why the tangential arc and the mock suns are more strongly coloured than the small halo. Finally, the diffraction theory makes it clear to us why the colours are at times most vivid in the small halo and at other times in the large one: the small halo is more strongly coloured when the faces of the prisms where the refraction takes place are broad, as with *plate-shaped* crystals; if, however, the faces are narrow, as with *pillar-shaped* crystals, the small halo is pale and the large one more vividly coloured.

The light of the small halo is polarised. In contradistinction to the rainbow, the vibrations in this case are stronger in directions at right angles to the ring than parallel to it. This is quite understandable since here there is no reflection at all and only two refractions. However, the effect is not nearly so pronounced as it was in the rainbow.

According to a popular belief, the small halo is a forerunner of rainy weather,[1] and when they say 'the larger the halo, the sooner we shall have rain,' they mean that *the small halo*, and *not the corona*, predicts rain. And indeed the cirro-stratus clouds are often the forerunners of a region of depression.

[1] *Quart. Journ.*, 1926, and elsewhere.

136. *The Parhelia or Mock Suns of the Small Halo* (Fig. 121, *b*)

These mock suns are two concentrations of light on the small halo at the same altitude as the sun. It often happens that only one of the two can be seen properly and sometimes the small halo is absent, whereas the parhelia are clearly visible. The intensity of these mock suns is usually very great; they are distinctly red on the inside, then yellow, changing into a bluish white.

On close observation one finds that in reality the parhelia stand a little way *outside* the small halo, the more so as the altitude of the sun is greater; when the sun is very high the difference may even amount to several degrees.

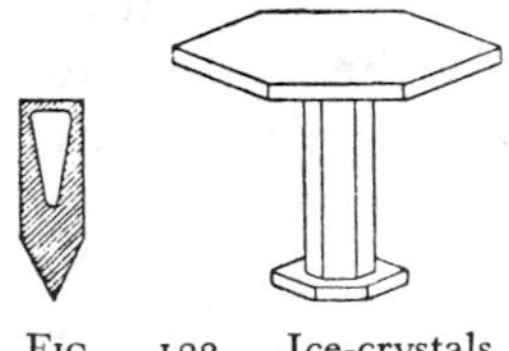

FIG. 123. Ice-crystals, which play a prominent part in the formation of mock suns.

The mock suns arise when the axes of a large number of the hexagonal ice-crystals are vertical. This is true for slowly descending 'umbrella-shapes' or for little ice-pillars concave at one end (Fig. 123). Through such prisms the rays of light no longer travel along the path of minimum deviation, because they do not lie in a plane perpendicular to the axis. When the sun's altitude is h, the 'relative minimum deviation' is in this case determined by the condition—

$$\frac{\sin \frac{1}{2}(A+D')}{\sin \frac{1}{2}A} = \sqrt{\frac{n^2 - \sin^2 h}{1 - \sin^2 h}}$$

so that the light behaves as if the refractive index were increased for the oblique rays (cf. § 135). From this equation, the following table is easily computed:

Altitude of the Sun	*Distance from parhelion to small halo*
0°	0°
10°	0°20′
20°	1°14′
30°	2°59′
40°	5°48′
50°	10°36′

This agrees very well with the observations. For altitudes of the sun greater than 40° we have, unfortunately, hardly any measurements, because then the phenomenon is apt to become indistinct; try to fill up this gap!

137. *The Horizontal Tangential Arcs to the Small Halo* (Fig. 121, *c*)

These arcs, which appear as an increase of brightness at the bottom and at the top of the small halo can be seen, under favourable circumstances, to be parts of a much larger

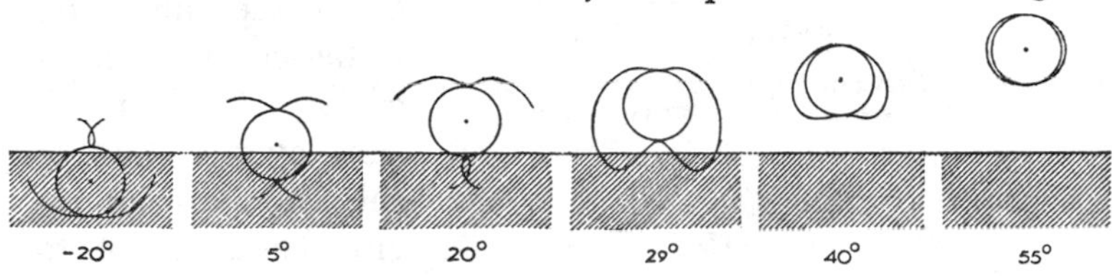

FIG. 124. Different forms of the circumscribed halo for increasing altitudes of the sun.

curve of light—the *circumscribed halo*. This very peculiar halo phenomenon is produced when the hexagonal crystals have their axis horizontal and oscillate slightly about that position; such a state of affairs will arise when the ice-crystals are rod-shaped instead of plate-shaped.

The circumscribed halo depends for its shape very much on the altitude of the sun (Fig. 124). When the latter is not high, all that can be seen is that the upper tangential bow is bent back in a downward direction on both sides; at greater altitudes, one can see an almost elliptical figure. The parts of the curves below the horizon have been found by calculation, and have been seen occasionally from a mountain, where the eye could be directed downwards (presumably this would be equally possible from a tower or an aeroplane).

138. *The Oblique Tangential Arcs to the Small Halo or 'the Oblique Arcs of Lowitz'* (Fig. 121, *d*)[1]

These are remarkable little arcs sloping downwards from the mock sun and touching the small halo: a very rare

[1] Visser, *Diss. Utrecht*, 1936.

phenomenon. It is possible to see it only when the sun is high and the mock suns consequently some distance from the small halo. The little arcs are produced when the minute vertical ice-prisms, from which the mock suns arise, vibrate slightly about the vertical. Often the only thing to be seen is an elongation of the mock sun, over 1° or 2°; the little arc is inclined at about 60° to the horizon. Once only was the arc fairly well-defined and long. Therefore it is always necessary to observe the mock suns carefully for possible traces of this phenomenon.

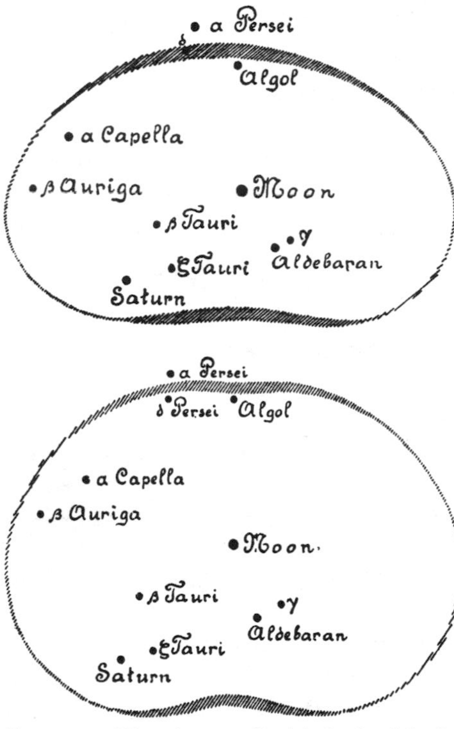

FIG. 125. The circumscribed halo fixed in its position relative to the stars in the vicinity of the moon. After Veenhuizen, *Onweders*, etc., **35**, 119, 1914. By kind permission of the Royal Dutch Meteorological Institute.

139. *Parry's Arc* (Fig. 121, *e*)

Seen very seldom! A slightly curved little arc, just above the small halo. It occurs when the six-sided prisms show a tendency not only to float with their axis in a horizontal direction, but also to bring one of their side-faces into a horizontal position.

140. *The Large Ring or the 46° Halo* (Fig. 121, *f*)

This appears fully twice as far from the sun as the small halo and has the same colouring, but its brightness is less and it is visible much more rarely. Accurate measurements of the radius of the inner edge are needed. This halo is produced in the same way as the 22° halo (the small ring) only this time by refraction in 90° ice-prisms orientated at random. As is shown in Fig. 126, the same type of ice-crystals can give rise to both the small and the large halo.

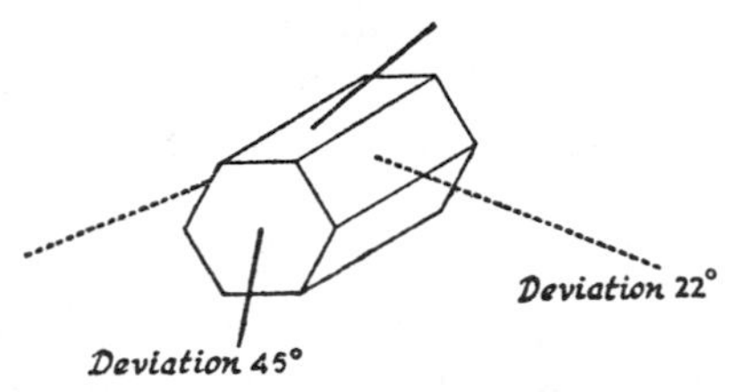

Fig. 126. In a hexagonal prism of ice the light-rays can suffer a minimum deviation of 22° and of 46°.

141. *The Mock Suns of the Large Halo* (Fig. 121, *g*)

These have seldom been observed, and no wonder; the refracting edges of 90° would have to be vertical in great numbers in order to produce it. That this should ever actually occur seems almost unthinkable, considering what the ordinary shapes of ice-crystals are like.

142. *The Lower Tangential Arcs to the Large Halo* (Fig. 121, *h*)

Also rare! They are due to a definite orientation, in which both the axis and a side-face are horizontal, while the light is refracted by those faces at right angles to each other. When the sun is *very* high, one can see the arc straighten and finally even become concave towards the sun.

143. *The Upper Tangential Arc to the Large Halo* (Fig. 121, *i*)

This arc can only occur when 90° prisms float in the air with refracting edges horizontal, the prisms oscillating round this position. Those among them which are suitably placed to give minimum deviation will then produce the tangential arc in question. Often an arc is observed which is very like this one but in reality it has another origin; it is the circum-zenithal arc of Bravais, not the true upper tangential arc.

144. *The Circumzenithal Arc* (Fig. 121, *i*)

One of the most beautiful halo phenomena! Of rather frequent occurrence, it is a vividly coloured arc parallel to the horizon and showing the colours of the rainbow. It is generally *a few degrees above* where one would expect the upper tangential arc to the large halo to appear.

To explain it, one must imagine crystals in the shape of plates or umbrellas (Fig. 123) floating in a stable position with axes vertical. A sunbeam will then be refracted in the 90° prism, but, generally speaking, this will not be in the position of minimum deviation. Fig. 127 shows that—

$$\sin i' = n \sin r' = n \cos r = n\sqrt{1 - \frac{\sin^2 i}{n^2}} = \sqrt{n^2 - \sin^2 i}$$

from which it follows at once that the angle of deviation is $i' + i - 90°$. For a solar altitude H=10°, this amounts to about 50°; for H=20° it has decreased to 46° (the minimum value); for H=30° it has increased again to 49·5°. This means that only when the sun is high or very low in the sky is it possible to distinguish the circumzenithal arc from the upper tangential arc (with an angle of deviation of 46°) to the great circle. According to theory, one should never see more than one-half of the circle; in practice this is reduced to one-third, yet the complete arc is said to have been observed once (Kern's halo).[1]

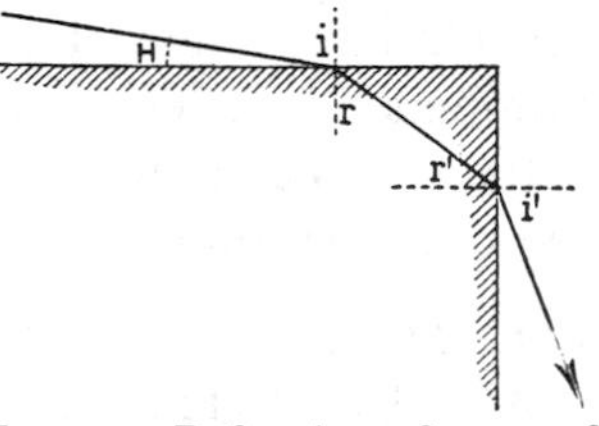

Fig. 127. Refraction of a ray of light by a 90° ice-prism.

If the tangential *and* the circumzenithal arcs should happen to be visible at the same time, *a gap* of a few degrees ought to be observed between them. And, indeed, it is on record that once a broad arc was seen, divided over its whole length by a dark band, appearing suddenly and disappearing after a short time.[2] But observations of this kind are bound to remain rare, for, in order to be possible at all, a group of horizontally floating plates and a group of irregularly directed plates must be present simultaneously.

145. *The Horizontal or Parhelic Circle* (Fig. 121, *k*)

This is a circle running at the same height as the sun, parallel to the horizon. Though one can at times follow its

[1] Observations by Lambert in 1838 after Pernter-Exner, p. 300; *M.W.R.*, **50**, 132, 1922.

[2] *M.W.R.*, 506, 1920.

course round the whole 360°, it is often difficult to see it in the neighbourhood of the sun, where the sky is rather bright. The fact of its being *uncoloured*, shows clearly that it is due to reflection and not to refraction; the reflecting planes in this case are the side-faces of ice-prisms floating with vertical axes.

A similar band of light can be seen when one looks at a source of light through a window wiped with a rather greasy cloth in one direction or reflected by finely ribbed glass. The band of light is always seen to be perpendicular to the ripples.

146. *Light-Pillars or Sun-Pillars*[1]

A vertical pillar of light or, rather, feather of light can be observed fairly often above the rising or setting sun, best of all when the sun is hidden behind a house, so that the eye is not dazzled. This pillar of light is in itself uncoloured, but when the sun is low and has become yellow, orange or red, the pillar naturally assumes the same tint. It is generally only about 5° high, seldom 15° or more. When the sun is high these pillars are rare; on the other hand, they can often be seen very well when the sun is actually below the horizon. Pillars *below* the sun occur only now and then; they are shorter than those above the sun.

Imagine a cloud of ice-flakes, all perfectly horizontal and falling very slowly. Under these conditions they reflect the incident rays of the sun; these reflected rays, however, will not reach our eye. But let these same ice-flakes become *slightly inclined* relative to their horizontal position at a small angle Δ, towards all points of the compass, and the reflected rays will now acquire all kinds of small deviations, and if the inclination remains smaller than $\frac{h}{2}$ ($h =$ sun's height) there will take shape a pillar below the sun, more or less in the same manner as path-like patches of light are formed in rippled surfaces of water (§ 14); when the inclination of the

[1] K. Stuchtey, *Ann. d. Phys.*, **59**, 33, 1919. Cf. references to § 14.

plates becomes larger than *h* we see not only a pillar below the sun, but also a fainter one above it.

This description, however, conflicts in two points with the observations. In the first place, the pillars would always have to be brighter below than above the sun; secondly, they could never be seen above the sun when it is fairly high, because the fluctuations of the ice-flakes round their horizontal positions are comparatively small (cf. § 148). Neither is true.

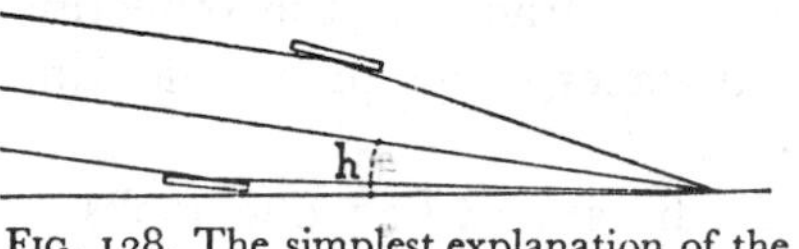

FIG. 128. The simplest explanation of the origin of light-pillars below and above the sun.

The pillar of light has been attributed to *repeated reflection*, but then the light phenomenon would be much weaker and the pillar become much broader than it usually appears to be, as can be shown mathematically. Another explanation was that it arose from *the curvature of the Earth*, which requires that in any one direction the observer should see flakes of noticeably different inclination. Finally it was supposed to arise from ice-plates rotating rapidly round a horizontal axis and for that reason assuming every possible orientation in space. This last supposition seems, indeed, to be one of the most likely, though its calculation has never been carried out to the end.

The pillars of light seemed such a simple phenomenon. Who would have thought that their explanation would incur so many difficulties?

147. *Crosses* (Plate IX, *b*)

When a vertical pillar and a part of the horizontal circle occur at the same time, we see a cross in the sky. To remark that the superstitious have made the most of this is superfluous!

On July 14th, 1865, the alpinist Whymper and his companions were the first to reach the top of the Matterhorn, but on the way back four of the men slipped and fell headlong down a precipice. Towards the evening Whymper saw

IX

(*a*) Strongly coloured primary bow; faint secondary bow. The contrast of light within and without the bow at the lower end and the supernumeraries under the primary bow are clearly shown. (*Copyright, A. Glark, King's College, Aberdeen*)

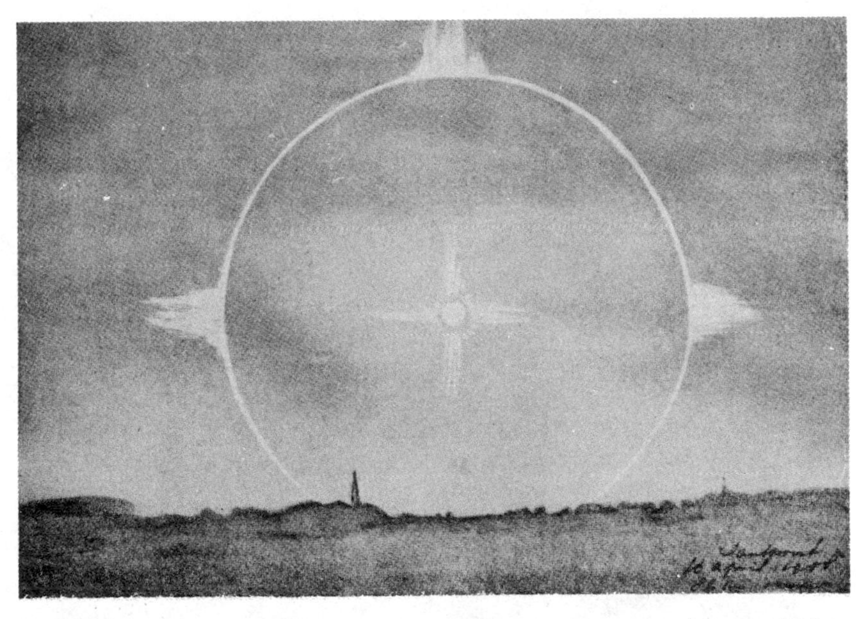

(*b*) Ring or Halo round the Moon, with mock moons, upper tangential arc and light cross. (*After a water colour by L. W. R. Wenckebach, by kind permission of the Royal Dutch Met. Inst.*)

an awe-inspiring circle of light with three crosses in the sky: 'the ghostly apparitions of light hung motionless; it was a strange and awesome sight, unique to me and indescribably imposing at such a moment.'

148. *The Sub-sun*

This is to be seen only from a mountain or an aeroplane. It is a somewhat oblong, uncoloured reflection; the sun reflected not in a surface of water but in a cloud! A cloud of ice-plates, in fact, which appear to float extremely calmly judging from the comparative sharpness of the image.

149. *Double Sun*

Sometimes we see a patch of light just above the sun, very rarely below it. The distance between the sun and this, its misty reflection, is generally not much more than 1° to 2°. In a few exceptional cases there have been two and even three of these images above the sun's disc. Probably this phenomenon is simply due to a local enhancing of the brightness of the pillar of light by unequal distribution of the clouds.

150. *Very Rare and Doubtful Halo Phenomena*

After the many different forms of haloes already discussed, the following short list is intended to give some idea of the amazing wealth of much more rare phenomena, which appear at the most unexpected moments with surprising clearness.[1]

Circles round the sun of 8°–9°, 16°–19°, 23°–32°; always be careful to screen off the sun when looking for these faintly luminous rings! They are due to refraction in pyramidal ice-crystals, fortuitously orientated. This is why they often occur several at a time.

A white circle of 90° radius round the sun. Very dubious.

The anthelion, a concentration of light lying on the horizontal circle opposite the sun, usually colourless and rather hazy.

[1] Numerous interesting observations in the periodical *Hemel en Dampkring* and in the publication of the Royal Dutch Meteorological Institute, *Onweders en Optische Verschijnselen.*

The mock suns on the 90° circle; at 33° from the sun; at 19° from the sun.

The white paranthelia on the horizontal circle at 120° from the sun; also at 40°(?), 84°–100°(?), 145°(?).

The mock sun below the horizon, visible as the reflection of the ordinary mock sun from an aeroplane or a mountain.

Light-pillars over the mock suns and over the anthelia.

Mock suns of the mock suns (a secondary halo phenomenon).

Mock suns located where the small circle and the vertical pillar meet the horizon.

Tangential arcs to the small circle at the position of the mock sun.

Oblique arcs through the sun and oblique arcs through the anthelion, usually white, but once observed coloured.

Arcs on the side opposite the sun, i.e. circles round the anthelion, with a radius of 33°, 35°, 38°.

Uncommon circumzenithal arcs at various heights.

An ellipse round the sun with its long axis of 10° vertical and its short axis of 8° horizontal.

The Bouguer's halo round the anti-solar point, with a radius of 35°–38°; this is often difficult to distinguish from a fog-bow, but the Bouguer's halo is entirely uncoloured, has no supernumerary bows and is generally accompanied by other halo phenomena.

151. *Oblique and distorted Halo Phenomena*

Pillars of light have sometimes been observed which were not vertical, but inclined as much as 20° with respect to the vertical ! The oblique pillar-like patches on undulating water were explained by the dominating direction of the wavelets; here we can obviously assume that the ice-crystals do not float horizontally, but are wafted in a slanting direction by certain air-currents, exactly how is not very easy to explain !

In the same way upper tangential arcs are known which touch the small circle at 10° to 12° from the top. The horizontal circle has occasionally been seen inclined. Once when the sun was 50° high, it appeared to be curved towards the horizon on both sides, and at 90° from the sun it was only 25° above the horizon. On another occasion this circle ran 1° to 2° below the sun! The mock sun of the small circle was once observed to be 40′ too high; this was seen particularly clearly because the sun was about to set.

More observations are needed, and special care should be taken to avoid all personal errors of judgment; make use of a plummet; take photos with a plummet held at some distance in front of the camera, so that it appears (somewhat blurred) on the plate.

152. *The Degree of Development of Halo Phenomena*

The regularity of natural phenomena is always exaggerated by inexperienced observers: they draw snow-crystals perfectly symmetrical, count seven colours in the rainbow, and see lightning as a zigzag! Similarly, there is a tendency to describe halo phenomena as being more complete than they actually are. Yet there is a vast difference between seeing half the circumference of the small halo and seeing all of it! 'The imperfection' of natural phenomena is also governed by fixed laws and is in its way only another regularity.

That is why it is important to note the degree of development of each halo phenomenon by making an estimate of its light-intensity as well as the extent of the part visible. By averaging the observations, the influence of any haphazard irregularity in the distribution of clouds can for the most part be eliminated. It is found as a rule that those parts whose intensity is greatest, are also developed the most frequently. A halo of particularly strong brightness is on an average also particularly extensive. A moderately thin layer of clouds is the most favourable for the development of haloes, very thin layers containing too few crystals, and very thick ones not transmitting enough light or else scattering it in all directions.

A very interesting fact is that the top part of the small circle is seen on an average about three times as often as the lower part. As a reason for this it has been suggested that the path of the rays through the layer of clouds is much longer for the lower part; though this may be an advantage as much as a disadvantage.

153. *Halo Phenomena close to the Eye*

Walking down a narrow street, an observer saw a halo round the moon, but noticed that part of it was projected

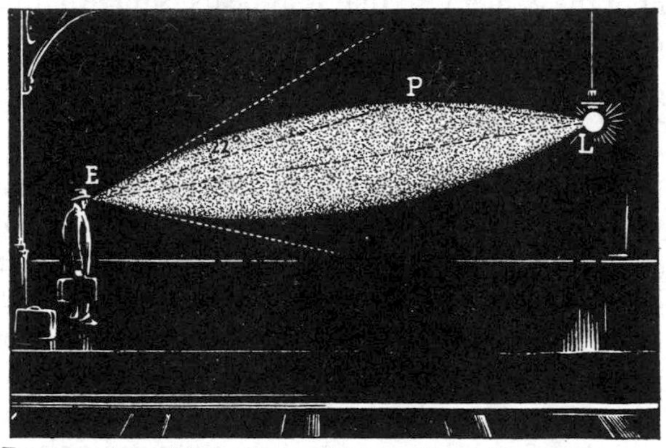

FIG. 129. A small halo observed in the immediate neighbourhood of the eye.

on a dark wall, and formed a whole with the remainder projected on the sky. Even when he screened off the moon with his hand, he could still see the halo: it could therefore *not* be a phenomenon in the eye itself, but apparently there were ice-crystals floating between his eye and the wall, only a few yards from the ground.

One very cold evening (17° F.) beautiful halo phenomena could be seen in the steam from a train in a railway-station. Near one of the lamps, where the cloud of steam was blown in every direction, a cigar-shaped surface of light could be seen, having one end near the eye, the other near the lamp (Fig. 129); all little crystals traversing this surface were lit up, but the space inside was quite dark; the cone tangential to the surface had an angle at the vertex of about 44°. It is at once clear that the cigar-shaped surface is simply the locus of all those points P such that the sum of the angles subtended by EP and PL at L and E respectively is 22°.

The remarkable part of this observation is its *three-dimensional* nature; this is only possible because the source of light is so near and the eyes, working together, see the individual points of light and estimate their distance stereoscopically.

That same evening, in a quieter part of the station, it was observed that the lamps there produced light-crosses. This phenomenon is not new: in Russia and Canada in the winter pillars of light can often be seen above distant lamps, which proves the presence of a floating mist of ice-crystals in the air.

The small halo, the mock suns, the upper tangential arc and the large halo have been seen at times in masses of whirling snow.

154. *Halo Phenomena on the Ground*

We have seen the rainbow projected on a horizontal plane as a dew-bow; in the same way one can sometimes see, on freshly fallen snow, the small and the large circle in the shape of hyperbolic arcs (Fig. 130), especially when the temperature is abnormally low (15° F. or lower), more often when there is hoar-frost.[1] In order to observe it, one should try to see it about half an hour or an hour at the most after sunrise or before sunset. The streak of light consists of a number of tiny separate crystals scintillating with the most wonderful colours; these are mostly red and gold-brown, but the tints are apparently not very

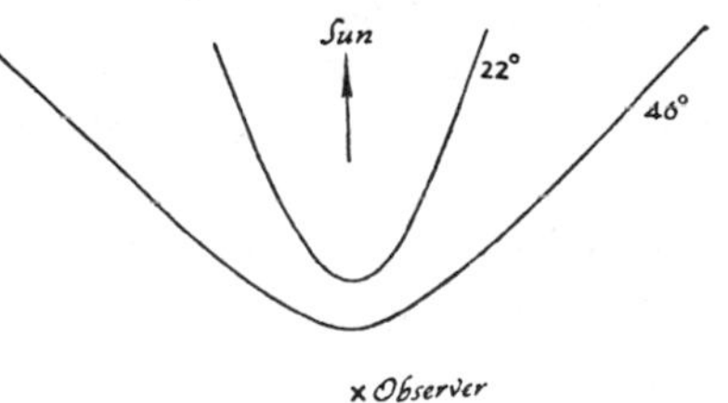

FIG. 130. The small and the large circle, appearing as hyperbolae on ground covered with freshly fallen snow.

[1] Listing, *Ann. d. Phys.*, **122**, 161, 1864. Meyer, *Das Wetter*, **42**, 137, 1925.

saturated. When one moves, the light phenomenon moves with one.

The angle contained by lines from the sun and eye to the crystal can be determined by simple measurements, and you can show that the light-rays are refracted through an angle of 22° (or 46° respectively). Examine the shapes of the crystals through a magnifying glass and then draw and measure them !

CORONAE

155. *Interference Colours in Oil-spots*

When the ground is wet after a shower of rain we can often see coloured patches on the dark asphalt of our roads, patches sometimes as large as 2 feet in diameter and made up of concentric coloured circles. On certain days and on certain roads these patches can be very beautiful, though as a rule they are merely blue-grey spots. They are evidently formed by drops of oil from passing cars; each drop spreads out into a very thin film and the combination of light reflected from the upper and the lower surfaces of the film produces *interference colours*—in other words, the famous 'Newton's rings,' identical with those in the fantastically lovely hues of soap bubbles. Their explanation is to be found in ordinary textbooks on science, but I must point out that we have here before our eyes a proof that light is a wave phenomenon.

The colours are enumerated in the following table, beginning at the outside of the spot and gradually approaching the centre; the thickness of the oil-film is in μ ($=\frac{1}{1000}$ mm.).

I.	μ	II.	μ
Black	0	Violet	0·190
Pale grey	0·080	Blue	0·210
Brown yellow	0·115	Green	0·270
Red	0·170	Yellow	0·305
		Red	0·340

III.	μ	IV.	μ
Violet	0·385	Grey-blue	0·595
Blue	0·400	Green	0·655
Green	0·455	Flesh-colour	0·695
Yellow	0·505		
Flesh-colour	0·525	V.	
		Pale blue-green	0·820

The layers of oil are thus thinnest at the edges and grow thicker towards the centre. Sometimes they reach no farther, even in the centre, than the first steps of the colour scale; at other times, they are so thick that, after the colours mentioned in our table, pink and green alternate with one another several times, growing steadily paler and turning into 'white of higher order' so that no more rings are to be seen in the middle.

Measure the diameter of a regularly shaped spot across the different colours and then draw to scale the transverse section of the layer of oil. If you repeat this after ten minutes you will see that the mound of oil has flattened out. Follow one definite colour as a function of time, and you will notice that the ring first expands and then contracts. Why? In the end all you see is a grey spot whose origin would never have occurred to you if you had not observed its formation. The best way is to stand and watch a single spot and measure every change in it. It will not require so

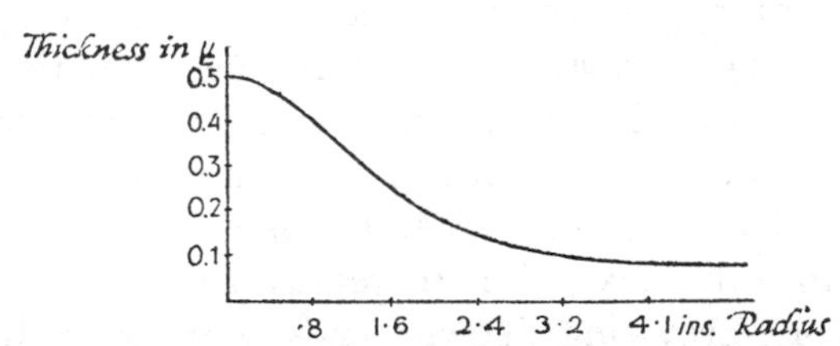

FIG. 131. Scale profile of a drop of oil on wet asphalt, measured by means of interference colours.

very much patience; perhaps not more than half an hour. Protect it from cyclists and pedestrians and pray that it may reach the end of its life before a car runs over it!

Observe the oil-spot obliquely; the colours become displaced as if the layer had become thinner. For if you look at it more slantingly the coloured rings seem to contract so that at any particular point the colour is replaced by the colour that belonged originally to a thinner ring. Try to account for this by calculating the difference in phase of the two interfering rays.

A little boy strokes it with his finger; the colours begin to change, but resume their positions with surprising rapidity; the rings have grown smaller, as a little of the oil has been taken away.

Sometimes we see finely shaped double spots which apparently belong together. There is nothing mysterious in this: they are simply a normal spot after a motor-car wheel has been over it!

We shall not be quite satisfied until we have made coloured rings ourselves. A drop of paraffin or a drop of turpentine poured on to a pond produces indescribably lovely colours! But if we use *oil* for this experiment we shall find a surprise in store for us! The oil does not spread out into a film, and we see nothing. The same thing happens on a wet road as on a surface of water. Are the spots in the road due to petrol perhaps, and not to oil? Again we are disappointed, for petrol produces only greyish white spots, apparently extremely thin and bearing no resemblance whatever to the magnificent rings of colour. Closer investigation has shown that only the *used, oxidised oil* dripping from the engine is capable of spreading on a wet surface.[1] The more complete the oxidation of the oil, the thinner the layer becomes.

Most oil-spots show radial bands. Each coloured ring merges into the next with fringes, as it were, and the outermost, white-grey ring likewise ends in fringes. By pouring petrol on to a wet road we can see how the spot it forms spreads and branches out on all sides, causing as we watch radial bands and fringes. The same phenomenon can frequently be seen in coloured films floating on dirty water.

[1] K. B. Blodgett, *J.O.S.A.*, **24**, 313, 1934.

It may be that complicated molecular forces are at work here.

Interference colours are present wherever a thin film is formed; for instance, on the thin layers of tar and paraffin floating on water; lines of constant colour are lines of constant thickness, and their distortions betray to us all the currents and eddies of the liquid. Beautiful colours can sometimes be seen on the tarnished copper of the funnels of locomotives. Is this because the copper has become hot and subsequently oxidised? Or has a layer of sulphide been deposited on the funnel from the atmosphere and combustion gases?

156. *Magnificent Colours on a Frozen Windowpane*

I once observed the following curious phenomenon. It was an exceptionally cold winter's night (14° F.) and the condensed vapour from the breath of my fellow-passengers was beginning to freeze on the windows of the railway compartment where I was sitting. Suddenly I noticed that the light of each lamp we passed gave rise to a wonderful display of colour, certain parts of the thin frozen layer being tinted sky-blue, others green or red. These colours remained more or less the same over an area of about 1 sq. cm. and all of them were visible only in transmitted and not in reflected light. The tints were so lovely and saturated that one realised at once that this was something very remarkable! The phenomenon lasted only a few minutes and then the layer of ice became a few millimetres thick and the colours disappeared.

I have since learnt that a phenomenon of this kind has already been described,[1] and in far more detail than was possible for me to notice in those few minutes. If you breathe on a very cold windowpane, it seems that your breath is frozen first into small hemispherical lumps of ice (*a*); after about half a minute cracks are formed in the layer, and

[1] Observed by Ch. F. Brooks, *M.W.R.*, **53**, 49, 1925; and by Schlottmann, *Met. Zs.*, **10**, 156, 1893.

the ice-particles gather into small groups (*b*) until, in the end, they form long needles, between which one can see the transparent ice. It is only in stage (*b*) that the colours appear, and this explains their short duration. Another typical peculiarity is that the observed lamp or source of light itself becomes coloured and is surrounded by a halo in the complementary colour; for instance in the daytime, the brighter parts of the snow-clad scenery seem rose-red, and the darker parts green.

The cause of this remarkable phenomenon is as yet unknown. The impression is that it occurs only when the weather is intensely cold. One of my younger readers informed me that at a temperature of 14° F. he could reproduce it at will by breathing on a windowpane. One is involuntarily reminded of colours of thin films or of the 'Christiansen's colours.' More experiments and observations are required to throw light on this matter. Can it be connected in any way with iridescent clouds (§ 166)?

157. *Interference Colours in Ferruginous Water*

The brown water of ditches running across heath-land, where the soil is ferruginous, is sometimes covered with a thin iridescent layer, the pale colours resembling mother-of-pearl. These colours are caused by the colloidal solution of iron-oxide present in the water arranging itself in small parallel plates, at distances of about $\frac{1}{4}\mu$ apart,[1] and this laminated membrane acts more or less like a Lippmann colour-photograph.

158. *Diffraction of Light*

It is night. In the distance a motor car comes whirring towards us through the darkness, its headlights casting a fierce glare on the broad road. A cyclist chances to cross in front of the dazzling light so that for a moment we stand in his shadow. And then suddenly the cyclist's silhouette is outlined by a strangely beautiful light apparently radiating from its edge. The same effect can be observed around

[1] Zocher, *Zs. f. allg. anorg. Chem.*, **49**, 203, 1925.

pedestrians and trees. It is a diffraction effect. Diffraction is the name given to the slight bending that a ray of light undergoes at the edge of an opaque screen, part of the wave-front penetrating the region where, according to geometrical optics, shadow would be expected. The light so deflected is fairly intense if the angle of deviation is small, but diminishes rapidly at larger angles, which accounts for the beauty of the effect when the cyclist is far away and the car far beyond that.

A similar phenomenon on a larger scale can be seen in mountainous country where the air is pure, when, standing in the shadow of a hill, you see its tree-clad upper part as a dark outline against the morning sky. When the sun is about to rise, those trees situated where the light is at its brightest are surrounded by a brilliant silvery white radiance.[1]

They say that in our country furze bushes in particular, seen against the sun, can produce an effect of this kind.

159. *Diffraction of Light by Small Scratches*

If you look at the sun through the window of a train you will see thousands of extremely fine scratches in the glass, all arranged concentrically round the sun. Through whatever part of the window we look, the figure is always the same, from which we may conclude that the glass is covered *all*

FIG. 132. Diffraction of light by small scratches on windowpanes.

over with small scratches in *all* directions, although we only notice those that are at right angles to the plane of incidence of the rays of light (cf. § 27). For every scratch spreads the light in a plane at right angles to its own direction and is therefore only visible to the observer in this plane.

[1] This phenomenon, superficially observed by Folie, was at that time a matter of much discussion. It can be found in *Rep. Brit. Ass.*, **42**, 45, 1872; later in *Nat.*, **47**, 364.

Where such very fine scratches are concerned, one can no longer speak of reflection or refraction, and it is better in this case to regard the deviation of the light-rays as *diffraction*. If you look carefully at one of these scratches, you will see that, in certain definite directions, it shows the most magnificent colours in every possible sequence; if you used a nicol, you would find that the light was strongly polarised when the direction of incidence and observation was oblique. All of these phenomena are very complicated and can be only partly explained by theoretical optics.[1]

160. *Coronae*

Thin, white, fleecy clouds glide slowly past the moon. Our eye is unconsciously attracted to this illuminated part of the sky, the centre of the nocturnal landscape. And each time another little cloud appears we see beautifully variegated rings of light round the softly radiant moon, rings having a diameter only a few times that of the moon itself.

Let us investigate the sequence of these colours carefully. Immediately next to the moon is a bluish border, merging into yellowish-white, and this in its turn has a brownish outermost edge. This *aureole* is the corona phenomenon in its simplest form, and by far the most frequent form of all. It becomes really remarkable only when surrounded by larger and more beautifully coloured rings. The sequence of these rings can be seen from the following table, which agrees almost exactly with the scale of Newton's interference colours, except that the meteorologists have decided on the limits between the various 'orders' in a way slightly different from that of the physicists, namely, so that each group ends with red. On very rare occasions three groups have been observed outside the aureole ('four-fold corona').

I. Aureole (bluish)—white—(yellowish)—brown-red.
II. Blue—green—(yellow)—red.
III. Blue—green—red.
IV. Blue—green—red.

[1] Rayleigh, *Phil. Mag.*, **14**, 350, 1907; *Papers*, v, 410.

It seems practically certain that the colour-gradations vary occasionally; those colours between brackets in the above table, are at one time present and at another absent. When investigating this changeability of the coronae the phases of the moon should be borne in mind, for they cause the diffraction pattern to be at some times hazier than at others.

The best way to estimate the radius of the coronae is to take as starting point the red border with which every order ends, because that colour stands out most sharply, and then to compare the size of the corona with the diameter of the moon (32′). The size of the coronae is found to vary considerably; the brown border of the aureole for instance, can have a radius of barely 1°, whereas at other times it is as much as 5°. The extreme values which have been recorded are 10′ and 13°.

Coronae round the sun can be seen very often, at least as often as round the moon, but they are not noticed so frequently because everyone naturally avoids looking into the dazzling light. And yet, owing to the intense luminosity of the sun, the coronae round this celestial body are usually the finest of all.

Observations can be facilitated by making use of one of the following suggestions:

(*a*) Observe the sun's reflection in calm water; this was how Newton made his famous observation of a corona round the sun.

(*b*) By way of a mirror use a piece of black polished marble glass, or the dense goggles worn by welders, or else an ordinary piece of glass backed by black varnish. One should hold these plates close to the eye in order to be able to survey a large field.

(*c*) Choose marble glass or welder's goggles, sufficiently transparent to allow you to observe the sun without being dazzled.

(*d*) See that the sun is screened off by the edge of a roof.

(*e*) Look into a garden-globe a few yards away, intercepting the sun's image with your head.

The aureole is faintly visible in almost every type of cloud. It is much stronger in alto-cumuli or strato-cumuli, which usually also show a faint indication of the second coloured ring. The most beautiful coronae of all, with delightfully pure tints, occur in thin cirro-cumuli and, apparently, in cirri.

At times small, feebly luminous coronae are to be seen even round Venus, Jupiter and the brighter stars.

161. *The Explanation of the Corona Phenomena*[1]

The coronae we see in the clouds are formed by diffraction of light by the drops of water in the clouds. The smaller the drops, the larger the coronae. In clouds where the drops are all of equal size, the coronae are well developed and their colours pure; in those clouds, however, in which drops of all sizes are mixed together, coronae of different sizes occur simultaneously, the one over-lapping the other. This is why finely developed corona phenomena occur only in very definite kinds of clouds, where circumstances causing the condensation of the water vapour are sufficiently uniform; and for the same reason the finer distinctions in the sequence of the tints will depend on the number of drops of various sizes, on the thickness of the clouds, etc.

The general line of argument in the theory is this:

(*a*) The diffraction by a moderately dense cloud of water drops, all of the same size, is essentially the same as that by one drop, only the intensity of the diffracted light is greater.

(*b*) The diffraction produced by a drop is the same as that produced by a small aperture in a screen (Babinet's principle).

(*c*) The diffraction by an aperture is calculated by taking the aperture as the starting point of vibrations (Huygens' principle), and by ascertaining how the waves from all parts of the opening enter the eye and interfere.

It is quite an easy matter to observe the resemblance

[1] G. C. Simpson, *Quart. Journ.*, **38**, 291, 1912; Ch. F. Brooks, *M.W.R.*, **53**, 49, 1925 (with references); Köhler, *Met. Zs.*, **40**, 257, 1923.

between the corona and the diffraction image of a circular hole. In front of a window on which the sun is shining, hang a piece of cardboard perforated in the middle, but with the perforation covered by a piece of silver paper pasted on to the cardboard. Prick a hole in the silver paper with a needle and then look at this brilliant point of light in the direction of the sun at a distance of about a yard, holding before your eye a second piece of silver paper, likewise provided with a fine needle hole. These holes should be made with the finest of needles, which should be rolled to and fro between the fingers while making the hole; the holes themselves should not be more than 0·5 mm. in diameter. This small hole, at which you look, will appear broadened into a disc, which is a miniature aureole, and surrounding this disc you will see a system of rings corresponding to the successive orders of the corona. The finer the hole in front of your eye, the larger the diffraction pattern.

The successive maxima and minima can be compared in every respect to the diffraction fringes occurring at a parallel slit, only their distances are somewhat different. The outermost red borders of the aureole and of the first order lie at $\delta = \frac{0 \cdot 00070}{a}$ and $\frac{0 \cdot 00127}{a}$ (a = diameter of the hole in millimetres; δ=angular distance reckoned from the centre).

We are therefore able to calculate from the coronae how large the drops are that form the clouds. If the radius δ of an aureole round the moon is four times the diameter of the moon itself, that is $\frac{4}{108}$ radians, then the clouds are formed of drops with a diameter of $\frac{108}{4} \times 0 \cdot 00070 = \frac{0 \cdot 076}{4} = 0 \cdot 019$ mm. This calculation is not quite exact because the sun or the moon are not mere points, but have a radius of 16′. The outermost border therefore becomes apparently too large, and these 16′ are, for that reason, often subtracted from the observed angle δ, before applying the formula, but it is very doubtful whether one is justified in doing this. The result is that you will find that the size of the drops in the clouds is from 0·01 to 0·02 mm.

It is probable that coronae can also be caused by clouds of *ice-needles* of equal thickness, which diffract the light in the same way as a slit does. For coronae with the finest colouring and of the best development are observed now and then in the thin, high cirrus clouds: and these clouds consist of ice-needles.

The thickness of the ice-needles can then be calculated just as simply as the size of the water drops: in the case of the corona mentioned above, where the brown border had a radius of four moon-diameters, the thickness of the ice-needles would be $\frac{0{\cdot}062}{4} = 0{\cdot}015$ mm.

It is very difficult, when observing a corona, to say whether it is caused by drops of water or by needles of ice. With ice-needles the distances between the successive dark minima are all exactly the same and equal to the distance between the centre and the first dark minimum, whereas with water drops the radius of the aureole is 20 per cent. larger than the widths of the orders following. Moreover, the light-intensity of the successive orders diminishes much more slowly with ice-needles than with water drops. But these distinctions are not easy to observe. The best measurements favour at one time one manner of formation, at another time the other, in both cases agreeing with what one might expect judging from the kind of clouds.

For the physicist the presence of a beautiful corona is not only an indication of the great uniformity of the water drops or the ice-needles in the cloud. He also concludes from it that the cloud was probably formed fairly recently; it is 'a young cloud.' For swarms of drops show a continual tendency to become of unequal size; those that happen to be a little smaller evaporate the quickest, whereas the larger ones grow the fastest at the expense of the small ones.

When cirro-cumuli or alto-cumuli (fleecy clouds) pass before the moon one can sometimes see very well how the coronae extend *asymmetrically* towards the edge of the cloud each time a new one drifts before the moon (Fig. 133). The drops are evidently smaller in the outer parts of these clouds than in the inner parts. It is indeed quite obvious that they have already begun to evaporate in these outer parts.

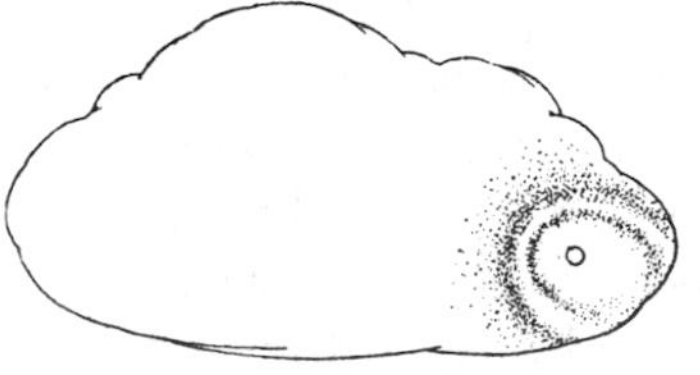

FIG. 133. Asymmetrical corona near the boundary of a small cloud.

162. *Coronae on Windowpanes*

If we walk past well-lit cafés on a winter night, we can often see that the lamps are surrounded by coloured rings, caused by the moisture on the windows. They are larger on some parts of the window than on others. Often we see only the aureole; sometimes, however, the coloured rings are surprisingly beautiful: it is as if some panes always show better ones than others do. The explanation is that the coronae are due to diffraction of the light by the minute drops of water on the window, and the more equal the drops are in size, the more lovely the coronae are. It is not unlikely that drops condense more evenly on some kinds of glass than on others.

These coronae bear a strong resemblance to the cloud coronae, but, after all, their manner of formation is the same. In the one case the diffracting drops are on the window, in the other they float as particles of a cloud high up in the air. And yet there is a difference between the coronae on the panes and those in the air, namely the source of light of the former is surrounded by a *dark field* instead of by a luminous aureole. This seems to be caused by the uniform arrangement of the drops which are formed at equal distances from each other, whereas the drops in a cloud are distributed irregularly.[1]

If we look through a window slantwise, we can see the shape of the corona become first elliptic, then parabolic and finally even hyperbolic. If the conditions were the same as

[1] Donle, *Ann. d. Phys.*, **34**, 814, 1888. K. Exner, *Sitzungsber. Akad. Wien* **76**, 522, 1877; **98**, 1130, 1889.

in the case of the dew-bow we should understand by this: 'the coronae, *as they are delineated on the windowpane,* are elliptical, etc.; but *seen from my eye* they lie on perfectly conical surfaces round the axis of the eye and lamp and they project themselves as circles.' Here, however, the conditions are different. In projection the coronae have actually become ellipses; they have suffered *a further extension* in a horizontal direction, evidently because every drop seen in that direction is foreshortened, i.e. elliptical. At the same time this proves that the diffracting particles are not spherical but hemispherical or spherical segments. For that direction, in which the projection of the drops is the smallest, the corona will be widest.

On blurred windows the coronae round the *reflection* of the sun can also be seen; strictly speaking, this phenomenon cannot be seen in the sky, but it differs only very slightly from a real corona.

Sprinkle a fine layer of lycopodium powder on a small piece of glass (it is a powder used by chemists to cover pills). Look through the glass at an electric lamp at least 10 yards away. You will see it surrounded by magnificent coronae. This powder is the only one that will produce this phenomenon, because the lycopodium spores, being all of about the same size, behave alike, whereas with irregular particles of matter larger and smaller coronae become confused. If you hold the glass in an inclined position, the coronae undergo *no* changes in projection, therein differing from those in the blurred window. The field surrounding the source of light is luminous and not dark, which is to be expected from the irregular separations between the lycopodium spores.

If you breathe on a windowpane from a distance of a foot or two and then examine and measure the coronae formed, you will observe that they do not increase in size as the condensed moisture evaporates; this shows that the drops become less convex, but are not reduced in circumference.

Watch the coronae in the clouds formed in the air by your

own breath on winter days: the brown edge has a radius of 7° to 9°.

Coronae are also sometimes formed on *frosted* windows, the brown border of the aureole having a radius of about 8°.

Measure the coronae formed by lycopodium powder, compute the dimensions of the grains and check your results by means of a microscope.

163. *Coronae of Light originating in the Eye*[1]

At night I can see a feeble circle of light round arc-lamps and other brilliant sources of light, contrasting strongly with a dark, black background; and also round the moon when the sky is clear, and round the blazing sun as it pierces the dense foliage of a tree. The diameter of the circle of light is approximately 6°. It is coloured blue inside, red on the outside; it must, therefore, be attributed to diffraction and not to refraction. The resemblance to the coronae in the clouds is striking, but there is a decided difference. If I stand where the moon is just screened off by the corner of a house, a 'cloud-corona' still remains visible, whereas an 'eye-corona' disappears entirely as soon as I screen off the source of light. Evidently it arises *in the eye itself* ('entoptic').

Can it be caused by small grains in the eye, all about the same size, which diffract the light in the same way as lycopodium powder or the water-drops in the clouds do? This is indeed true with some people.

But with other observers the coronae seem to grow stronger and clearer, if the eye is carefully exposed to vapours of osmic acid. In these cases the cells of the cornea protrude like so many little lumps, sufficiently uniform in size to cause a corona by diffraction. One observer gives the following dimensions for such a corona: red border of the aureole, radius $= 1^\circ 23'$; the blue-green ring, $3^\circ 46'$; red ring, $4^\circ 22'$.

A third type of entoptic corona is one that I observe myself and which apparently is the most common of all.

[1] A. Gullstrand in Helmholtz, *Physiologische Optik.*

The fact that I can see, sometimes for weeks at a stretch, certain *sectors* of this corona with extraordinary clarity, proves that quite another explanation must be given of it than of the cloud-coronae; it is difficult to see how this could arise from diffraction by small particles. Take a piece of paper with an aperture 2 mm. in diameter and hold it before the pupil of your eye, first in the very centre and then more and more towards the edge of the pupil until only two parts of the corona remain, namely, the parts to the left and to the right of the source of light, the hole being held in front of the lower part of the pupil. If the hole is held to the right or left of the pupil only those parts below and above the source can be seen. From this we conclude that the corona in question is due to diffraction by radially arranged filaments, probably in the crystalline lens, since that would explain all the details of the experiment. The use of the small hole is a reliable method for distinguishing this third type of entoptic corona from the two preceding ones. For if the diffracting centres are grains and not filaments, the screening off will only make the corona feebler, and this to the same extent along its entire circumference.

There are times when the corona is practically invisible to me, unless I look upwards and to the side, or unless I am very tired. At other times I can see it continually.

Experiences of this kind help us to decide with greater accuracy in what part of our eye the corona is formed. It appears at night the moment I look at a street-lamp, but disappears in a few seconds. I have noticed that this is connected with the contraction of the pupils when the eye, after its adaptation to darkness, suddenly encounters a strong light. This is why anyone waking up in the middle of the night and becoming suddenly aware of a burning candle or lamp, sees such a bright corona around it.[1] It seems probable that the corona is formed in the outermost parts of the crystalline lens and therefore disappears immediately the pupil contracts.

[1] Cf. a similar observation by Descartes in Goethe's *Theory of Colours*.

In the case of these diffraction phenomena by grains or filaments in the eye, the connection between the angle of diffraction and the size of the diffracting particles is more complicated than usual.

Examine and measure the entoptic corona in the light of a yellow sodium-lamp, such as those along many roads.

164. *Green and Blue Sun*[1]

One observer informs us that, when he was looking at the sun through a column of steam issuing from the funnel of an engine, the sun appeared bright green during three definite puffs, though all the other puffs had no particular effect at all. I, too, once noticed a similar effect during the departure of a local train. The engine (a fairly old-fashioned one) emitted clouds of steam, which again and again darkened the low-lying sun for a moment as they rose into the air. As one such cloud gradually dispersed and disappeared, there came a moment when the sun could be seen again, its colour being sometimes light green, sometimes pale blue and even light green merging into pale blue or *vice versâ*. After a fraction of a second the light was so fierce and the cloud so thin that nothing more was to be seen.

Phenomena like these occur when the drops of water of which the steam consists are very small, between 1μ and 5μ. It is then no longer correct to describe the way in which they affect the light by imagining the water-drops to be replaced by small apertures or opaque discs which diffract the light. An approximate idea of the phenomena can be obtained by examining the united effect of diffracted light, light reflected on the surface and directly transmitted light.[2]

Green, pale blue and azure blue colours of the sun and moon have been repeatedly observed to last for hours at a time without the presence of steam and were seen best of all during the years following the famous eruption of Krakatoa

[1] *Nat.*, **37**, 440, 1888; *Quart. Journ.*, **61**, 177, 1935.

[2] The recent work of H. Blumer has proved, however, how careful one must be with these approximations.

(1883).[1] We know that, at the time, tremendous quantities of extremely fine volcanic dust were hurled into the highest layers of the atmosphere and that it took years for the dust to settle, in the meantime spreading far and wide over the world, and causing the most magnificent sunrises and sunsets everywhere. One can imagine the passing dust-clouds on certain days to consist of very fine grains all of the same size; this would explain the very striking colours of the sun. Blue coloration of the sun has been observed during sand storms.

Under the same group of phenomena can be included an abnormal corona that was once observed in the mist:[2] a vivid yellow-green aureole was encircled by a broad red ring which in its turn was surrounded by a blue one, while green rings were also present. The explanation can certainly be found in the smallness of the fog-drops.

The rarity of these phenomena is indicated by the popular expression, 'once in a blue moon.'

165. *The Glory* (Plate I, frontispiece)

If we happen to be on the top of a hill when the sun is low we can sometimes see our own shadow outlined against a layer of mist, in which case the head of the shadow will be surrounded by a glory having the same vivid colours as those shown by coronae round the sun and moon. On one occasion a fivefold glory of this kind was observed. Note, however, that although everyone sees his own shadow and also the shadows of those around him if they are near enough to him and if the mist is far enough away, the *glory can be seen by each one round his own head only* !

When flying towards the sun through a cloud, an aviator often sees a corona in front of him and at the same time a glory behind him.[3]

As yet no satisfactory explanation of the glory is known. The resemblance to the corona has led many to believe that

[1] Kiessling, *Met. Zs.*, **1**, 117. 1884. *Nature*, 1883.
[2] H. Köhler, *Met. Zs.*, **46**, 164, 1929.
[3] Douglas, *Met. Mag.*, **56**, 67, 1921.

the cloud of drops of water scatters the light of the sun somehow, in the same direction as that from which it came, and that the returning rays are then diffracted by other drops, in a similar way to the diffraction of the direct rays in the corona. However, the most likely explanation seems to be that the glory is formed by the backward scattering itself.[1]

The radius of the glory changes frequently: some parts of the mist evidently consist of larger drops than others. When the mist has just arisen, the glory is very large and the calculated size of the drops it contains is not more than 6 μ.

The glory is often surrounded by a mist-bow; always, if the distance from our eye to the mist is more than 50 yards. It is remarkable that the mist-bow seems much farther away from us than the glory[2]—surely a psychological effect.

The simultaneous occurrence of these two phenomena gives us a splendid opportunity of ascertaining whether the usual theory of the corona can be applied to the glory, for, according to § 161, the size of the drops can be calculated from the radius of the glory, and also from the radius of the mist-bow (§ 128). The question now is whether the results will agree. The agreement is not quite satisfactory; the mist-bow always shows higher values than the glory, sometimes as much as three times greater. Although it is true that the formula for the mist-bow requires great accuracy of measurement, yet such striking and systematic deviations as these ought not to occur.

The results are also unsatisfactory if one calculates the radius of the drops from two different rings of the glory, one being measured immediately after the other; the further out are the rings used, the smaller the values found. In other words, the diameters of the successive rings are not quite what one would expect from the theory of the corona. The whole explanation is obviously not yet as it should be.

[1] B. Ray, *Proc. Ind. Ass.*, **8**, 23, 1923. *Nat.*, **111**, 83, 1923.
[2] Tyndall, *Phil. Mag.*, **17**, 244, 1883.

Such thou art, as when
The woodman winding westward up the glen
At wintry dawn, where o'er the sheep-track's maze
The viewless snow-mist weaves a glistening haze,
Sees full before him, gliding without tread,
An image with a glory round its head;
The enamoured rustic worships its fair hues,
Nor knows he makes the shadow, he pursues!

S. T. Coleridge: *Constancy to an Ideal Object*

166. *Iridescent Clouds*[1] (Plate X)

Those not accustomed to studying the heavens will be surprised to learn that clouds can often show the most glorious and the purest colours, such as green, purple-red, blue. . . . These colours are in no way related to twilight phenomena and appear both when the sun is low and when it is high. They are distributed irregularly over the clouds in the form of coloured edges, spots and bands; some observers maintain that they have a 'metallic' lustre; what do they mean? Our feelings at the sight of such lovely clouds are of intense delight, which is difficult to describe, but which is certainly due, to no small extent, to the purity of the colours, their delicate mingling and their radiant light. We cannot take our eyes off this exquisite sight.

Iridescent clouds like these appear at all seasons of the year, but especially in the autumn. They appear close to the sun, and within distances of 2° from the sun they are mostly dazzling white. If one uses a piece of dark glass they can be seen, most frequently of all, at distances from 3° to 10°; and with the eye unprotected, from 10° to 30°, purple and red being the colours which occur the most, growing paler as the distance increases. Iridescent clouds have been seen by very few observers at still greater distances (up to 50°) and even round about the anti-solar point (Brooks, *loc cit.*).

The intensity of the light is often so fierce as to be almost intolerable for many observers. Always stand in the shade

[1] Ch. F. Brooks, *M.W.R.*, **53**, 49, 1925.

X

Iridescent clouds. The sun was behind the building, the clouds on the right were ribbed and the iridescence was particularly strong. (*From Stanhope Eyre*, *Eder's Jahrbuch*, **14**, 317, 1900)

of a house or a tree or use one of the protective measures given in § 160.

After staring at the iridescent clouds for a long time, without applying one of these methods, I occasionally discovered that afterwards the colours purple and green danced before my eyes, these being the colours that remain as an after-image of all such fierce impressions of light (§ 90). And, as it happens, these are the most predominant colours of the iridescent clouds, so that I almost wondered whether the whole phenomenon was not a consequence of tired eyes. However, this is most certainly not so, for two different observers see the colours in the same way, which remain visible when the light is tempered in one of the ways indicated above; and, finally, the iridescence is often seen on clouds of comparatively feeble luminosity.

Tints in the clouds can practically always be seen if the clouding of the sky is fragmentary, and are absent only when the sun is very low, or when the sky is absolutely white instead of blue. We can distinguish three groups.

(i.) Those with the loveliest iridescence of all are the brilliant white alto-stratus clouds, which look almost like cirrus clouds, only they are more flaky and float at a lower level; that they are not very high is seen when the shadows of high cumuli are cast on them at times when the sun 'draws water' or when at sunset they are not illuminated as long as the real cirri.

(ii.) Cirro-cumuli and alto-cumuli also show lovely iridescent hues; in both these groups the colours are arranged in stripes, bands and 'eyes.' Iridescence is seen particularly when the clouds change shape rapidly, shortly before and after a storm.

(iii.) Cumuli, cumulo-nimbi, and cumulo-strati show iridescence only at the edges, but the light there is so blinding that it is hardly possible to observe it without using a black mirror or something of that kind; a cumulus on the point of dissolving as it passes over the sun is a grand sight! For the rest, the question remains whether the colours in this third group can be considered to be 'genuinely' iridescent and whether they can be counted as belonging to the same phenomenon as groups (i) and (ii).

The arrangement of the colours should give us valuable indications as to the way in which they arise. At first sight the distribution of the colours seems very irregular, but after a time we begin to discover some kind of law. The distribution of the colours in clouds at some distance from the sun is apparently determined by the structure of the cloud: certain streaks show the same colour all over or a purple-red edge is seen round the cloud, etc. If the clouds are nearer the sun, then the main factor is the distance. One notices, for instance, that the clouds begin to show iridescent colours each time they reach a certain part of the sky; or that the colours are arranged in more or less irregular rings round the sun.

We are probably justified in assuming that iridescent clouds are nothing but parts of coronae. Which colour appears at a given point depends on the product:

Size of the particles × angular distance to the sun (cf. § 161).

It then becomes quite clear that, when the angular distance is great, the dimensions of the cloud are negligible and that all parts can be considered as being equally far from the sun; the colour is then determined by the size of the particles, which can differ noticeably between the edges of a cloud and the centre (§ 161). With small angular distance, however, the variations of this factor will play the main part. According to this supposition, iridescent clouds are characterised by drops of very uniform size in each definite part of the cloud, but different in different parts of it. Not everyone agrees with this theory, but in my opinion, there is not a better one.

The iridescent clouds at distances greater than 30° from the sun are extremely interesting. What had been taken for iridescence was often simply a fragment of a halo. But there are certain instances where the observation is indisputable, as I know from my own experience. In such instances one would have to imagine extremely minute particles (2μ), or little feathery ice-crystals, forming a kind of diffraction grating.

The light of iridescent clouds is *not* polarised.

Though less frequently than round the sun, iridescent clouds have been observed round the moon, and are then paler in colour, evidently owing to the very faint luminosity.

On one single occasion iridescence has been observed on the artificial clouds from an aeroplane writing advertisements in the sky!

167. *Mother-of-pearl Clouds*

These are a very rare and remarkable kind of iridescent clouds, on a much larger scale than the usual forms; whole banks of clouds sometimes as iridescent as the scales of fishes, and sometimes full of pure tints and lovely colours. They are particularly beautiful just before sunset at distances of 10° to 20° from the sun. Their salient feature is that they remain visible for as long as two hours after the sun has set, a fact which points to their great height.[1] This was determined recently by more accurate methods to be sixteen miles, while the ordinary clouds are never higher than eight miles. When the mother-of-pearl clouds grow dark, they do so fairly suddenly, in about four minutes, the space of time needed for the sun's disc to sink below the horizon, so that it seems very likely that their illumination is due, not to the twilight, but to the direct sun.

The arrangement of the colours depends almost entirely on the part of the cloud. Sometimes these clouds are striped, undulating, cirrus-like, sometimes the entire cloud bank is almost one colour, with spectral colours along the edges or in oblong horizontal rows, between which one can see the sky, a strange opal-coloured background. The colours sometimes remain constant, at other times they

[1] Their height follows from the time their illumination lasts; accurate computations in Mohn, *Met. Zs.*, **10**, 82, 1893, also in Jesse, *Met. Zs.*, **3**, 1886, etc.

change gradually; they disappear as soon as the distance from the sun to the cloud exceeds 40°. The whole scene is indescribably lovely and majestic.

If the clouds are examined through a nicol the colours will be seen to change on rotating the nicol. On one occasion a halo was observed in these mother-of-pearl clouds, which indicates that they probably consist of ice-crystals (cf. § 134). They are formed most often just after a depression has passed, when the sky is very clear. They are usually visible in Oslo in the winter, when a deep depression lies to the north or east, or while a storm is raging over the Atlantic Ocean and a warm, dry wind is blowing (*föhnwind*); for the sky is very clear at such times and the highest layers are observable.

An exceptionally lovely development of mother-of-pearl clouds was observed on May 19th, 1910, the day when the earth passed through the tail of Halley's Comet. It would seem as if there were some connection between these two phenomena.[1]

For ultra-cirri and luminous night-clouds, cf. §§ 198, 199.

HEILIGENSCHEIN

168. *The Heiligenschein on Bedewed Grass*[2] (Plate XI)

In the early morning, when the sun is still low and casts our long shadow on the dewy grass, we perceive a remarkable aureole of light, uncoloured, lying near and above the shadow of our head. No, it is not an optical illusion, nor a contrast phenomenon, for when the same shadow falls on a gravel path you no longer see this aureole of light.

This phenomenon is at its best when the length of the shadow is at least 15 yards, and when it falls on short grass or clover, greyish-white from the heavy dew. Under such circumstances the Heiligenschein is very pronounced. It is less distinct in the middle of the day after a shower, or at

[1] Slocum, *J.R.A.S., Can.*, **28**, 145, 1934, with a beautiful photo.
[2] *Quart. Journ.*, **39**, 157, 1913; E. Maey, *Met. Zs.*, **39**, 229, 1922.

XI

The *Heiligenschein* on meadows covered with dew

night in the light of strong electric lamps. If there is any doubt about this phenomenon, the best way to be certain is: (i.) Survey the whole lawn, and note how the light increases near your shadow. (ii.) Take a few steps: you will see the glow of light go with you, and places where the grass was not particularly bright become illuminated as your shadow approaches. (iii.) Compare your shadow with that of other people; you will see the Heiligenschein surrounding only your own head. This may lead you to philosophise! When Benvenuto Cellini, the famous Italian artist of the sixteenth century noticed it, he thought that the shimmer of light was a sign of his own genius!

What is the explanation of this curious phenomenon? The dewdrops are certainly essential, for when the dew has once evaporated the Heiligenschein practically vanishes; it can be brought about once more by sprinkling water on the grass. Drops of water sprinkled on a white sheet or a piece of white paper sparkle clearly when the shadow of our head falls close to them.

Fill a spherical glass flask with water and hold it in the sun's rays; this represents a drop of dew on a large scale. Hold a sheet of paper behind it to represent the blade of grass on which the dewdrop is formed. If we observe this flask from a direction forming only a small angle with the direction of incidence we shall see that it is brightly illuminated, provided the paper is held a short distance away from it, more or less at the focus.

This leads us to assume that each of the dewdrops forms an image of the sun on the blade of grass supporting it, and that rays are emitted from the image along almost the same path as the path of the incident rays, i.e. in the general direction of the sun (Fig. 134, *a*). This would explain why the drops seem to emit light from within, in the same way as the eyes of a cat. It is also an excellent explanation of why one can see so much light coming from the grass in directions close to that of the anti-solar point, and why the intensity of the light diminishes rapidly when one looks away from it. But why, then, is this light not *green*?

There must be other factors at work. If we look at our flask again we shall see that the front of it reflects as well as the back. A simple calculation shows that the brightness of the reflection from the back is about half that of the light re-emitted by the blades, and of the reflection from the front about an eighth.

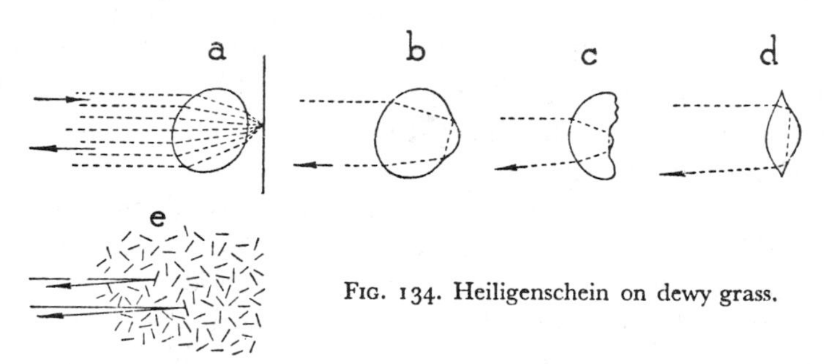

FIG. 134. Heiligenschein on dewy grass.

But a much stronger light comes from the neck and the flat base of the flask, light that has suffered total reflection. And in our dewdrops this is very likely to be the most important factor, since the drops are irregularly distorted (Fig. 134, *b*, *c*, *d*), especially on hairy, white-woolly plants, and the light, being totally reflected at various points, is as fierce and white as when it arrived from the sun. This second group of reflected rays shows no decided preference for reflection in the direction of incidence. But the following ingenious observation has been made: only those blades of grass on which the sun's light actually falls re-emit light, and these are naturally unscreened in the direction of the sun by other blades, whereas in most of the other directions there is no clear opening in front of them (Fig. 134, *e*). This explains why an observer always sees more light when looking more or less in the direction of incidence. This curiously simple principle (put forward by Seeliger and Richarz) has already been applied in astronomy to explain the distribution of light in the rings of Saturn which are known to be composed of small stones.

The combination of the different light effects just mentioned appears to be a sufficient explanation of the whiteness, as well as of the direction, of the light of the Heiligenschein.

169. *The Heiligenschein on Surfaces without Dew*

This phenomenon is much more difficult to observe, and the methods described in § 168 will come in very useful. It has been seen on stubble fields, on short grass and even on rough soil; I have seen it clearly and unmistakably, when the sun was low, on a well-kept lawn, all the blades of grass being upright and of equal length, and still more clearly on clumps of the grass *Molinia coerulea.*

If the observer is standing at some distance from the lawn, say a few hundred yards, his shadow is so hazy as to be almost indistinguishable (cf. § 2) and all that strikes one is the Heiligenschein itself, as a patch about 2° in diameter (that is about four times the diameter of the moon) somewhat elongated in the direction towards us.[1]

The explanation is the same as the one given by Winterfeld for the Heiligenschein on dewy grass (cf. § 168) and which we may formulate as follows: the sun illuminates most of the stalks through the spaces between the foremost rows; anyone looking more or less in the direction of the sun's rays will see all the small illuminated surfaces; if he looks more sideways he will see many blades in the shade, the average brightness, therefore, being less.

A strong Heiligenschein can often be seen on the white chenopodium. This plant is covered with a farinaceous coating of globular cells, which evidently act in the same way as the dewdrops and are very strongly developed on certain varieties of this species.[2]

170. *Heiligenschein round the Shadow of a Balloon*[3]

Anyone enjoying a flight in a balloon should watch the shadow of the basket passing over the country below. It is

[1] *Nat.*, **90**, 621, 1913.
[2] v. Lommel, *Ann. d. Phys.*, 1874, Jubelband 10.
[3] C. Flammarion, *L'Atmosphère*, p. 232 (1888); *Met. Zs.*, **30**, 501, 1914; *Deutsche Luftfahrer Zeitung*, **17**, 83, 1913.

nearly always surrounded by an aureole of light, and that this is no subjective contrast phenomenon is proved by the fact that it becomes more pronounced on fields and meadows covered with dew; on cornfields it changes into a vertical pillar of light, parallel to the direction of the stalks of corn. This is a particularly beautiful form of Heiligenschein for, owing to the great distance from the balloon to the earth, we see everything beneath us at an extremely small angle with the incident rays of the sun. If the shadow floats over banks of clouds, there is a chance of seeing a magnificent shadow phenomenon with coloured glory-rings (§§ 128, 165).

Can a Heiligenschein be seen round the shadow of an aeroplane?

XI

Light and Colour of the Sky

171. *Scattering of Light by Smoke*

We will begin our observations concerning the scattering of light by taking a walk beside a busy canal. Many of the passing boats have oil or petrol engines which emit a fine smoke that looks blue against a dark background. If, however, the smoke is seen against the light background of the sky it does not appear to be blue at all, but yellow. It is clear then that blueness is not an inherent property of the smoke in the way that the blue of blue glass is, but the colour of the smoke depends on the manner in which it is illuminated, which is different in the two cases. The explanation is as follows:

Against a dark background, the smoke is illuminated by rays from the sun falling on it obliquely from all directions except from behind; these rays are scattered by the smoke in every direction and some of the scattered rays enter our eyes and make the smoke visible. The particles which make up the smoke scatter blue light much more than red or yellow: therefore we see the smoke as blue. On the other hand, when the background is bright we see the smoke by transmitted light and it appears yellow because the blue in the incident white light has been scattered in all directions, very little can reach our eyes, and only the yellow and red remain to be transmitted and give colour to the smoke.

'Years gone by, I used to see something similar to this in Killarney, when on windless days the columns of smoke rose above the roofs of the cottages. The lower part of each column was shown up by a dark background of pines, the top part by the light background of clouds. The former was blue because it was mainly seen through dispersed light, the latter was reddish because it was seen through transmitted light' (J. Tyndall).

This same phenomenon of blue in scattered light and of yellow in transmitted light, can be seen very clearly in the smoke issuing from the exhaust of Diesel engines at the moment when they are accelerated to set the train going, and in Diesel motor lorries and 'buses. Or, again in the smoke of dry leaves, weeds and rubbish of bonfires in the autumn, and in the smoke from our own chimney when we have a wood fire.

In all these cases the smoke consists of extraordinarily tiny drops of tar-like moisture, whereas the combustion of ordinary coal produces much coarser flakes of soot. And it is the size of the scattering particles in comparison with the wave-length λ of the light (0·0006 mm.) which decides the colour of the smoke. When the particles are not larger than one-tenth or two-tenths of a wave-length the scattering is proportional to $\frac{1}{\lambda^4}$, and it therefore increases rapidly towards the violet end of the spectrum; the scattering by such small particles, whatever they may consist of, always shows a beautiful blue-violet light. But with larger particles this increase of the scattering towards the violet is much less pronounced, and the scattering is then proportional to $\frac{1}{\lambda^2}$. When the particles are very large, the dependence of the scattering on the wave-length is no longer noticeable and the scattered light is white. By 'very large' we again mean very large in comparison with the wave-length as, for example, 0·01 mm.!

This explains why the smoke of a cigar or cigarette is blue when blown immediately into the air, but becomes white if it has been kept in the mouth first. The particles of smoke in the latter case are covered by a coat of water and become much larger.

The steam of an engine is bluish close to the exhaust-opening of the safety-valve, and white higher up, owing to the increased condensation and the larger drops. Notice the difference between the *smoke* and the *steam* of an engine, in incident as well as in transmitted light, and take care never to confuse them !

We have so far considered scattering in comparatively rarefied clouds of smoke, but in very dense smoke the phenomena are more complicated, for then the light undergoes repeated scattering from one particle to another. Watch the smoke rising from dry leaves on a bonfire and you will see that the edges of the smoke column are a lovely blue, like that in the smoke of all wood fires, but the parts towards the centre, where the smoke is thickest, are almost white. It is easy to prove that the light, which reaches our eye after being scattered by layers of sufficient thickness, will always be white, however blue the scattering by each particle may be. For all the light falling on the cloud of smoke must ultimately leave it if there is scattering only and no absorption (§ 176).

Our chimney smoke and the smoke from factories is generally black in incident light, however thick and opaque the column of smoke, which shows that the soot flakes not only scatter light, but also strongly absorb it. Thin layers of smoke of this kind make the sky seem brown when you look through them and yet the colour of the smoke in scattered light can hardly be called bluish. The brown colour must therefore be attributed to absorption by the smoke particles. This agrees with the fact that the absorption of carbon increases rapidly from the red to the violet of the spectrum; this characteristic is exemplified in the blood-red colour of the sun when seen through the smoke of a house on fire !

172. *The Blue Sky*[1]

Above the clouds the sky is always blue !

H. DRACHMANN.

In unending beauty the blue sky spans the Earth. It is as if this blue were fathomless, as if its very depth were palpable. The variety of its tints is infinite; it changes from day to day, from one point of the sky to the other.

What can be the cause of this wonderful blue ? Not light emitted by the atmosphere itself, for then it would shine at night, too. Nor a source of blue light somewhere behind it, for at night we can see the beauty of the dark background on which the atmosphere is visible to us. It must, therefore, be inherent in the atmosphere itself. And yet it is no ordinary absorption-colour, since the sun and the moon are by no means blue, but yellow rather. It must then, surely, be a case similar to that of very fine smoke. This leads us to assume that the light of the sky is simply scattered sunlight ! We know that the scattering by small particles is greater, the nearer we are to the violet end of the spectrum. The colour of the sky is, indeed, largely composed of violet (to which our eye is not very sensitive) and further of a fair amount of blue, a little green and very little yellow and red; the sum of all these colours is sky-blue.

What, now, are those particles of matter that scatter light in the atmosphere ? In the summer, after a long period of drought, the air is filled with innumerable particles of sand and clay borne by the wind and clouding our view of the distant landscape, and it is at times like these that the sky seems less blue and more whitish. But after a few heavy showers, when the rain has washed away the dust, the air becomes clear and transparent, the sky a deep and saturated blue. Whenever high cirrus-clouds appear, filling the air with ice crystals, the lovely blue disappears and changes

[1] The famous Swiss geologist, A. Heim, has written a splendid book called *Luftfarben* (Zürich, 1912), in which he describes in a popular and enthusiastic way the colours of the sky and the twilight phenomena. The coloured reproductions of water-colours are superb.

into a much whiter colour. Therefore it can be neither the actual grains of dust nor the small particles of water and ice that cause the blue scattering that colours the heavenly vault. The only possibility is that the *molecules of air themselves* act as scattering centres—very weak ones certainly, but strong enough to cause a noticeable brightness in a layer many miles deep, and with a decided preference for the violet and the blue rays (the $\frac{1}{\lambda^4}$ law).

The sunlight, as we see it, lacks the blue and violet rays which the air has scattered. The sun, therefore, takes on a pale yellow hue, which becomes more pronounced the lower the sun is, its rays having then to traverse a longer path through the air. This colour changes gradually into orange and then into red, the red peculiar to the setting sun.

Rayleigh's famous law of scattering for particles smaller than 0·1 of the wave-length of light is expressed by:

$$s = \text{const} . \frac{(n-1)^2}{N. \lambda^4}$$

where s denotes the scattering per unit-volume, N the number of particles per cm.³, n the index of refraction.

173A. *Aerial Perspective*[1]

A distant wood forms an excellent dark background against which to observe atmospheric scattering, and the more distant it is the hazier and bluer it seems. The thick layer of air between us and the wood, illuminated laterally by the sun's rays, scatters light which is superposed on the background, just as the light of a veil covers the objects behind it. In this way the contrasts between the light and dark parts are softened, making the background *more uniform* and also *more blue*. Our estimate of the distance of groups of trees is involuntarily influenced by the extent of this *aerial perspective*. A tree standing 100 yards away has a

[1] Heim, *Luftfarben* (cf. § 172); Vaughan Cornish, *Geogr. Journ.*, **67**, 506, 1926, from which paper especially the end of § 173 has been taken.

more bluish tint than one close to us. The green of meadows is made blue-green (and afterwards blue) by the distance amazingly quickly. Distant hills are often a lovely blue, like that so much used by sixteenth-century painters, such as van Eyck and Memling in their background landscapes. The dunes along the coast, with their luxuriant vegetation, undulating like waves, crest after crest, further and further away, also show these 'blue' horizons beautifully. Owing to this aerial perspective, every tint approaches the same blueness, and merges harmoniously into the others; only the red of houses and the green of meadows quite near to us are conspicuous and break the colour harmony. Observe this for yourself in the landscape.

On the other hand we can try to find complementary changes of colour in a bright background. In mountainous country one would choose the snow-clad mountains; in flat country one looks at the rows of cumulus clouds, dazzling white when close by, and becoming gradually *yellowish* further back in the landscape.

Yet the scattered blue light on a dark background is much more distinct than the yellow colouring of the bright parts. In the former case, darkness is replaced by a small quantity of light, and in the latter there is only a small change in an already considerable brightness: the *relative* difference is much smaller (§ 64).

In the wide horizons of the flatter parts of the country aerial perspective develops in all its glory, and owing to the constant changes in the degree of moisture the blue scattering of air molecules and the stronger but greyer light of the hazy sky, predominate in turn.

Sometimes between two showers of rain a wedge of high pressure passes over us, and the air is very transparent and pure. Shadows and colours in the foreground become distinct, and the dark parts of the background change to purplish-blue.

On a hazy day the foreground is less richly coloured, verging more on grey. The undulations in the ground in the middle distance stand out more because the hollows are seen through a thicker curtain of mist than the heights (*see*, however, § 91), and finally the view in the very far distance becomes much less clear.

In beautiful summer weather when the barometer is high there are many particles of dust in the air, the sky is very bright but not very blue, so that the contrasts of light and shade are less pronounced, and, moreover, the observer is continually being partly blinded by the brightness of the sky.

Moonlit scenery is at its best when there is absolutely no mistiness, for this makes the light weaker, the contrasts less striking, and the scene is likely to turn into a monotonous grey.

It is due to aerial perspective that the sailor sees the coast loom blue and ethereal in the distance, a contrast to the darker blue of the waves, whose more powerful shapes develop in the foreground of the scene. The distant land seems to him a place of peace, an enchanted kingdom. . . .

173B. *Light and Colour in Mountain Regions. The Landscape seen from an Aeroplane*

The wonderful charm that mountain scenery has for those used to living in flat countries must, in the first place, be ascribed to the greater purity of the air, rather than to the height as such. There is no smoke from factories or big towns, consequently fewer large particles of dust, more saturated colours and a more striking perspective in the air. The original beauty of the hues in the landscape, more and more spoilt in our parts by the effects of industrial development, can still be enjoyed in its full splendour in the mountains. Moreover, as a consequence of the greater height, the air is appreciably more rare, so that its scattering power is diminished. At heights of more than 10,000 feet the inexperienced traveller again and again makes the same mistakes in his estimation of distances. Without knowing it, he is apt to attribute the slight scattering to the nearness of the scenery he is looking at. From the mountains we can see how the air lower down, strongly illuminated by the sun, covers the valleys as with a veil, whereas there is nothing to interfere with people in the valleys seeing the brightly illuminated tops of the mountains.

At heights of more than 13,000 feet the sky looks blue-black, the sun and the moon have a fierce, almost white colour, instead of the warm yellowish hue they usually show us. The bright snowfields are dazzling, the shadows sharp and dark. Observing these harsh contrasts, we learn to appreciate to their full extent the mellowness and harmony of the scenery in flat countries.

Seen from an aeroplane the optical effect is again different. While flying low the light from the landscape underneath has only to travel a short distance through the scattering layer of air to reach our eye. The haze hiding all colours as long as we were on firm ground, has practically disappeared altogether, and the hues are now displayed for the first time in their full warmth and saturation. This explains the charm of the scenery experienced by everyone who has had the opportunity of making a trip by air. At greater heights, the effects become more and more like those observed in the Alps.

174. *Experiments performed with a Nigrometer*[1]

'Nigrometer' is a learned name for a very simple instrument. A cardboard cylinder of the kind used for sending drawings by post, 20 inches long, with a diameter of, say, 1 inch, is provided with a little lid at both ends. In one of these lids a hole is cut with a diameter of $\frac{1}{4}$ inch, and in the other lid one of $\frac{1}{8}$ inch. A cap of black paper is then rolled round both ends of the cylinder and the apparatus is ready for use.

When looking through this apparatus, the smaller of the two apertures should be held before the eye, which sees the other aperture illuminated against an almost completely

[1] R. Wood, *Phil. Mag.*, 1920. **39**, 423, 1920.

dark background. Direct the cylinder towards a window some distance away and you will see the dark window-opening look distinctly bluish in colour, which is the scattered light of the sunlit air between you and the window-opening. Go nearer to the window: the nearer you get, the weaker the bluish light becomes, the shorter the scattering

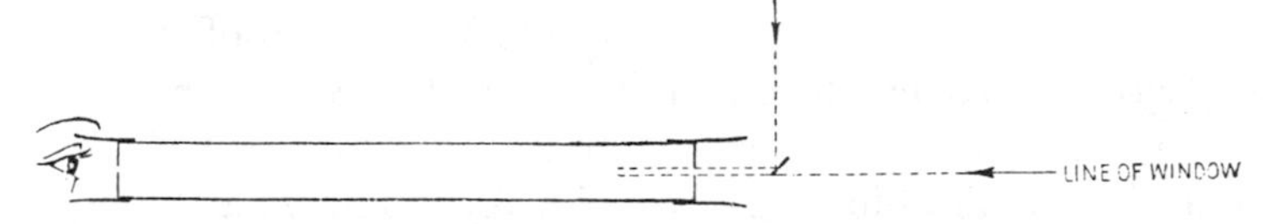

FIG. 135. Observing with the nigrometer: measuring the atmospheric scattering.

column of air. For short distances it is better to direct the nigrometer towards a box, black inside, with a small opening, which forms an almost perfect 'black body.'

We will now determine what column of air scatters to the same extent as the entire depth of the atmosphere. Hold a piece of glass, blackened at the back (for example, a piece of a heavily fogged photographic plate) in front of half of the aperture, at an angle of 45° with the axis of the cylinder. If you can, choose your direction of observation in such a way that the reflected light comes from a part of the sky about 60° away from thc sun. Through the uncovered half of the aperture our dark window opening is visible. Now, how far must you go back in order to see both halves of the aperture equally strongly illuminated ? When the weather is sunny and bright, you will find the required distance to be about 350 yards; when it is sunny, but slightly hazy, you will find the distance to be perhaps only 140 yards.

By the reflection, the piece of glass reduces the light to 5 per cent. of its original intensity (§ 52). The amount of scattering by the sky at 60° from the sun is, therefore, equal to that of a column of air of 350×20 yards $=4$ miles, roughly. Now, if we were able so to compress the atmosphere as to make its density over its entire height equal to the density at the earth's surface, its 'equivalent height' would be 5·5 miles. For the total weight of the atmosphere is $1{\cdot}033 \times 10^3$ gm. per square centimetre and 1 $cm.^3$ of air weighs 0·001293 gm., so that we obtain for the 'equivalent height':

$$\frac{1{\cdot}033 \times 10^3}{0{\cdot}001293} = 8{\cdot}8 \times 10^5 \text{ cm. or } 5{\cdot}5 \text{ miles.}$$

The agreement with our optical determination turns out to be not at all bad! We may consider it as a proof that the scattering particles, to which the aerial perspective at the earth's surface is due, are of the same nature as those responsible for the blue colour of the sky. The fact that our result, 4 miles, falls somewhat short of 5·5 miles, may be taken to prove that owing to their higher content of dust, the lower layers of air scatter more strongly than the upper layers. Besides, our determination is in all respects a very rough one: you can hardly expect it to yield more than the right order of magnitude.

The nigrometer shows us each colour in the landscape separately, free from any influence of the contrast with its surroundings. To some observers all the tints under this condition seem more yellow than usual, the sea and sky seem almost white, and when a cloud passes by the blue appears again.[1]

This may not perhaps be experienced by everyone, but nevertheless the nigrometer is undeniably a very valuable and interesting apparatus.

175. *The Cyanometer* (instrument for measuring the blueness of the sky)

Mix zinc-white and bistre with Prussian blue or cobalt blue in different proportions. These mixtures do not fade. By painting them on strips of cardboard and numbering them we shall be well equipped for measuring the colour of the sky. This method is still often applied when travelling, and the composition of the light of the different numbers on the scale can be examined colorimetrically

[1] J. S. Haldane, *The Philosophy of a Biologist* (Oxford, 1935), p. 52.

later on. Similar scales have been made for practical purposes, and can be bought ready made.

Care should be taken when using these blueness-scales to stand with one's back to the sun and to let it shine on the scale.

176. *Distribution of Light over the Sky*

Study on a bright day the distribution of light over the sky, using, if you have one, a cyanometer, or our useful nigrometer. Above all study your surroundings intently. Use a little mirror to compare one part of the sky with another (Plate XIII; p. 304), and draw lines of constant brightness (isophotes) and blueness in a diagram like the one in Fig. 136; repeat this for different altitudes of the sun.

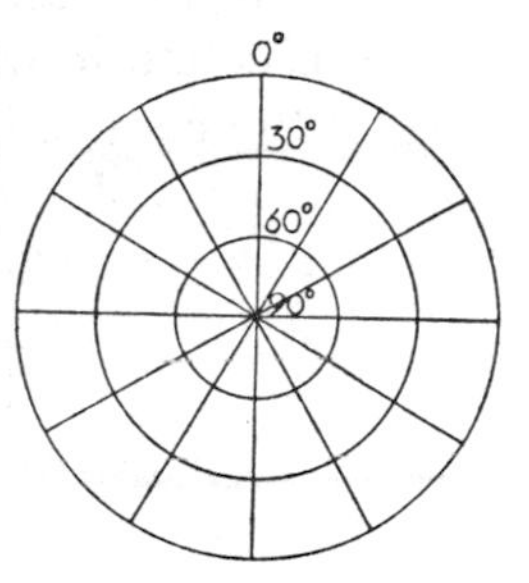

FIG. 136. Diagram for drawing the lines of constant brightness and blueness of the sky.

'In time the practised eye sees the course of the isophotes as if they were painted in blue on the background of the sky.'—C. Dorno.

The darkest point always lies in the vertical circle through the sun at a distance of about 95° from the sun when it is low, and at a distance of about 65° when it is high. Through this point passes the 'line of darkness' dividing the sky into two: a bright region surrounding the sun, and another bright region opposite it. The shape and size of these regions vary according to the altitude of the sun. We can consider this distribution of light to be due to the combination of the three following phenomena:

1. The brightness increases rapidly close to the sun and even becomes dazzling, its colour approaching white more and more (you should stand in the shadow of a building near the edge of the shadow).

2. At a distance of 90° from the sun, the sky tends to become darkest and bluest but

3. there is still another effect. The light intensity increases

from the zenith to the horizon whilst at the same time the colour changes to white. This effect combines with the two just given.

We can measure the first phenomenon very well with the nigrometer. We cover one half of the field with a piece of glass, backed by a layer of black paint, which will reflect that part of the sky close to the sun, and we turn the other half in a direction 40°–50° from the sun. By varying the direction a few degrees one way or the other, we can easily find a direction in which both halves of the field are equally bright; the change in brightness produced by such a movement is particularly marked in the half of the field illuminated by the reflection of the bright part of the sky. The fact that such an adjustment is possible leads us to the conclusion that at this point close to the sun the brightness must be at least twenty times stronger than at a distance of 45° from the sun. This exceedingly strong scattering at small angles with the incident light is to be ascribed to coarser particles floating in the air, both dust and drops. This agrees with the fact that the colour near the sun is less blue, but whiter and even yellowish like the sun itself, for these large particles scatter all colours to about the same extent.

The second effect is a consequence of the law of scattering itself. At an angle of 90° the scattering must be almost twice as weak as at the anti-solar point; moreover, the coarser particles hardly, if at all, scatter the light at such a large angle. What we see is therefore only the saturated deep blue scattered by the molecules of air themselves.

The third effect arises chiefly from the great thickness of the layer of air between our eye and the horizon. Although every particle of the air scatters the violet and blue rays by preference, these, in their turn, are weakened most in their long path from the scattering particle to our eye. When there is a very thick stratum of air, these two effects just counteract one another.

Suppose an element of volume at a distance x from our eye, to scatter the fraction sdx. This amount of light is weakened in the ratio e^{-sx} before reaching our eye. The light received from an infinitely thick layer would therefore consist of the sum of similar contributions arriving from all the elements dx, i.e. $\int_0^\infty se^{-sx}\,dx$, which is equal to 1. This is evidently independent of s, i.e. of the colour. The sky close to the horizon must therefore show the same brightness and colour as a white screen illuminated by the sun.

It is quite possible, also, that the layers close to the ground contain more floating particles of dust, which make the scattering of light more intense and the colour whiter, even when the layer of air cannot be considered to be of infinite thickness.

The darkest parts of the sky are always the most blue, and the colour there is the most saturated. This means that no clouds occur having particles smaller than 0·0001 mm., for these would increase the light-intensity locally and yet leave the blue unchanged.

Ruskin mentions the blue sky as being the finest example of a uniform gradation of colour.[1] He advises us to study a part of the sky after sunset, mirrored in a windowpane, or in its natural frame of trees and houses. Try to imagine that you are looking at a painting and admire the evenness and the delicacy of the transitions.

177. *The Variability of the Colour of the Blue Sky*

The colour of the blue sky changes daily in proportion to the quantities of dust and water-drops present in the air; the cyanometer is indispensable for comparisons of this kind. The deepest blue is seen during the temporary clearing up of the weather between two showers of rain, in the wedges of high pressure. On the other hand, the sky becomes whitish the moment the cirrus clouds appear, however thin they may be, or in the dust-laden air of summer (§§ 172, 173).

[1] Ruskin. *See* note on p. 21.

Compare the blue of the sky with the skies of Italy during your holiday. Compare the blue of the tropical skies with the blue sky in England.

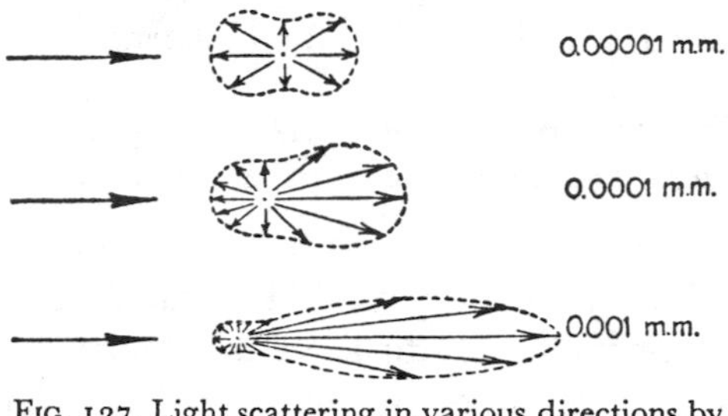

FIG. 137. Light scattering in various directions by grains of increasing size.

Compare the blue at different times of the day. The sky is at its bluest during sunrise or sunset,[1] which is readily understood, since a point near the zenith is then 90° from the sun and 90° from the horizon (cf. § 176).

Small particles scatter preferentially violet and blue light, and this they do with almost equal intensity in all directions.

Large particles scatter all colours equally strongly (white light) and do this mostly at small angles (Fig. 137).

178. *When is the Colour of the Distant Sky Orange and when is it Green?*

We have seen that when the sky is cloudless the horizon has the same colour as a sheet of white paper illuminated directly by the sun. It is therefore clear that towards sunset, when everything is flushed by the sun's warm orange glow, the same colour must appear all along the horizon too.

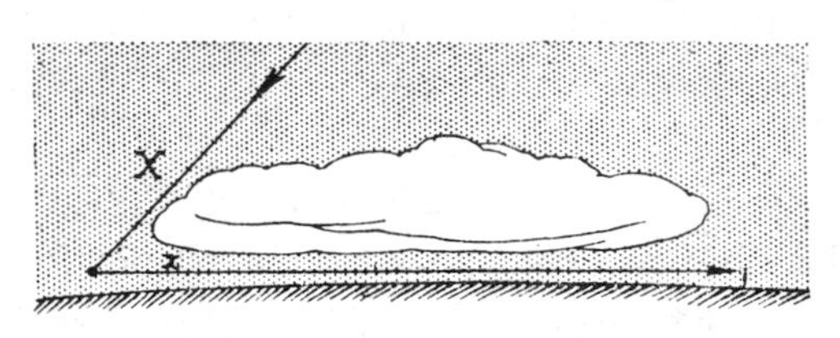

FIG. 138. When a large part of the landscape is covered by a heavy bank of clouds, the horizon sometimes shows a warm orange colour.

But there are times when the distant horizon becomes orange long before the moment of sunset. A heavy dark

[1] *Phys. Rev.*, **26**, 497, 1908.

bank of clouds stretches across the whole landscape and there remains, far away near the horizon, only one low patch where the sun is shining (Fig. 138). At such times this small portion of the heavens has a surprisingly warm orange colour, throwing up the dark, black silhouettes of distant farms, and made all the more impressive by the darkness of the rest of the landscape.

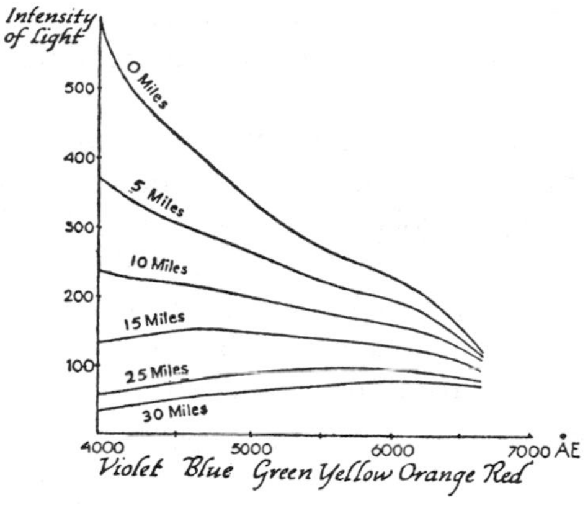

Fig. 139. Composition of light coming from a small volume of air, lying at various distances from our eye.[1]

The explanation is as follows: Consider a volume of air at a distance x illuminated by sunlight which has travelled over a path of length X in the atmosphere. Assuming that a small fraction s of light is scattered per kilometre of path traversed, the intensity at x will be proportional to e^{-sX}. The molecules of air at x scatter towards our eye a fraction of the incident light proportional to s, so that if the intensity at x were unity the fraction reaching our eye would be se^{-sx}. But the intensity at x is proportional to e^{-sX}; hence the amount of light actually entering our eye is proportional to $se^{-sX} \times e^{-sx}$ or $se^{-s(X+x)}$. This expression yields a maximum for moderate values of s, but for small and large values of s, it becomes practically zero. To put it another way: light of longer wave-lengths is not scattered to a great extent by the air through which it passes; light of shorter wave-lengths, on the other hand, is weakened considerably while travelling over the long distance through the atmosphere. The graph (Fig. 139) shows how the light is

[1] For calculating this figure, we used the coefficients of the scattering which have been observed for the atmosphere as a whole. Properly speaking, we ought to have used those belonging to its lowest layer.

composed which reaches our eye from volume-elements of air for which $X+x$ equals 0, 5, 10, 15, 25, 30 miles. The maximum, that is, the colour in the light reaching us having maximum intensity, moves more and more from blue towards red, according as the illuminated part of the air is farther away from us. When $(X+x)=20$ miles the colour is practically green, but at 30 miles it has changed into orange.

This also explains the origin of the lovely green seen in the colour of the sky at times, for instance, after a fall of snow. It follows from Fig. 139 that in that tint the green component predominates only slightly over the other colours, so that the green colour will be only slightly saturated, as is in fact observed.

The green and yellow components in the light from the horizon are in reality always there, though when the air is cloudless they mix with the blue of the nearer particles to produce white. The exceptional colour effects appear as soon as a shadow falls on part of the light-path; and whenever different openings occur in the clouds covering the sky very different shades of colour become possible.

179. *Colour of the Sky during Eclipses of the Sun*

A *partial* eclipse of the sun provides us with an opportunity of seeing how the colour of the sky is changed by the shadow of the moon and how the colour differs on the side from which the shadow comes, from that on the side towards which the shadow moves.

A *total* eclipse of the sun, occurring alas far too seldom, shows much more magnificent colours. The side of the sky from which the shadow approaches is dark purple, as if a thunderstorm were gathering. During totality the sky in the distance is a warm orange colour because the parts of the atmosphere there, being outside the zone of totality, are still illuminated by the rays of the sun and are now seen straight across an unilluminated part of the atmosphere (cf. § 178).

180. *Objections to the Theory concerning the Blue Colour of the Sky*

It is an interesting fact that to this day there are scientists who do not consider the problem of the blue sky as being definitely solved.[1] They have two objections and both can be checked by means of our simple apparatus. On very exceptional days, occurring perhaps not even once a year, the sky is beautifully blue *right down to the horizon.* Observations on days like these should be carefully recorded and described with the help of the normal colours on a cyanometer, for, according to the theory of scattering, such a phenomenon is impossible: with layers of such thickness the air ought to appear white (cf. § 176).

Under abnormal conditions mountains can be visible at distances of 100 miles and more. According to the theory of scattering, this, too, is impossible.

When one considers, however, that these observations are only qualitative, that several corrections have to be applied to them before they can be used, and that up to now the theory given to explain these complicated phenomena is only approximate, one cannot be very convinced by these objections when balanced against the enormous number of experimental and theoretical confirmations.

181. *Polarisation of the Light from the Blue Sky* (cf. § 182)

The light from the blue sky is polarised to a rather high degree. The effects are especially clear when the sun is low. They can be observed with the help of a nicol or, more simply,[2] by using a piece of glass with a dark backing. If a ray of light strikes this glass at an angle of incidence of about 60° with the normal (the 'angle of polarisation'), the light reflected is almost completely polarised and the vibrations reflected are those at right angles to the plane of incidence.

We will now see how the sky immediately above us is reflected in the glass, which should be held about 8 inches

[1] J. Duclaux and R. Gindre, *Bull. Obs. Lyon*, **11**, 5, 1929.

[2] It is possible to buy polarising film, a newly invented device called 'polaroid.'

above the level of our eye, so that reflection occurs as nearly as possible at the angle of polarisation (Fig. 140, *a*). If you turn successively to each point of the compass, at the same time holding the glass so that it always reflects the *same point* above your head, you will see that the reflected image is *bright* when you face towards or directly away from

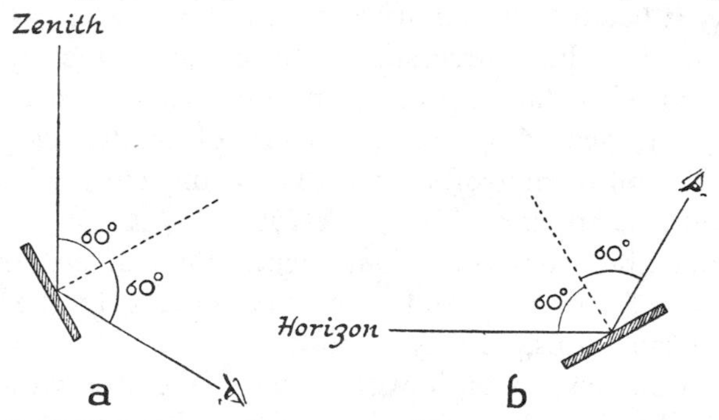

FIG. 140. Investigation of the polarisation of the light from the sky. (*a*) Near the zenith. (*b*) Near the horizon.

the sun, but *dark* when you stand at right angles to these directions. This proves that the light from the zenith vibrates perpendicularly to the plane containing the sun, the zenith and your eye. Such is indeed the general rule when light is scattered by minute particles.

We will next examine the reflection of the part of the sky near the horizon, continuing to hold the glass in such a way that the angles of incidence and of reflection are equal to the angle of polarisation (Fig. 140, *b*). We see that the image is bright on the side of and opposite to the sun and dark in the direction 90° away. That it is brighter on the side of the sun is no matter for surprise, but the other three points of the compass appear to the eye, without the reflecting glass, pretty well equally bright, so that the difference we observe in the reflected light, is a genuine polarisation phenomenon. The sky near the horizon opposite the sun sends light to us that is only slightly polarised, whereas the polarisation in the directions at right angles is strong and the vibrations are directed vertically—

that is, perpendicular to the plane containing the sun, the point observed and our eye.

One wonders whether Nature herself ever arranges tests of this kind for us. Surely the reflection of the sky in placid water should enable us to see a darker region there as well; we look at the surface of the water at an angle of incidence rather greater than 50° and then turn to all four points of the compass. When the sun is low, the water on the north and south sides should appear noticeably darker than on the east and west sides. My own experience is that this experiment is sometimes a success, but not often; the sky is usually not sufficiently uniform in brightness or the water not sufficiently smooth.

A more convincing fact is that sometimes small clouds, hardly visible in the air, can be seen more clearly in the reflection in water because their light, not being polarised, is weakened to a less degree than the light from the sky, which is polarised. The same effect is, naturally, still more striking when sky and clouds are seen through a nicol or reflected in a piece of dark glass. One should look preferably when the sun is low in the west or east, at a small cloud at a height of 20° to 40° to the south or to the north, where the light in the sky is polarised strongest. The direction of the vibrations is almost perpendicular to the line connecting this part of the sky and the sun, that is they are vertical; so that, in a piece of glass lying on the table in front of us, we see the light from this point in the sky considerably weakened and the little cloud becomes clearer.

The proper instrument for examining the polarisation of the sky is Savart's polariscope, a simple and yet very sensitive little apparatus. Considering however that only a few Nature-lovers are so fortunate as to possess one, and these phenomena constitute an entirely separate field of meteorological optics, we confine ourselves to mentioning some literature on this subject.[1] For those who are sufficiently

[1] Fr. Busch and Chr. Jensen, *Tatsachen und Theorien der atmosphärischen Polarisation* (Hamburg, 1911); Plassmann, *Ann. d. Hydr.*, **40**, 478, 1912.

interested to make a series of systematic observations, this is an exceedingly delightful and many-sided subject.

The polarisation of the sky is easy to observe with the aid of a nicol simply by rotating it about its axis. The following method is a very sensitive one but applicable only in the twilight. Select a star so faint as to be only just perceptible, and try to ascertain whether, on looking at it through a nicol, its visibility is greater in some positions of the nicol than in others. This method is based on the same principle as that used for our observation of small clouds described above. The light of the star is not polarised and the darker its background the clearer the star; the change in visibility indicates, therefore, a change in the brightness of the background, and so polarisation.

It is for this reason that in the daytime the nicol increases the visibility of distant objects, provided it is rotated in such a way that it cuts off the light scattered from the sky.[1] White pillars in the distance, lighthouses, seagulls, etc., become more distinct in comparison with the background, though only on bright days; on misty days the light from the grey sky is not noticeably polarised. The effect of the nicol is generally most pronounced at 90° from the sun.

Examine with your dark glass the polarisation of a number of points in the blue sky and try to obtain a general survey of them. Would it be possible to observe the regions of abnormal direction of polarisation above the sun as well as above the anti-solar point? And what happens if we reflect the blue sky in a garden-globe and then observe it with our dark glass at the angle of polarisation?

182. *Haidinger's Brush*

Many a laboratory physicist is astonished and inclined to disbelieve us when we tell him we are able to see with our naked eye, unaided by any instrument, that the light from the sky is polarised! It does, however, require a certain

[1] H. N. Russell, *Science*, **63**, 616, 1917.

amount of practice. A beginning should be made with completely polarised light by studying the reflection of the sky in a glass surface at the angle of polarisation (§ 181). After one has observed the reflection of the uniformly blue sky for a minute or two, a kind of marble effect will begin to appear. This is shortly followed in the direction in which the eye is looking by that remarkable figure known as 'Haidinger's brush,' a figure resembling more or less the one shown in Fig. 141. It is a yellowish brush with a small blue cloud on either side. The yellow brush lies in the plane of incidence of the light reflected in the glass —in other words, the yellow brush always lies *perpendicular* to the direction of the light-vibration.

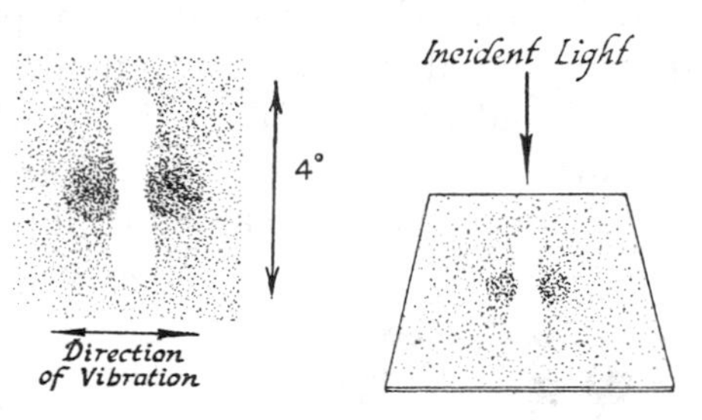

FIG. 141. Haidinger's brush, a remarkable figure that can be seen in the blue sky and is an indication of the polarisation. (The light brush is yellowish, the clouds at each side are blue.)

It disappears in a few seconds, but if you fix your eyes on a point close to it on the glass surface you will see it again ![1] The figure is not easily distinguished from its surroundings and presumably the task consists in learning to pick out this faint contrast from among the unavoidable irregularities of the background. One should practise a few times daily for a few minutes at a time. After a day or two one is able to distinguish Haidinger's brush fairly easily when looking at the blue sky, although the light emitted by it is only partly polarised. I can see it particularly clearly in the twilight if I stare at the zenith; the whole sky seems to be covered by a network, as it were, and everywhere I look I see this characteristic figure. It is very pleasing to be able to determine the direction of polarisation without an instrument in this way, and even to obtain an estimate of its degree.

[1] If you have a nicol at your disposal, then look through it at a white cloud—or at an evenly illuminated surface and try to distinguish the figure by the fact that it revolves when the nicol is rotated.

The yellow brush is generally to be found pointing towards the sun if one prolongs it as the arc of a great circle, which shows that the scattered light vibrates, in general, perpendicularly to the plane containing the sun, the molecule of air and the eye.

Haidinger's brush can be seen still more clearly in the reflection of the sky in a garden-globe, the head of the observer screening off the image of the sun (cf. § 11).

In this case a small region can also be seen near the sun in which the yellow brush does not point in the direction of the sun, but at right angles to it; the boundary between the normal and region of deviation is to be seen as a kind of shadow.

Haidinger's brush is caused by the dichroism of the yellow spot of our retina. That all observers do not, apparently, see this remarkable figure in the same way no doubt depends on the difference in shape and structure of this yellow spot. For instance, some cannot see the blue part of the figure; some see the yellow region continuous; others the blue (Fig. 142). The two following assertions are opposed to one another:

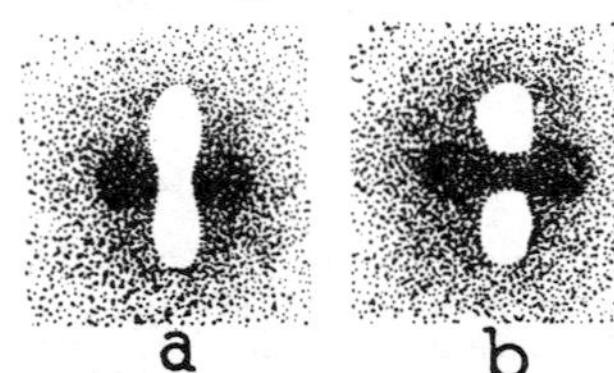

FIG. 142. We do not always see the brush of Haidinger in the same way: (*a*) here yellow is continuous; (*b*) here blue is continuous.

(*a*) One's first impression is that yellow is continuous; when the eye is fatigued from gazing too long at it the image changes and blue is seen to be continuous.[1]

(*b*) The continuous colour is always the one perpendicular to the line joining the eyes. If, therefore, you look at a fixed point of the blue sky and turn your head through 90°, you will see first the one and then the other colour to be continuous.[2] The transitory nature of the figure makes it difficult to form an opinion about this.

[1] Hardinger, *Ann. d. Phys.*, **67**, 435, 1846.

[2] Brewster, *Ann. d. Phys.*, **107**, 346, 1859. Apparently in agreement, A. Hoffmann, *Wetter*, 34, 133, 1917.

Haidinger's brush can be seen with much greater clearness if one holds a green or blue glass in front of one's eye, whereas it disappears when a red or yellow glass is used.[1]

183. *Scattering of Light by Fog and Mist*

A thin early morning haze, with the sun shining through it, is delightfully exhilarating and lends a poetic touch to the most prosaic scenery. A denser mist obstructs the view in the distance, but covers the trees and houses near to us with a mistiness such as we are accustomed to seeing only on objects far away; at the same time we are struck by the large angle subtended by them, which in its turn conveys to us the impression of their being extraordinarily high. By the combination of these impressions, which often is quite subconscious, the mist gives a palatial stateliness to large buildings, and elevates the tops of towers into the clouds.[2]

The colours of objects seen through the mist are generally unchanged. The sun though much less bright, is still white, and there is no noticeable difference in colour between the street-lamps in the distance and those close by. Yet there are other cases as well—for instance, when the sun, at a considerable height above the horizon, shines *red* through the fog. Everything depends, of course, on the size of the drops of mist; the source of light appears reddish when the drops are so small as to approach the wave-length of light and therefore scatter chiefly the blue and violet rays, while the yellow and red rays are scattered to a smaller degree (§ 171).

The mist itself is white at such times, decidedly whiter than the faded orange sun, for it is illuminated by both the scattered and the transmitted rays. A heavy mist like this is not bluish; the scattered light amounts to perhaps 99 per cent of the incident light, and therefore is bound to be almost white as a whole, even though each element of volume may show a preference for scattering blue.

Comparatively large drops, like those constituting mist, scatter most of the light forward at small angles to

[1] Stokes, *Papers*, **5**. [2] Vaughan Cornish, *Geogr. Journ.*, **67**, 506, 1926.

the original direction of incidence (§ 177). This explains why a thin mist can be seen so much more clearly in directions approximately that of the sun. The splendid photographs of sunny mist in a wood are taken against the light, the camera pointing a little away from the sun.

The most striking thing about heavier mist is the 'solidity' of the shadows (Fig. 143). On approaching a tree whose trunk is illuminated by the sun, you will see a great deal of light in the directions AO and BO, because in those directions there lie a number of drops of mist which by scattering

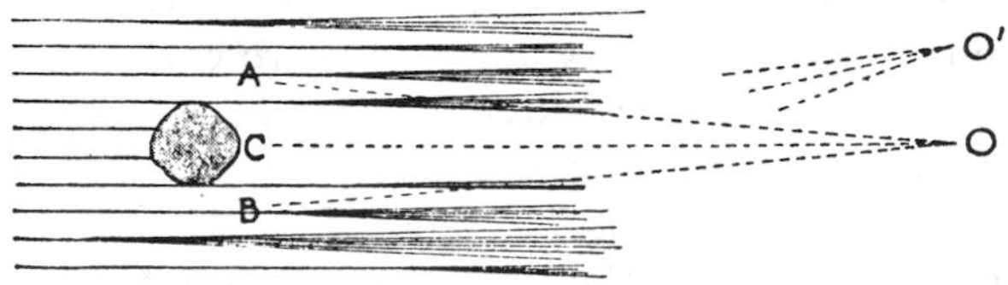

FIG. 143. How shadows arise behind an object in the mist.

light make the air appear self-luminous. Along CO you see much less light because you are looking through air that is not illuminated. Now, if you move your eye a little to one side, as far as O′, the light and the dark regions of the mist overlap, and the shadow becomes indistinct; moreover, we see hardly any light arriving from the directions AO′ and BO′, because at such fairly large angles the scattering has become negligible (§ 177).

In this way a shadow hangs in space behind each branch and behind each post, and you see nothing of these shadows until you are almost in them. Still stranger is this scene at night, when every street-lamp and the head-lamps of every car cause the mist to become luminous and to cast shadows behind every object which, however, are only visible from behind. A walk through the mist is a real pleasure, optically speaking !

Sometimes you can see the streaks of shadow when looking across them, for example when the sun's rays shine obliquely over the roofs of the houses and you look more or less along the line of the shadow, traced faintly in the air.

Backward scattering by the mist is far more difficult to observe. The mist must be formed of very fine drops and yet be dense; there must be a dazzling source of light behind us, and a dark background in front of us. Sometimes we can see our shadow projected on the mist simply by standing before an open window on a misty night, with a strong light behind us. Note that the shadow is not cast on the ground, for it remains even if the lamp is a little lower than your head. Let your eyes grow accustomed to the darkness outside and protect them with your hands from light from the side (Fig. 144). The shadow of your arms on the mist is very elongated, and that of your body conical and huge. All the shadow streaks converge towards the shadow of your head, which is also the anti-point of the lamp. A glow of light surrounds this point, most noticeable when you move slightly to and fro. This wonderful picture is nothing less than the 'spectre of the Brocken,' which is so impressive when seen with mist and sun on a high mountain-top.

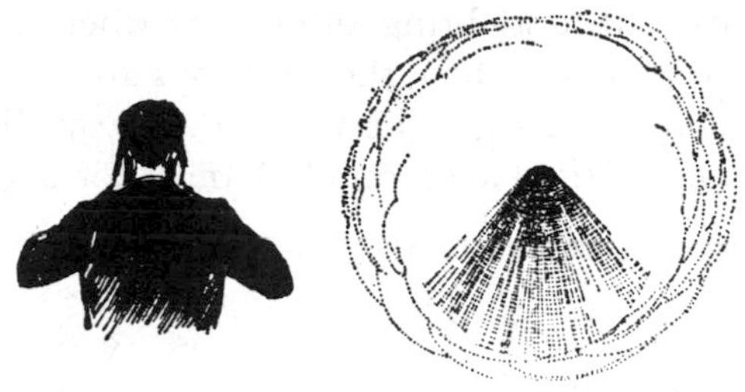

FIG. 144. The spectre of the Brocken as mist.

The great size of this phenomenon is due to the shadow not lying in one plane but extending over a depth of perhaps tens of yards.

A cyclist with the glaring lights from the head-lamps of a motorcar shining on him from behind, sometimes sees his own shadow in the fog magnified to an enormous size. The light from the lamp of another cyclist, if directed towards the head of the first, produces the same phenomenon.

The glow of light and the shadows traced on it arise from the backward scattering of a small fraction of the light by the drops of mist; all those beams which appear to converge *towards* the shadow of our eye are in reality *parallel* (or nearly so) (cf. §§ 191, 217).

184. *Visibility of Drops of Rain and Water*

It is worth while during a shower to observe in which direction the falling rain is most easily seen. The drops are not visible against the bright sky nor against the ground, but they are against houses and trees. Evidently they can be seen only when they deflect the light from its path and bring brightness where formerly it was dark. Apparently, then, the light rays are deflected chiefly through *fairly small angles* between 0 and 45°. The more the brightness of the background changes for a given small deflection

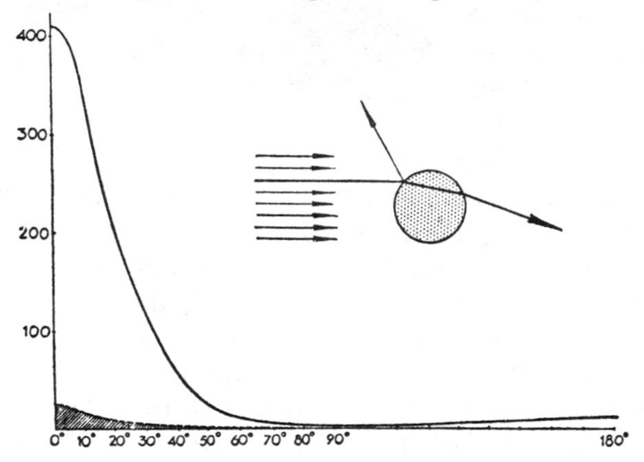

FIG. 145. Sunlight, sparkling in raindrops, is refracted and reflected in every direction. Distribution of light at various angles of deflection. (Shaded area indicates the share of the reflected rays.)

of the light, the more clearly the drops will be seen. If the sun is shining while it rains the drops in the vicinity of the sun are seen to sparkle brighter than ever; this is due to the enormous difference between the brightness of the sun and of the sky, so that every refracting drop is conspicuous.

You can nearly always see them shining like pearls against a dark background; against a light sky they seldom appear dark. This is an application of the general principle that the eye is sensitive to the *ratio* of light-intensities, not their *difference* (§ 64). If light of intensity 100 falls on a drop and the intensity scattered by the drop is 10, then it will show up very well against a dark background of intensity 5, since the ratio of intensities is 2:1. On the other hand, the diminution in intensity of the transmitted light from 100 to 90 means that the intensity ratio of the drop seen against the sky is only 10:9, which is scarcely perceptible. But if they are close to us—for instance, large drops off our umbrellas—they look dark as they fall, and during a heavy shower we can see dark parallel streaks against the light background of a gap in the sombre rain-clouds. Similar phenomena can be observed in fountains and in the jet of water from a garden spray.

By applying the ordinary optical laws it is easy to calculate the contributions to the resulting distribution of light, which are due respectively to the rays reflected at the surface of the drop and to the rays that have passed through the drops, after refraction (Fig. 145). It appears that the latter play by far the larger part and that they do indeed cause the light to deviate only through fairly small angles, just as direct observation had led us to conclude.

185. *Visibility of Particles floating in the Air*

The foregoing description of the visibility of water drops can be applied more or less to everything floating in the air. Clouds of dust are seen much better in the direction towards the sun than away from it. When the weather is sunny a slight haziness along the horizon, to a height of about 3° above it, can often be seen when we look towards the sun; at a distance of not more than about half a mile colours in the scenery are no longer clearly distinguishable, and the spires of distant churches cannot be seen. If we look away from the sun the same haziness along the horizon is seen to be darker. The difference between the light hazy band near the sun and the dark one opposite can be seen particularly clearly when, in a balloon or climbing a mountain, you reach the top surface of the haze. The point of transition lies about 80° from the sun, where the brightness of the layer is practically equal to that of the sky.

When night begins to fall, the rising moon is coloured a deep red, but changes with surprising rapidity into yellow-white.

If, while there is a slight mist, one stands in the shadow of a chimney, one can see the sun surrounded by an aureole of light which was not perceptible so long as one was dazzled by the sun's glare. At times there is a red edge to the aureole. A similar, though feebler, light-effect, caused by dust and tiny drops of water, can also be seen even when there is *no* mist (§ 197).

Swarms of insects resemble dancing sparks of light when they are on the same side of the observer as the sun, whereas on the opposite side they are barely visible. The beards on ears of rye waving high in the air shine with a lovely golden purple when seen against the rays of the setting sun. Dry leaves, stones, and twigs, all shine whenever they are seen towards the sun, and hardly or not at all in the opposite direction.

These observations confirm that light is diffracted only through small angles at the edges of a screen. The same is true of the reflection, refraction or diffraction by small globes, if they are not too small (§§ 177, 184). Objects of irregular shape act in much the same way as small screens and spheres of about the same size.

186. *Searchlights*

The beam of a searchlight furnishes matter for various interesting observations. We must remember, first of all, that the beam would not be visible at all but for the particles of dust and drops of water in the air, which are illuminated by it. The brightness of the beam is therefore a criterion of the purity of the air.

It seems strange that the beam ends as suddenly as it does, even when the sky is very clear and there is no cloud to act as a 'screen.' The explanation is that an observer at O sees light arriving along AO, BO, CO and so on, from all the points along the beam. But however long the beam he can never see any point on it in a direction beyond OD parallel to LC. This direction represents the 'end' of the beam for the observer and therefore fixes accurately its direction in space. The fact that he receives an appreciable amount of light from the remote parts of the beam must be attributed to the obliquity with which at those distances his

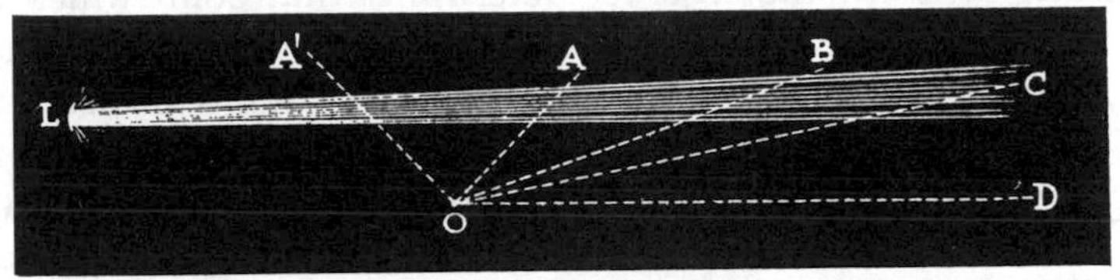

FIG. 146. The track of a beam from a searchlight appears to end suddenly in a very definite direction.

line of vision traverses the beam, and therefore the thick layer of scattering particles along that line; on the other hand, in the direction OA, he looks through the illuminated air for a short distance only.

Go and stand close to the beam and compare the light-intensity in directions of 45° and 135°. You will find that the forward scattering along A'O is much stronger than the backward scattering along AO. Yet the amount of scattering material in the line of vision is the same in both cases and we may take it that the diameter of the beam in the direction of A differs so slightly from that of A' that the difference can safely be neglected. Obviously the explanation is to be found in the unsymmetrical scattering of the dust particles, for these are fairly large, and therefore scatter most in a forward direction (§ 177). A more reliable way to carry out this experiment is to stand near a lighthouse and compare intensities of the beam when it shines obliquely in your direction and when it shines obliquely away from you.

A few experiments of this kind can be done with the beam of a really good electric torch, provided the night is dark enough. One can even point out to other people some particular star, so clearly defined is the 'end' of the beam.[1]

[1] Davis, *Science*, **76**, 274, 1933.

187. *Visibility*[1]

Visibility is measured in a stretch of open country in which a series of well-known landmarks can be chosen at increasing distances from the observer; suitable landmarks are factory chimneys or church spires in far off villages, of which the distance can be obtained from a good map. Then each day the observer determines that point which is only just visible, the distance of the point being defined as the 'visibility.' If the number of points at his disposal is insufficient, he can estimate the visibility according to his general impression on a scale from 0 to 10. Obviously the result is an extremely intricate combination of several factors, in particular of the drops of water and particles of dust in the air by which a false light is spread over the darker parts. Let us suppose that an object reflects an amount of light A, the air in front of it B, and the air behind it C, and, moreover, let us suppose that after being diminished by their passage through the atmosphere the amounts a, b, c of A, B, C enter our eye. The visibility of the distant object is then determined by $\frac{a+b}{b+c}$, and this is also the fraction on which the 'visibility,' as defined by the distance measured above, depends. This explains why the visibility is not governed by atmospheric conditions only, but also depends to a certain extent on the position of the sun. In order to reduce the effect of the sun to a minimum, it has been agreed that the landmarks or reference points should be dark objects about 10 ft. high, projecting against the sky and subtending an angle between 0·3° and 1·0°. It is interesting that, when these requirements are fulfilled, the visibility may be proved to be nearly independent of the position of the sun and of the kind of landmark chosen.

At night one can choose lamps the distance of which is known, or one can determine the lowest altitude in degrees at which a star of the first magnitude becomes visible. These

[1] W. E. Knowles Middleton, *Visibility in Meteorology* (Toronto, 1935), with numerous references to the extensive literature.

determinations will naturally not be in perfect agreement with the determinations made by day, for the quantity measured is, strictly speaking, not the same.

Observations have been made and their results worked out statistically by innumerable observers. The main factor determining the visibility is undoubtedly the quantity of dust carried by the wind; moisture condenses round the dust nuclei and the drops thus formed scatter the light. From this it is clear that both the amount of dust and the humidity of the air have a great influence. Visibility is at its best in sunny weather, when we are in the wedge-shaped maxima of atmospheric pressure, as they appear on weather charts, between two depressions, which bring us fresh 'polar air' containing very few dust nuclei. These weather conditions are usually of short duration. On the other hand, visibility becomes bad when regions of high pressure remain unchanged for a long time in the same place, so that the dust gradually sinks down to the lower layers of the air.

For those living on the coast it is interesting to compare visibility when the wind blows from the sea, and when it blows from the land. This, however, should always be done under the same conditions of humidity—that is, when a wet and a dry bulb thermometer give nearly the same readings.

In a small town in Scotland the visibility was found to be six or nine times greater when the wind blew from the mountains than when it had passed over a densely populated region. The influence of moisture is evident from the fact that the visibility was four times as great when the psychrometric difference was 8° as when it was 2°. One can picture this very clearly to oneself by drawing lines on a map in the direction from which the wind comes, and making the length of the lines proportional to the visibility distance. This should be done for various values of the humidity, and in this way a set of curves will be obtained showing the average transparency of the air from different sources.

Visibility becomes suddenly greatly improved when a strike breaks out !

Statistics show, moreover, that visibility is better when there is a strong wind, and in the summer months (March to October) than in the winter. Generally it is better in the afternoon than in the morning, because during the day the ascendent currents of air have carried high in the air the dust which was floating in the lower layers. After a long period of rain or snow, the dust is nearly entirely precipitated and the visibility is often excellent.

It is remarkable how much farther we can see into a shower of rain than we can into the cloud or bank of mist from which it falls. The reason for this will be apparent from the following argument (which is admittedly very rough):

Let us denote by V the volume of water present per unit volume of air, and let us divide V into drops with a diameter d and, therefore, with a volume of about d^3. The number of drops will be $\frac{V}{d^3}$ and, since each drop screens off an area of (about) d^2, the whole area, blocked by the drops, will be $\frac{Vd^2}{d^3} = \frac{V}{d}$. Therefore the smaller the drops, the less transparent their aggregate. In the case of a heavy mist, V is of the order 10^{-6}, and with pouring rain, curiously enough, it is about the same. However, the mist drops have a diameter of the order 0·01 mm.; the rain-drops of 0·5 mm. Let us now consider a column with a cross-section of 1 cm.^{2} and length l. In order to stop about half of the light, we must have $\frac{Vl}{d} = 0{\cdot}5$. This gives for mist $l = 5$ m. $=5{\cdot}5$ yards, and for rain 250 m.$=$280 yards. This is the right order of magnitude. The great extent to which the result depends on whether the drops of water are fine or coarse, is very evident from this example. Occasionally it happens that the visibility diminishes substantially during a heavy shower, when the drops, on hitting the earth, are splashed into much finer drops and we look through them close to the ground. This, in its turn, fits in with our reasoning.

188. *How the Sun 'Draws Water'*

It is a lovely fresh autumn morning; the bright sunshine penetrates the foliage of the trees. At a distance, we can see how beautifully parallel the pencils of rays seem in the misty air. But, on drawing nearer, it seems as if they are no longer parallel, but radiate from one single point—the sun.

The same phenomenon on a large scale is also familiar to us. When the sun is hidden behind loose and heavy clouds, and the air is filled with a fine mist, groups of these sunbeams can often be seen darting from the sun through the openings in the clouds, showing a path of light through the mist, thanks to the scattering by the drops of which it is constituted. All these beams are in reality *parallel* (their prolongations do pass through the sun, but this is so far away that I am quite justified in saying 'parallel'). Their perspective gives us the impression that they diverge from one point, their 'vanishing point' being the sun, in the same way that railway lines appear to run towards each other in the distance (Plate XV, *a*; p. 334).

Depending on the shifting of the clouds, some of these beams become stronger or weaker, or move from one place to another, etc. Sometimes the whole landscape is filled with them; or, again, the sun is hidden by one solitary cloud casting a dark shadow. Shadow-beams of this kind are often seen in mountainous countries, cast by ridges and peaks of mountains in front of the low-lying sun.[1]

Light-beams can also arise from the moon but with such feeble intensity that they are only visible when the scattering in the atmosphere is strong. This very rare phenomenon conveys an impression of ominous gloom.

Why are the bundles of rays only visible at small distances from the sun, seldom as far as 90° from it, for instance? (Cf. §§ 177, 183, 184.)

[1] Vaughan Cornish, *Scenery and the Sense of Sight* (Cambridge, 1935).

189. *Twilight Colours*[1]

The ordinary man's ideal sunset is one draped in purple golden clouds, glowing with an inner glow of a deep warm colour. He tries with childish delight to find a camel or a lion depicted in it, or a flaming palace and a fantastic sea of fire. The physicist, however, tries to begin his observations with the sunset in its simplest form and prefers a perfectly cloudless and bright sky. He studies the fine ranges of colour, evanescent, tender tints, transitions from the blue of day into the dark depths of night, which are perceptible only after a certain amount of practice, but return ever and again in more or less the same order, their development forming a grand drama of Nature—the drama of the departing sun.

To what is it due, this sense of infinite calm emanating from these light phenomena? Compare them with the rainbow, arousing feelings of cheerfulness and joy. This twilight atmosphere is surely due to the broad arches of interflowing colour, lying so flat across the sky as to be almost horizontal. The horizontal line, wherever it may be in the architecture of the landscape, brings rest and peace.

A serious study of the colours of twilight will provide us with information concerning the condition of the highest layers of the atmosphere far above the regions where the clouds are formed, layers of which we know hardly anything but what we gather from their scattering effect on the light. The best months for beginning this study are October and November. The distinctness of the phenomena varies from day to day, their colours often being robbed of their glory by dust and haze, and more especially by the smoke of our towns. For this reason, the studies should be repeated again and again!

In order to see the fine twilight colours properly, the eye must be perfectly rested. However brief our glance may be of the sun before it sets, we shall be too dazzled for some

[1] The extensive literature is condensed in: P. Gruner and H. Kleinert, *Die Dämmerungserscheinungen* (Hamburg, 1917).

time to be able to continue our observations satisfactorily. If we intend to observe the eastern sky we must not look too long at the very bright sky in the west. Each time we have rested our eyes for a moment by going indoors, or by glancing at a book, we realise how much richer the colours of the twilight phenomena are, and how much farther they reach than we first thought. My advice is, therefore: begin by following the development of the twilight in general and, after that, study the peculiar beauty of each part of the sky.

Compare different parts of the sky with each other frequently, by means of a little mirror held at arm's length, in this way projecting on that part of the sky at which you are looking a part from quite a different direction.

You may perhaps experience some difficulty in seeing any *shape* at all in colour phenomena, merging so completely into one another. Yet the secret is quite simple. You draw imaginary lines of equal *brightness* or equal *hue* along the sky; these are the lines mentioned again and again in descriptions, as, for instance, when it is said that twilight phenomena develop usually in the shape of coloured arcs.

The following is a description of a typical sunset in our part of the world on a clear evening (Fig. 147). The minus sign given with the sun's altitude denotes its depth *below* the horizon.

Sun's Altitude 5°*; Half an Hour before Sunset.* The colour of the sky near the horizon changes into warm yellow or yellow-red, a colour entirely different from the usual whitish-blue seen there by day. The *horizontal stripes* below the sun become faintly visible as a long, yellowish band of colour. (By 'stripes' we only mean that the lines of equal hue run horizontally, and not that there are sharp boundary lines.) Above them, concentrically around the sun, a large, very luminous, whitish patch of light, the *bright glow*, often bordered by a faintly indicated *brown ring*.

If there are white clouds near the eastern horizon they assume a soft red hue, and the sky above shows the upper part of the *counter-twilight*, a coloured border of 6° to 12° in height, showing transitions to orange, yellow, green and blue.

Sun's Altitude, 0°; Sunset. But do not think that the twilight phenomena are over now! The interesting part is only just beginning. In the *west*: along the horizon lies the colour-bank of horizontal stripes, the colours from below upwards being, white-yellow, yellow, and green. Above it the magnificent bright glow, transparent and white, encircled by the brown ring, its height reaching as far as 50°. In the *east*: the *earth-shadow* begins to rise almost at the same moment as the sun begins to set. It is a very striking bluish-grey segment, shifting gradually across the purple layer, and as a rule it cannot be followed further than about 6° above the horizon. Occasionally it seems as if a trace of the earth-shadow is visible a long time before the sun begins to set, but this is simply a layer of dust or mist. Above the earth-shadow, the counter-twilight in all its glory. Higher, the *bright reflection* of the light in the west, a widespread diffuse illumination.

Sun's Altitude, −1° to −2°; 10 Minutes after Sunset. In the *west*: the horizontal stripes (from below upwards) become brown, orange, yellow. The bright glow with its brown ring still reaches a height of 40°. In the *east*: the earth-shadow rises higher and higher, and everything within it is now of a dull, uniform hue, more or less green-blue (a subjective contrast colour! Cf. § 95). The counter-twilight develops its border of colours, from below upwards, violet, crimson, orange, yellow, green, blue, and above that the bright reflection.

Sun's Altitude, −2° to −3°; 15 to 20 Minutes after Sunset. In the *west*: now begins the most interesting of all twilight phenomena. At the top of the bright glow of light at about 25° above the horizon, a rose-red spot appears. It gets quickly larger and larger but at the same time its imaginary centre slides downwards so that it develops into a segment which becomes flatter and flatter. This *purple light* radiates

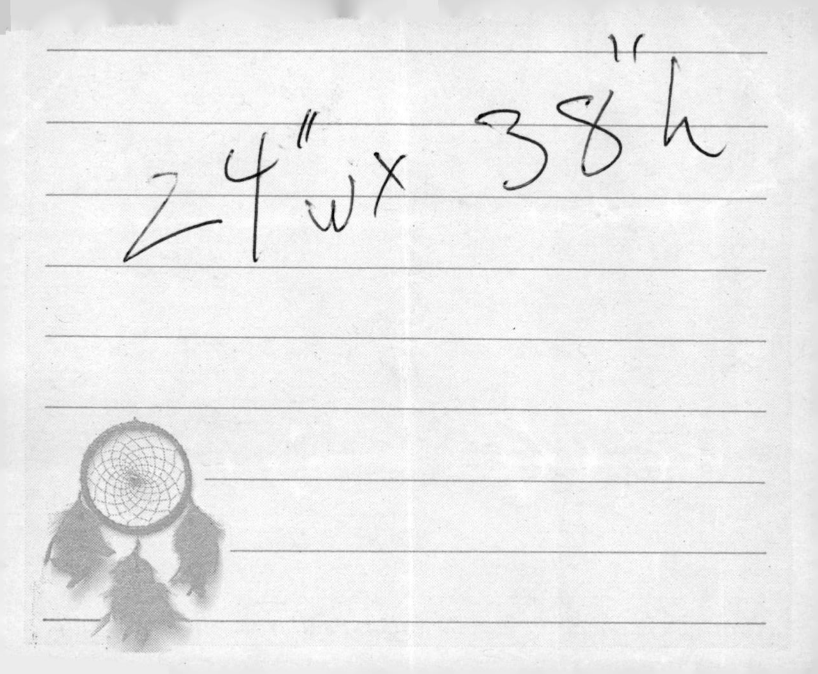
24"w x 38"h

colours of a wonderful soft transparency, more pink and salmon-coloured than true 'purple.' The colour of the horizontal stripes has grown more dim. In the *east*:

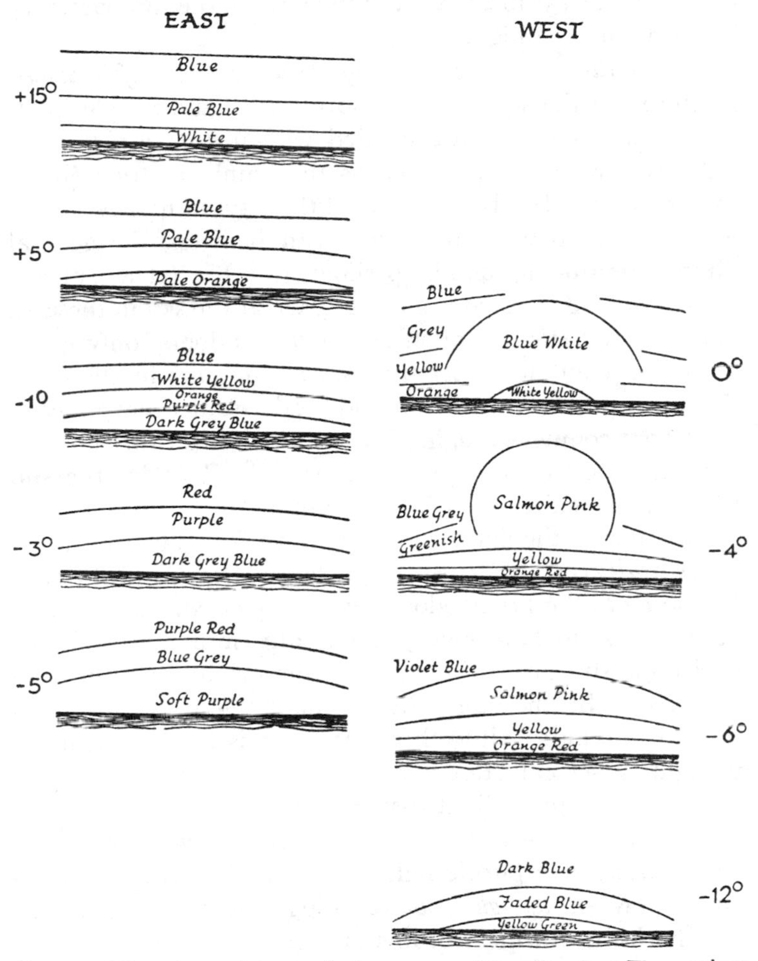

FIG. 147. The colours of the sky during sunset, with a clear sky. The numbers give the height of the sun above and below the horizon.

the earth-shadow is higher still. The upper counter-twilight reaches its strongest development. Above that is the bright reflection.

Sun's Altitude, $-3°$ *to* $-4°$*; 20 to 30 Minutes after Sunset.* In the *west*: the bright glow is still 5° to 10° high. The purple light is more strongly developed. The greatest light-intensity is at 15° to 20° above the horizon, the top boundary being about 40° high.

Sun's Altitude, $-4°$ *to* $-5°$*; 30 to 35 Minutes after Sunset.* In the *west*: the strongest development of the purple light. Buildings facing west are flooded with a purple glow; the soil has a warm tint, as well as the trunks of trees (birch trees especially !). In the heart of the city, in narrow streets from where no western horizon can be seen, the general illumination of the buildings shows clearly that the purple light is shining. Be careful not to gaze at the sky in the west too long and stay as much as possible indoors, only going out occasionally for observations. In the *east*: in the earth shadow a pale meat-red coloured border sometimes appears, the lowest counter-twilight; it is due to the fact that the east is illuminated by the purple light instead of by the sun itself. It is seldom seen in our climate.

The stars of the first magnitude have become visible.

Sun's Altitude, $-5°$ *to* $-6°$*; 35 to 40 Minutes after Sunset.* In the *west*: the bright glow has disappeared. The purple light begins to fade away, apparently mingling with the horizontal stripes, for these are getting brighter and orange-coloured. In the *east*: the boundary line of the earth shadow has faded altogether. If there is a lower counter-twilight, a second faint earth shadow can be seen at the moment the purple light disappears.

Sun's Altitude, $-6°$ *to* $-7°$*; 45 to 60 Minutes after Sunset.* In the *west*: The purple light disappears, leaving a bluish-white glow, the *twilight glow*, reaching a height of 15° to 20°. The horizontal stripes become orange, yellow, and greenish in that order. The disappearance of the purple-light gives us the impression that the illumination of the landscape is decreasing rapidly; reading becomes difficult, the 'civil twilight' is over.

Sun's Altitude, $-9°$. In the *west*: The twilight glow still reaches 7° to 10°. In the *east*: The lower counter-twilight has vanished; only one last, very feeble reflection remains.

The darkest point of the sky is now near the zenith, a little towards the west.

Sun's Altitude, −12°. In the *west*: The horizontal stripes are considerably weakened and are now a faded green. The green-blue twilight glow is still 6° high.

Sun's Altitude, −15°. In the *west*: Twilight glow still 3° to 4° high.

Sun's Altitude, −17°. In the *west*: The twilight glow has disappeared.

Stars of the fifth magnitude are now becoming visible. This moment can be determined fairly precisely and changes according to the season of the year and from day to day. The 'astronomical twilight' is over.

Remarks on the purple light. The intensity of the purple light varies very much from one day to another. The presence of very thin veils of clouds, floating high up in the air, can intensify it to a great extent, and its development is often strikingly beautiful when the weather has cleared up again after a long series of rainy days. It is on an average stronger in late summer or in the autumn than in spring or summer. It is polarised to a small extent only, whereas in the surrounding parts of the sky polarisation is particularly strong. The experiment with Haidinger's brush is sufficient to prove this difference (§ 182).

Its development during the course of the twilight is not always as we have outlined it. It may arise in one of the following ways:

(i) From the brown border encircling the bright glow. (ii) From the bright glow itself, which passes from yellow to rose and purple. (iii) From the counter-twilight, which spreads more or less invisibly over the zenith, and on reaching the west, becomes visible there. (iv) From delicate cirrus clouds, illuminated by the sun after it has set. (v) From a purple patch formed at the top of the bright glow and spreading from there. This is the type described in our text, but it is not of frequent occurrence.

Never, if you can help it, miss seeing the sunset and the dawn.
RUSKIN—*Modern Painters.*

190. *Measurements of Twilight Phenomena*

The earth-shadow is quite easy to measure (cf. the methods, § 235). Make a graph, in which its height is plotted against the time. At first the earth-shadow rises at about the same rate as the sun sinks, later, twice and

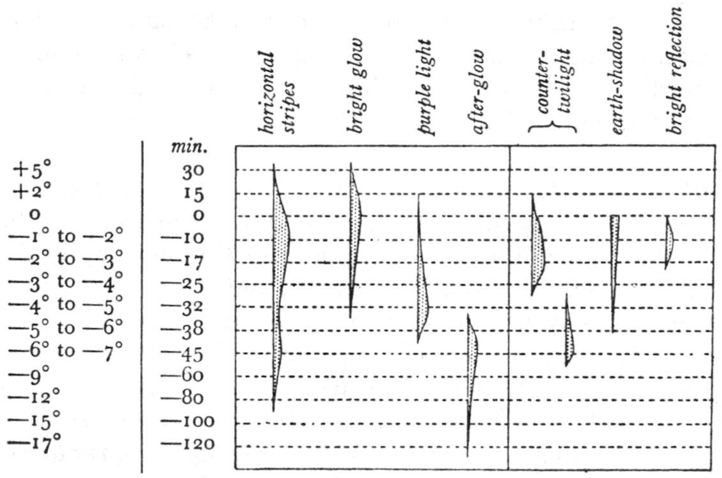

FIG. 148. Concise table, showing the development of the different twilight phenomena.

even three times as fast.[1] The height above the horizon at which the earth-shadow vanishes gives one an idea of the purity of the air. It is very sensitive to the slightest trace of turbidity: the more particles of dust in the atmosphere, the sooner the shadow becomes invisible.

Measurements of the bright glow and the purple light are more difficult. Apart from its being desirable to rest the eye from time to time, it should be borne in mind that

[1] For the theoretical explanation of the velocity at which the earth's shadow rises, see Pernter-Exner.

every dark silhouette against the background of the sky is bound to show a contrast effect, and is for this reason to be avoided. It is quite surprising how the line that we had estimated to be the boundary of the purple light can be influenced simply by holding out a pencil or a flat strip of wood. The best method is to compare its height with trees or towers in the landscape.

It is worth while to mention here that measurements of the brightness of the sky have shown that the purple light by no means consists in an *increase in brightness*, but in a *slower decrease in brightness* of a certain part of the sky compared with the parts round it. In this way a maximum of '*relative*' brightness is produced, and this leads to the visual impression of fresh radiation being developed there. Similarly, the modified colour must be ascribed to a slower decrease of the intensity of certain definite wavelengths than of other wavelengths.

After the purple light has faded away, the movement of the afterglow becomes interesting. Its topmost boundary is, in reality, the last stage of the earth-shadow, which has passed the zenith and now appears on the western side. It descends rapidly at first, and then slower and slower.

191. *Crepuscular Rays*

Twilight phenomena are remarkably beautiful when clouds, hidden by the western horizon, spread their stripes of shadow over the evening sky like a huge fan. They radiate from the imaginary point below the horizon where the sun is, exactly in the same way as the sunbeams 'drawing water'; only this time the sky is very clear and now we can see how the dark beams are outlined especially in the purple light, their blue-green colour forming a particularly good contrast, the more so because there is also a subjective colour contrast contributed by the eye. The crepuscular rays show us how the sky would appear if the purple scattering were absent, and we notice now, for the first time, exactly how far the purple light extends. They can be observed not only in the west, where the sun is setting, but also, occasionally, in the *eastern* sky, against the purple background of the counter-twilight where they converge towards the anti-solar point.

Therefore whenever crepuscular rays are observed the eastern sky should be included in the observations. Accurate observation teaches us that the crepuscular rays in the east and in the west correspond exactly in pairs, and are apparently the same rays, which in reality run round the whole celestial vault, but of which the ends are the parts we see best. It is even possible at times to follow these stripes all the way round, like huge arcs converging at their extremities. These familiar stripes, however, we know to be really parallel, their arched shape being due to optical illusion (§ 108).

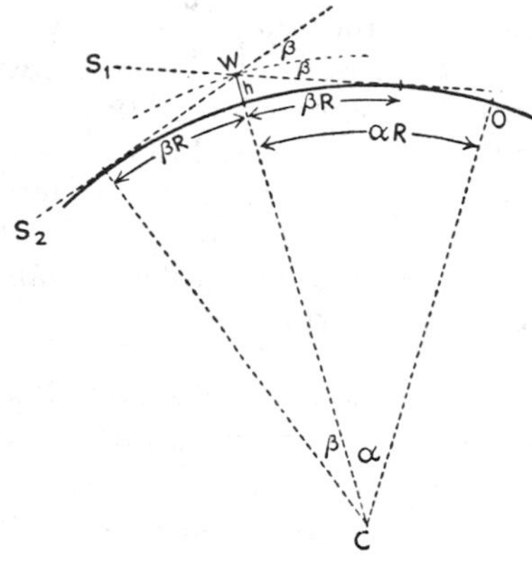

FIG. 149. How to estimate the distance of the clouds causing the crepuscular rays.

These crepuscular rays are visible only where scattering particles float in the air. With the sunbeams 'drawing water,' they were outlined against the light mist; with the purple light, against the far finer particles of dust causing this twilight phenomenon. In twilights without purple light the crepuscular rays are absent, and they never appear outlined against the greenish parts of the sky. On the other hand, they can remain visible long after the purple light has melted away into the horizontal stripes; this is, indeed, a proof that the former of these light phenomena is always present, contributing appreciably to the light of the western sky.

The crepuscular rays can more readily be seen near their vanishing point than in directions at right angles away from it, in the same way that twilight phenomena in general are more pronounced in the eastern and western sky than in between. And this, again, follows from the law of scattering (cf. § 183).

We can also estimate how far the shadow-casting cloud is from us. If the cloud rested on the earth it would produce crepuscular rays the moment the rays of the sun became tangential to the earth. Then if the crepuscular ray became visible the instant the sun is at an angle α below the horizon we should know that the distance between the cloud and our eye is α R (R=radius of the earth). If, however, the cloud were to occupy the position W, at a height h, its distance to the observer, as is shown in Fig. 149, could have any value between R $(\alpha-\beta)$ and R $(\alpha+\beta)$, according as the sun's position is in directions between S_1 and S_2. Here $\cos\beta=\dfrac{R}{R+h}$ or approximately $\beta=\sqrt{\dfrac{2h}{R}}$.

Suppose now a crepuscular ray to be observed half an hour after sunset, that is when $\alpha=4°$, for the sun's position below the horizon. The kind of clouds that cause this phenomenon can safely be assumed never to be at a greater height than three miles, that is, β can, at the most, be equal to $\sqrt{\dfrac{2\times3}{4{,}000}}=\dfrac{1}{25}$ rad. (roughly), or 2·3°. For this value of β, $\alpha-\beta$ and $\alpha+\beta$ will amount to 1·7°=0·03 rad. and 6·3°=0·11 rad. respectively, and the distance of the cloud can, therefore, have any value between 120 and 450 miles. This result makes it clear why sometimes crepuscular rays are visible, although to all appearances the sky is perfectly cloudless.

192. *The Explanation of Twilight Phenomena* (Fig. 150)

We follow in our imagination the course of the sun's rays when it is close to the horizon. They travel a long distance through the atmosphere and their colour becomes more and more red, as the molecules of air scatter the violet, blue and green rays. In this way the setting sun acquires its copper-red colour. Once hidden below the horizon its rays continue to illuminate the layers of air above our heads. The lower layers are the denser and they scatter most, whereas the upper layers become increasingly more rarefied, for that reason scattering less and less. If we are at O_1 and look upwards along O_1A, the layer of air is not very deep and, moreover, the molecules do not scatter much at an angle of 90°. Close to the zenith, therefore, the sky will be dark. Looking along O_1B and O_1C, on the other hand, our eye will receive a considerable amount of scattered light, because our gaze now travels a long way through the illuminated layer. The light arriving from B will be stronger

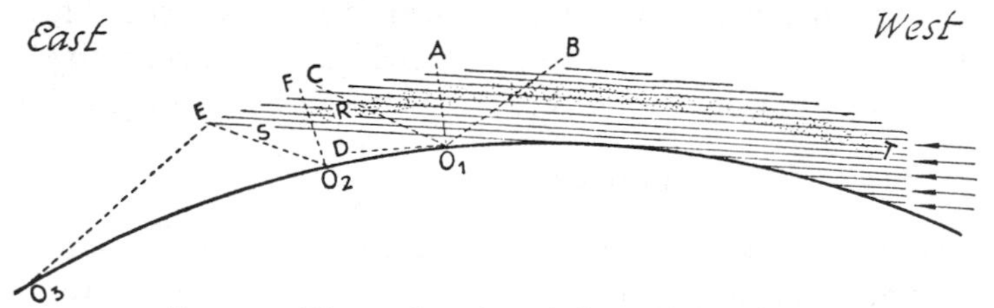

FIG. 150. The explanation of the twilight colours.

because, apart from the scattering effect of the air, we also receive the rays scattered over *small* angles by little drops and the coarser particles of dust. Here we find the explanation of the origin of the *horizontal bands*, of which the direction corresponds with the layer-like grouping of the larger particles. At the same time this furnishes the explanation of the *counter-twilight* in the direction O_1C and why its colour changes from blue through green and yellow into red; for as we lower our gaze, it travels through layers, that are so dense and extended that finally the only light by which they are illuminated is red. Still lower, along O_1D, our gaze meets the *earth-shadow*, so that we should receive no light at all from D, but for the fact that objects lying in that direction are illuminated by a faint diffuse light from all parts of the sky, making all contrasts vanish. After some time we are at O_2, where we can no longer see the red border of the counter-twilight, for we are now looking in a direction making a larger angle with the rays of the sun and our gaze no longer grazes the plane separating the illuminated and the non-illuminated parts of the air. The amount of light arriving along the ray from E is insufficient, while the steeper ray from F conveys equal amounts of blue, yellow and red. The boundary of the illuminated part of the atmosphere in this way becomes increasingly indistinct and dull.

Still later, the slope of the illuminated twilight layers has become so much steeper, that we no longer see any red colouring in the western sky. We must imagine our observer to be, by this time, at O_3. The boundary E of the illuminated atmosphere, which had first climbed the eastern sky as the boundary of the earth shadow, has risen higher and higher and then passed the zenith (without our seeing it do so) to appear to us again in the western sky; for the direction of our gaze towards E once more makes a small angle with the separating plane between the illuminated and non-illuminated parts. Moreover, the scattering through small angles by the larger particles comes into play again, and the general illumination of the scenery is so much weaker now, that one is struck even by a fairly faint brightness. That is why E is found as the upper limit of the twilight-glow.

Finally, it remains to explain the purple light. It can only be explained on the assumption that there exists a layer ST of extremely fine particles of dust between 6 and 12 miles up in the air.[1] The beam of light by which we see this layer illuminated, arrives from the sun when it is already below the horizon. The colour of the lower parts of this beam will be an intense red, since these rays have traversed the longer and denser layers of air. The part SR of the layer will, therefore, contribute the greater part of the purple light. The striking feature here is that the scattering by SR is only seen at O_2 and not at O_1 (where it ought to be visible in the eastern sky). From this we may conclude that the scattering particles are considerably larger than the molecules of air, and scatter chiefly in a *forward* direction (cf. § 183). Whenever in the evening we

[1] P. Gruner, *Helvetica Phys. Acta*, **5**, 351, 1932.

see the purple light appear, we must take it as an indication that we have now entered the cone of forward scattering of the dust layer.

193. *Are there any Differences between Dawn and Dusk?*

If any they are so small that it is not possible to mention any really typical differences. One important thing, however, is that the eye is completely rested in the morning and sees the light-intensity increase continuously, so that it is more sensitive to dawn phenomena than to dusk phenomena. The latter have generally a greater richness of colour on account of the greater humidity of the air, and because the air is a little more turbulent, and contains more particles of dust than in the morning.

194. *'It is Darkest before the Dawn'*

Denning, the famous observer of shooting-stars, believes in the literal truth of this English proverb. Just before day begins to dawn he feels somewhat nervous, and things which he was sure he could see before perfectly well now seem to disappear.

The measurements of the illumination do indeed show irregular fluctuations at times, but they are too variable and too small to have any real meaning. The first brightness of dawn possibly disturbs the adaptation of the eye, though it is still too feeble and too limited in extent to illuminate the surroundings perceptibly.

195. *Morning and Evening Red as Weather Forecasts*

When it is evening, ye say, It will be fair weather: for the sky is red.

And in the morning, It will be foul weather to-day: for the sky is red and lowring. O ye hypocrites, ye can discern the face of the sky; but can ye not discern the signs of the times?

MATT. xvi. 2–3.

This ancient and universal rule, as modern statistics prove, is in the majority of cases actually fulfilled. Each case has its own individual explanation. The horizontal stripes are red only when the air contains dust or water droplets; in the morning there is not much dust and a red colour must then be due to water. With high pressure and fine weather in the evening the sky is clear, and the purple light visible.

A pale yellow, dull, and drizzling, sky in the west when the sun is setting is looked upon as a forerunner of storm and rain.

196. *Disturbances in the Normal Course of the Twilight*

The twilight phenomena are an extremely fine reagent for testing the purity of the high layers of air. The abnormally colourful sunrises and sunsets in the years 1883–6, were a direct consequence of the presence of finely distributed volcanic ash, ejected high into the air during the eruption of the volcano Krakatoa in the Dutch Indies, and spread in the course of a few months all over the world. But before that time, and afterwards too, small optical disturbances have taken place repeatedly which could usually be traced to volcanic eruptions: 1831, Pantellaria, near Sicily; 1902–4, Mont Pelée; 1907–9, Sjadutka, on Kamchatka; 1912–14, Katmai in Alaska. After every violent eruption of Vesuvius or Etna we may expect abnormal twilights, though it usually takes more than a week for the finely divided ashes to reach as far as here.

It seems very probable that a strong development of spots and prominences on the sun causes disturbances in the twilight phenomena, because the electrons, ions, and atoms ejected by the sun may be the cause of ionisation in our atmosphere. On this theory maxima ought to occur about 1938 and 1949.

A third cause of disturbance was discovered when the Earth passed through the tail of Halley's Comet on May 18th and 19th, 1910. The magnificent twilight phenomena seemed to be an indication that particles of dust from the comet had entered our atmosphere (§ 167). Equally striking phenomena were seen in 1908, when the Earth was struck by a huge meteorite landing in the desert wastes of North Siberia.

The chief optical phenomena indicating the occurrence of a period of disturbance are the following:

(*a*) The 'Bishop's Ring.' For the whole day the sun is in the centre of a shining, bluish-white disc, encircled by a red-brown ring. The brightest part of the ring has a radius of the order of 15°. When the sun is very low, this 'Bishop's ring' becomes a kind of triangle with a horizontal base. The fact that cirrus clouds can be seen passing in front of the ring proves that it occurs very high in the atmosphere.

(*b*) A similar copper-red ring can also be seen at times round the anti-solar point, its radius being about 25°.

(*c*) The blue of the sky is turbid and whitish; when the sun is low it is a dull red, owing to the layer of haze it has to shine through. Stars of the sixth and even of the fifth magnitude are no longer visible.

(*d*) Abnormally few haloes.

(*e*) Abnormally clear nights.

(*f*) Abnormally strong, fiery, purple light.

(*g*) Second purple light. This is a change in the course of the twilight. When the purple light has declined and the sun is 7° to 8° below the horizon, a feeble red-violet glow appears where the purple light had risen, develops in a similar manner and declines when the sun is 10° to 11° below the horizon.

(*h*) Ultra-cirrus clouds (cf. § 198).

(*i*) Luminous night clouds.

(*j*) The moon has a greenish hue.

Even the uninitiated are struck by the more pronounced of these phenomena. But it requires a great deal of practice to be able to observe the fine distinctions, which prevent the possibility of there ever being two sunsets alike, and which are, at the same time, a very sensitive means of identifying the slightest disturbance in the optical phenomena.

197. *The Glow of Light round the Sun*[1]

If we stand facing the sun in such a way that the sun itself is screened off by the edge of a roof, we will see a radiance spreading out on all sides round the sun, and diminishing gradually as the distance from the sun increases. It can also be seen very clearly in a garden-globe a few yards away, if you cover the sun's image with your head. Some observers maintain that it consists of 2 parts: (*a*) a silver-white disc with a radius of about 2° to 5°, rather changeable, occurring mostly in the afternoon; (*b*) a much larger glow of light with a radius certainly as great as 30° to 40°, hardly ever absent and changing at twilight into the 'bright glow.' Others distinguish a yellowish white aureole of 0·25° to 2° radius, a bluish white 'corona' of 2° to 5°, a 'central disc' of 15° to 23°, an 'inner disc' of 10° to 40°, an 'outer disc' of 25° to 70°. The dimensions depend greatly on the altitude of the sun and vary from day to day. For instance it appears that when the sun is very low—less than 2° above the horizon—it is accompanied by a kind of aureole having dull yellow rays, which disappears when the sun is lower than 1° above the horizon.

An accurate photometric investigation of the light round the sun has seldom been carried out. In all probability what appears to be a *ring* is simply a rather slower decrease of the light-intensity, which otherwise diminishes gradually as the distance from the sun increases. This scattered light is no doubt caused by diffraction of the sunlight by particles of dust, drops of water, or grains of ice, all of which scatter mainly through small angles (§ 183). Owing to their being of every size, these aureoles and coronae are superposed on one another, so that one can hardly speak of colours. The varying brightness and distribution of light in this glow is a criterion of the purity of the air, and it certainly is very well worth our while to continue to observe it. They at once betray the occurrence of optical disturbances in the atmosphere and are closely related to the twilight phenomena.

[1] J. Maurer, *Met. Zs.*, **32** and **33**, 1915–16.

Whenever there are volcanic ashes floating in the air an indistinct brown-red ring appears as the circumference of the glow of light, the ring of Bishop (§ 196).

198. *Twilight Cirri or Ultra-cirrus Clouds*

In rare circumstances the sky may appear cloudless just before sunset and then after a while show very fragile cloud undulations, lying low on the western horizon, and bluish grey in colour. A very remarkable fact is that they are supposed to be visible only about the time the sun is setting, and then again when its altitude is $-3°$ and $-7°$; for this would prove that they must be illuminated from certain definite directions. This observation has, however, been made too seldom to attach any general importance to it. The appearance of ultra-cirrus clouds is usually attended by particularly colourful sunsets and optical disturbances (§ 196), so that we can safely conclude that they consist of volcanic matter. They are so thin as not to be visible by day, but appear at dusk, evidently owing to their being brightly illuminated on a dark background. Considering that they were visible up to 10° above the horizon when the sun's altitude was $-7°$, their height cannot have been much more than seven miles; which proves that they float in the lowest layer of the stratosphere.

199. *Luminous Night-clouds*[1] (Plate XII)

These are very thin clouds, much higher than all other kinds, but they have also been observed under normal atmospheric conditions. Strangely enough, they have been seen only between latitudes 45° and 60° north, and the same south, especially from the middle of May to the middle of August. In our latitudes, look for them especially at the end of June.

As long as the sun has not yet set the sky seems perfectly clear. About a quarter of an hour after sunset the luminous clouds begin to appear in the shape either of delicate

[1] R. Süring, *Naturwiss*, **23**, 555, 1935; *Die Wolken* (Leipzig, 1936), p. 30.

Luminous night clouds. (*After C. Störmer, Vidensk. Akad. Oslo Avh. I*, 1933, No. 2, Plate IX)

feathers or ribbing or bands; they are clearest of all an hour or more after sunset. They stand out bright against the background of the after-glow (§ 189), whereas ordinary cirrus clouds are *dark*. It is evident, therefore, that they are still bathed in sunlight, and so must be high in the stratosphere; properly speaking they do not emit light themselves. Their bluish white light can be observed for hours, but the later it gets the smaller the illuminated surface of the layer and the lower it is above the horizon; at midnight it reaches the minimum, after which, apparently, it becomes brighter than before. These clouds are seldom seen higher than 10° above the horizon.

Their mysterious silver-white splendour, changing into golden-yellow near the horizon is very imposing. The particles of dust constituting these clouds must evidently be very fine, scattering as they do chiefly blue light, as appears from the fact that they are visible through blue glass but not through red. One can understand from this why they are not coloured by the red glow of the twilight; for only rays traversing our atmosphere high enough not to be reddened will be scattered by the night-clouds. Some observers maintain that their light is not polarised, while others speak of very strong polarisation (with the vibrations perpendicular to the plane containing the sun, the cloud and the earth, i.e. the same as in the blue sky and in various scattering processes). Is it possible that the luminous night-clouds consist sometimes of larger and sometimes of smaller particles?

Their height can be determined from observations of the upper limit of the illuminated part, preferably carried out for different positions of the sun below the horizon. In one instance it was found that the height α above the horizon of the upper limit was equal to 10°, 5° and 3°, when the sun was a depth $\beta = 12°$, 13° and 14° below the horizon. For the height of the night-clouds, one easily deduces the formula $h = \frac{R}{4} \beta^2 \left(\frac{2\alpha + \beta}{\alpha + \beta}\right)^2$, where

R is the earth's radius, while α and β are expressed in radians.

The height obtained in this way must be slightly increased because rays nearly tangential to the earth are not scattered. A more accurate method consists in taking photographs from two places. The general result is that their heights vary in the majority of cases between fifty and sixty miles. Once their height is known one can also find the true size of the ribbing delineated in these clouds. The distance from one rib to the next is, on an average, four to six miles.

The importance of the night-clouds is enhanced by the fact that they are our only source of information concerning the currents in the uppermost layers of our atmosphere (apart from a few, very rare, observations of meteors). If no photographs can be taken, one can determine the velocity of the clouds by means of the cloud-mirror; mostly they arrive from the north-east at a velocity between 40 and 80 yards/sec., at times from west-north-west at 30 yards/sec.; occasionally exceptional speeds up to 300 yards/sec. have been measured.

The theory formerly generally accepted was that the mysterious light phenomenon of luminous night-clouds was caused by volcanic dust thrown by violent eruptions very high into the atmosphere. However, it has now been observed so often that we are forced to assume another cause as well, viz. the very fine dust in the Universe around us which is brought into our atmosphere by shooting stars and meteorites and also perhaps comets passing near the Earth leaving fairly large quantities of cosmic dust in their wake. The large meteorite that fell in 1908 in Siberia was immediately followed by the appearance of very striking night-clouds. In other cases the origin of the dust is more probably volcanic.[1]

For photographing these clouds, a camera of large aperture is advisable. Using a lens of $f/3$, the exposure times were 16 seconds, 35 seconds, 72 seconds and 122

[1] Cf. the discussion, R. Süring, *Die Wolken* (Leipzig, 1936), pp. 30–6.

seconds, when the sun was respectively 9°, 12°, 14° and 15° below the horizon.

200. *Nocturnal Twilight and Nocturnal Light Phenomena*

If we wish to study the faintest forms of twilight phenomena, we should begin at night, and, while our eyes are perfectly rested, observe the first stages of dawn. We choose a moonless night with a perfectly cloudless sky in May or August to September, and a spot as far as possible from human habitations. It will not be easy to break our usual daily routine and begin at midnight with a few hours of observation out of doors. But once this difficulty is overcome we shall be abundantly rewarded by the sight of the magnificent scene unfolded before us. The glory of a starlit sky is beyond the imagination of the ordinary town dweller. The extent to which our eyes are capable of adapting themselves to the dark is quite astonishing, and it is also remarkable how many more stars we can see after waiting for an hour than when we first go outside. One almost thinks that the whole sky is luminiferous. This is a favourable moment for observing very feeble light phenomena, some of which can be seen fairly clearly, while others are more often invisible.

First of all, we will probably see here and there a faint glow low down on the horizon. This is the reflection of the lights from distant towns and villages. It is clearer on some nights than others, according to the cloudiness, mistiness or clearness of the sky. These factors can easily be taken into account if the observations are always carried out from the same place.

Right across the sky, like a ribbon, runs the *Milky Way*, consisting of large and small clouds of light interspersed with dark spaces. Those who have never observed a starlit sky before will be surprised at the brightness of some of its parts.

The sky in the background becomes clearer towards the horizon, which is bordered all the way round by 'earth-light,' reaching a maximum brightness at a height of about

15°. This is a kind of continual faint aurora in our atmosphere. The more oblique our glance, the longer the distance we look through the luminous layer, and the brighter, therefore, the earth-light. The fact that it decreases in luminosity near the horizon is to be ascribed to the weakening effect of the air.

Sometimes broad bright stripes are seen.[1] Twice a year they occur apparently with special frequency—namely, in August and December—and are supposed to be due to cosmic dust penetrating our atmosphere.

A few times a year the *Northern Lights* are seen in our country, at least in years of great sunspot activity, e.g. in 1938, and, probably, in 1949. They appear in the sky to the north as arcs, bundles of rays, etc.; the rays often move very rapidly, increasing and decreasing in length. Beware of confusing them with searchlights moving somewhere in the distance!

All along the 'zodiac' there is an increased brightness in the sky from the *zodiacal light*, becoming remarkably strong close to the sun, diminishing rapidly towards the anti-solar point. It resembles an oblique pyramid of light rising from the horizon, in the spring in the west after sunset, in the autumn in the east before sunrise (cf. § 201).

Independently of all these phenomena, the sky, as a background, possesses a positive brightness: your outstretched hand, the silhouettes of trees and buildings, stand out darkly against it. Fifty per cent. of this brightness is due to the sum of all the millions of invisible weak stars, 5 per cent. to the scattering of starlight by the atmosphere of the Earth, and the rest to earth-light. The brightness of the night-sky is by no means always of the same intensity, attaining on some nights a value four times the normal one, even when there is no moon. You can see the time by your watch and distinguish large letters. This variability is to be ascribed to variation in the intensity of the earth-light,

[1] C. Hoffmeister, *Die Sterne*, **11**, 257, 1931. *Sitzungsber. Akad. München*, 129, 1934.

which in its turn must be explained by the variable intensity of the currents of electrons and ions shot off by the sun into our atmosphere.

Finally, we turn to the observation of the *nocturnal twilight* phenomena. Examine the border of earth-light along the north side of the sky. Here the border rises by about 10° in a gradual slope, the maximum lying somewhere above the point at which the sun, now of course invisible, lies below

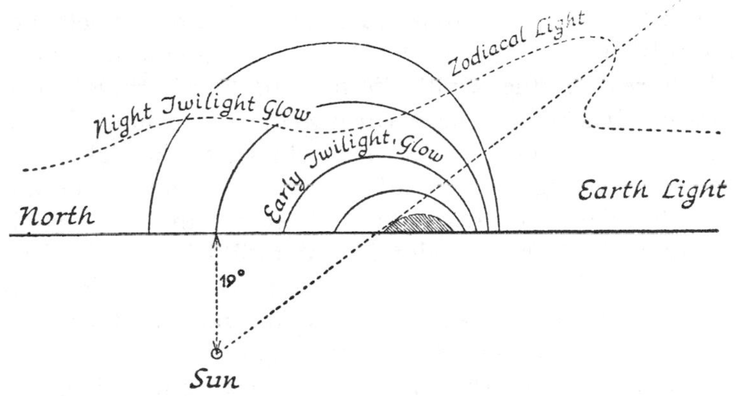

FIG. 151. Nocturnal twilight.
After Gruner and Kleinert, *Die Dämmerungserscheinungen* (Hamburg, 1927), p. 7.

the horizon. This is the *night twilight glow*. It can always be recognised by the fact that it invariably moves eastwards with the sun as the night advances. Its altitude above the sun is of the order of 40°; in the most favourable circumstances (in Greenland) it is observable to the height of 55° above the sun. It is clear, therefore, that in our climate the nights are never completely dark in the summer; the twilight lasts in reality the whole night long. Only in the winter is our sky perfectly dark. One can also understand why the tropical starlit sky is such a dense black, since the sun in that part of the world descends so steeply and so far below the horizon. There are cases in which the night twilight is abnormally strong.

Two and a half or three hours before sunrise the twilight

glow becomes asymmetrical, rising in the east and descending from there more steeply, to assume after a while the shape of a cone of light sloping upwards—the zodiacal light—with its axis having practically the same inclination as the ecliptic (§ 201).

About two and a half hours before sunrise, while the sun is still 20° below the horizon, a very faint bluish light appears at the base of the zodiacal light, a little to the right of the sun; it is observable only with difficulty, and rises slowly upwards at the same time, spreading towards the left, i.e. towards the sun (Fig. 151). This is the *early twilight glow*, which reaches the zenith in half an hour's time. The twilight arcs usually lie vertically above the sun. If the early twilight glow seems to have shifted to the right, it is because its brightness is added to the brightness of the zodiacal light on the right. However, the stronger it becomes, the more it predominates, until it has regained its normal position above the sun. For the rest, it continues to accompany the sun in its daily motion, moving, therefore, slowly more and more towards the right.

The feebler stars (fifth magnitude) have now faded, but the stronger ones are still perceptible; the main features of the countryside can already be discerned. In the western sky the counter-glow has grown very pronounced. The yellow twilight glow now begins to make its appearance, fading away at the top into a green-blue tint. The twilight proper has begun; the sun's altitude is −17° to −16° (*see*, further, § 186).

In other seasons of the year the course of the phenomena is the same, but the sun's altitude is different. In the middle of June, for instance, the sun descends no farther than 10°–15° below the horizon, so that all kinds of phenomena that only occur when the sun's position is much lower, are not seen.

201. *The Zodiacal Light*[1]

When the evening twilight has drawn to a close, or when the morning dusk is about to begin, we can see, in some

[1] Fr. Schmid, *Das Zodiakallicht* (Hamburg, 1928). W. Brunner, *Publ. Sternw. Zürich*, 1935.

months of the year, the softly radiant zodiacal light rising obliquely in a rounded pyramid. The steeper it rises the better we can observe it. The most favourable times are in January, February and March in the evening in the western sky, and in October, November and December in the early morning in the eastern sky (not so favourably as in the evening sky).

In June and July nothing of it is to be seen in our latitudes, because then the sun does not descend far enough below the horizon, and the zodiacal light cannot be distinguished from the lingering twilight phenomena.

In order to determine its position, we must begin by finding the zodiac itself—that is, the great circle running through the constellations of the *Ram*, *Bull*, *Twins*, *Crab*, *Lion*, *Virgin*, *Scales*, *Scorpion*, *Archer*, *Capricorn*, *Waterman*, *Fishes*.

This is the path we 'see' the sun cover in the course of the year. We cannot, of course, see the constellations at the actual moment the sun is in them, but as soon as it has declined and darkness has set in, the remaining part of the zodiac becomes visible. A kind of luminous mist extends along that circle, at its brightest and broadest near the sun, growing narrower as it runs out in both directions. On the one side of the sun is the part of zodiacal light that we see in the early morning; on the other side that observed in the evening. During the winter an experienced observer can see the zodiacal light in the evening as well as in the morning for six consecutive months.

The light itself is feeble, of the same order as the Milky Way, but not so 'granular,' and milkier. Practice is needed to see it. There must, of course, be no moon, and every lamp, even in the distance, is a hindrance, while luminous planets, like Venus and Jupiter, can also be troublesome. The vicinity of large towns should always be avoided; the best place from which to observe is an elevated spot with an open view in all directions.

One should begin by drawing on a star-chart the outline

of the zodiacal light relative to stars that are easily recognised, and afterwards to trace lines of equal luminosity. The part in the middle is the brightest, the brightness diminishing gradually towards the top and the edges, but more abruptly on the south than on the north side, so that the greatest brightness is shifted towards the south relative to the axis of symmetry of the weaker parts. By means of this kind of rough drawing, one is able to estimate the breadth of the light phenomenon, which, measured at right angles to its axis, amounts to about 40°, 20° and 10° at distances of 30°, 90°, 150° from the sun.

One will be well repaid if one takes the trouble to devote a whole night to observing the zodiacal light, and admiring the beautiful variations of the changing scene. About two hours after sunset, when the sun's position is −17°, a very faint, wedge-shaped cone of light becomes visible, rising obliquely, towards the south-west. When the sun's position is −20°, the sky has become so much darker that an enormous pyramid of light is observable. This western zodiacal light becomes, in the course of the night, more upright and spreads wider and wider; its position relative to the stars remains on the whole the same. A slight shifting is just perceptible, the stars that were a little to the south shift their positions later on more to the north of the zodiacal light. The best time for observing this curious phenomenon is in the first half of the winter.

Gradually the west zodiacal light begins to decline and the east zodiacal light appears in the east. It is now nearly midnight, the favourable moment for finding the famous *Gegenschein* or 'counter-glow,' one of the most difficult phenomena to observe, which we can hope to see only on clear winter nights, when the sky is very dark. At the antisolar point (p. 170), that is, practically in the south, an extremely faint bridge of light is observed, connecting the tops of the east and west zodiacal lights. Later on in the course of the night the east zodiacal light can be seen moving with the stars, but at the same time shifting slightly; the stars seem to move from the north to the south side of the pyramid. Once more it is as if the zodiacal light accompanies the daily rotation of the sky, but lags behind very slightly relative to the stars.

Day approaches; when the sun's position is -20° or -19°, it seems as if the base of the pyramid of the east zodiacal light were broader and brighter. When it reaches -19° to -17°, the early twilight glow appears.

The zodiacal light is a much more complicated phenomenon than was originally thought. It arises chiefly as the combined effect of two causes. (i) A tremendous disc or ring of cosmic dust surrounding the sun and scattering its light; we see this cloud of dust illuminated by the sun and increasing gradually in brightness the nearer the direction of our gaze approaches the sun. (ii) The night twilight (§ 200), which can be regarded as the very feeble light scattered by the highest layers in the atmosphere, and as such constitutes the last stage of the evening twilight. It is possible that to this must be added the emission of self-generated light, when these strongly ionised atmospheric layers come within the shadow of the Earth, and the ions are able to recombine. The brightness of this light also increases nearer the sun, but this increase is much quicker than that of the cosmic constituent of the light; its lines of constant brightness span the sun like arches, as do all real twilight phenomena; the zodiac has no effect on them (Fig. 151).

The combination of phenomena (i) and (ii) forms the 'typical' light-pyramid of the zodiacal light, and from the changing position of the horizon and zodiac one can understand why this light phenomenon shifts to a certain extent in the course of the night and of the year, the shift depending also on the geographical position of the place of observation. To this there must also be added the glow known as 'earth light,' showing its maximum light-intensity at about 15° above the horizon. Finally, there is the extinction of light by the atmosphere of the Earth, which causes the glow near the horizon to be increasingly weakened.

If the cloud of cosmic matter forming the zodiacal light were to lie in a ring outside the Earth's orbit, this ring ought to shine just as brightly on the side near the *Gegenschein* as on the other side, at Z. (Fig. 152.) The fact that this is by no means true definitely proves that the cloud of dust is found mainly *within* the Earth's orbit. The zodiacal light is

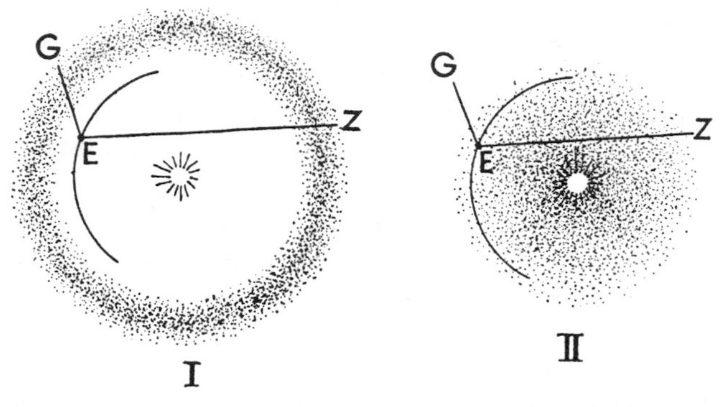

FIG. 152. Two theories of the origin of the zodiacal light. I. Ring of cosmic dust outside the Earth's orbit. II. Disc of cosmic dust round the sun.

light scattered at small angles; the *Gegenschein* or counter-glow is the light scattered back by the outer parts of the disc, and is always very much feebler when the dust particles are not very small (§ 183).

It has been maintained that the zodiacal light grows periodically brighter and weaker every two or three minutes and that these changes coincide with disturbances of the magnetic needle. It is said, also, that the light is particularly strong during magnetic storms. Before accepting these observations, we would do well to determine their reality by letting at least two persons carry out observations simultaneously and independently, and also to make quite sure that they are not due to veils of clouds or shadows of clouds.

A very interesting observation might be performed some time during a total eclipse of the sun, when there is a possibility of the shadow-cone of the moon being seen passing through the layers of dust which scatter the zodiacal light.

These observations would have to be carried out after the sun had set.

There seems to be such a thing as lunar zodiacal light, too, which appears just before the moon rises and after it has set, but this light is at least as difficult to observe as the *Gegenschein*.

202. *Eclipses of the Moon*[1]

Eclipses of the moon are caused by the shadow of the Earth falling upon the moon. Wouldn't it be worth while to see what this shadow looks like? Viewing it in this way, a moon eclipse is really a means of learning something about our own Earth!

No two moon eclipses look the same. It very seldom occurs that the moon is so completely eclipsed as not to be visible at all in the night sky. The colouring in the centre of the shadow is generally a faded copper red, surrounded by colours of increasing brightness. We distinguish:

The Dark Inner Part

0–30′, reddish black; towards the outer edge a brighter brownish orange;
30′–41′, grey border of fairly uniform brightness;
41′–42′, transition border.

The Bright Outer Part

At times this is observed to consist of rings coloured successively bright sea-green, pale golden, copper, peach-blossom pink (from the inner part outwards).

These colours, and the way they change, lead us to suppose that we are not now dealing with an ordinary shadow. Indeed, closer investigation shows that it is quite impossible for the shadow of the Earth's globe to cause an eclipse of the moon, because the curvature of the rays in our atmosphere makes the rays bend more or less round the

[1] Freeman, *Nat.*, **15**, 398, 1877. W. J. Fisher, *Smiths. Misc. Coll.*, **76**, No. 9, 1924; and *Science*, 70, 404, 1929.

earth ! 'The shadow of the Earth' is in this case nothing but the light beam, that has traversed the lower layers of our atmosphere up to a height of about five miles, and has become dark red in colour on the way. This takes place in the same way as the change in colour of the sun's rays that reach us during twilight through a dense atmospheric layer; only the colour is still more dull, owing to the fact that the distance travelled by the rays of light is now *double*. The colour of the central parts of the shadow of the Earth is therefore an indication of the degree of transparency of our atmosphere. It is not a mere coincidence that the moon, when eclipsed, should seem extremely dark at times when our atmosphere contains great quantities of dust from volcanic eruptions. The moon eclipses are also, on an average, darker when the moon is in the northern part of the shadow of the Earth than when in the southern part, so that apparently there is more volcanic and desert dust in our northern than in our southern hemisphere.

A simple way to judge the brightness of a moon eclipse is to use the really remarkable fact that when the light-intensity is small our eye can no longer distinguish any fine detail: for instance, while the large headlines in our newspaper are still readable in the twilight, we can no longer read the ordinary print. Similarly, we must notice now whether the large plains (the so called 'seas') of the moon's surface, which are usually visible as grey spots, remain visible during the eclipse (*a*) with the naked eye; (*b*) with a small telescope of 3 to 10 inches objective-diameter; (*c*) with a larger telescope.

These three ways of observing will suffice to help us to form a rough division of the moon's eclipses into light, medium and dark. A systematic comparison of notes made on these lines during a number of years is bound to furnish material for numerous remarkable conclusions. (Note that the telescope does not make the image of an illuminated plane seem brighter, but that it is only the optical magnification which accounts for the increased visibility !)

203. *Ash-grey Light*

When the new moon has just appeared, we can see at the side of her slender crescent the rest of the moon's surface, faintly illuminated (Fig. 80). This 'ash-grey' light comes from the Earth, which shines like a large, bright source of light on to the moon. The remarkable thing here is that the ash-grey light is not always equally strong. At times it is almost invisible, at others almost milk-white and so bright that the darker spots usually visible on the moon's surface can be distinguished. The changes in the strength of the ash-grey light are attributed to the fact that the half of the Earth facing the moon contains at times many oceans, and at other times many continents, and is at times more clouded, and at other times clearer. In this way a glance at the ash-grey light may give us a comprehensive impression of the conditions on one hemisphere of the Earth! In this respect the study of the ash-grey light belongs, properly speaking, to the physical science of the terrestrial sphere.

Estimate the strength of the ash-grey light on a scale from 1 to 10 on a number of days (1=invisible, 5=fairly visible, 10=exceptionally bright). You will very soon notice that the visibility is very dependent on the phases of the moon, because its bright crescent blinds us as it gets broader. A comparison of the visibility of the ash-grey light on various days, therefore, has meaning only when it is made for equal phases. On the other hand, the moon's height above the horizon appears to influence the visibility only to a very slight degree.

XII

Light and Colour in the Landscape

204. *The Colours of the Sun, Moon and Stars*

It is difficult to judge the colour of the sun, owing to its dazzling brightness. Personally, however, I should say it is decidedly yellow, and this, combined with the light from the blue sky, forms the mixture that we call 'white,' the colour of a sheet of paper, when the sun is shining and the sky is clear. Estimations of this kind give rise to difficulties, owing to a certain vagueness in the notion 'white.' Generally speaking, we are inclined to call the predominating colour in our surroundings white or nearly white (cf. § 95).

On a cloudy or misty day the rays from sun and sky are already intermingled owing to innumerable reflections and refractions by drops of water, and the colour of the sky is therefore a compound white. If we consider that the blue light from the sky is in reality scattered light that was first contained in the light from the sun, we must conclude that the sun, seen outside the atmosphere, would also be practically white.

We know already that the *orange* or *red* colours of the setting sun are accounted for by the rapidly increasing length of the path traversed by its rays before they reach our eye; gradually the more refrangible rays become almost completely scattered and only the dark red ones remain (§ 172).

In a few rare cases the sun, when high, shines *copper-red* through the mist, i.e. when the mist-drops are very small and therefore scatter the shorter waves more particularly (§ 183).

In other cases it is *bluish*, and this is said to occur most commonly when the clouds have an orange-coloured edge. It is possible that colour contrasts play a part here or that inexperienced observers confuse the colours of clouds in the immediate neighbourhood of the sun with the colour of the sun's disc itself. Entirely different is the phenomenon of the *blue* sun, when seen through a dense cloud consisting of drops very uniform in size (§ 164).

The moon by day is a striking *pure white,* because then the intense blue scattered by the sky is added to the moon's own yellowish light. Also when it rises and sets by day it is practically colourless, dull and only slightly yellowish. It grows gradually yellower as the sun sets and the blue light of the sky disappears; at a certain moment it becomes a beautiful *pure yellow,* though the colour probably seems stronger to us because of the psychological contrast with the still faintly blue background. As twilight draws to its close, the colour returns to *yellow-white*; very likely because the surroundings become darker, so that the moonlight appears very bright to us, and so owing to a curious peculiarity of our eye, tends to white like all other very bright sources of light (§ 77).

The moon remains for the rest of the night a light yellowish colour, exactly like the sun by day. The colour becomes most nearly white on very clear winter nights, when the moon is very high; but near the horizon it shows the same *orange* and *red* colours as the setting sun; that the impression made on our eye by the colours of the moon is rather different is due to the much feebler intensity of the light.

The full moon in the middle of the blue earth-shadow has a lovely *bronze-yellow* colour, no doubt caused by complementary contrast with its surroundings. When encircled by small clouds of a vivid purple-red, its tint becomes almost *green-yellow*; if these clouds become salmon-pink it changes almost to *blue-green.* These contrast colours are still clearer in the moon's crescent than in the full moon.

Not to be confused with the colour of the moon is the colour of the landscape by moonlight, which is commonly considered to be blue, or green-blue. No doubt this effect is to a great extent due to the contrast with our orange-coloured artificial light, which makes the blue of the sky, illuminated by the moon, all the more striking.[1]

In order to obtain a preliminary idea of the differences in colour shown by the stars, let us look closely at the large square of the constellation Orion. We notice that the colour of Betelgeuse, the bright star α at the top to the left, is a striking yellow, or even orange, compared with the three other stars (Fig. 62). Close to this constellation we see another orange-coloured star, Aldebaran in the Bull.

The next important thing is not to be satisfied with this first and very easy distinction of colour, but to try to trace finer differences of tints. This is a severe test for our sense of colour, but a great deal can be done by practice. Since the differences in the colours of the stars are due to their different temperatures, we can understand their showing the same succession of colours as a glowing body cooling down gradually: that is, from *white*, via *yellow* and *orange* to *red*. It has not been definitely settled whether the hottest stars should be described as blue or as white, different observers not being of the same opinion, as to what 'white' really is. Some are influenced more than others by the feebly illuminated background of the sky, which seems bluish to us and which we are accustomed to consider as colourless simply because it is the average colour of the scenery at night.

The following scale gives an idea of the different colours occurring among the stars, together with the numbers by which they are usually denoted and a few examples. Colour estimates made independently by skilled observers often turn out to be a whole class above or below the average given. The estimates in the examples given here were made by observers who did not see blue as such, and therefore recognised no negative values.

[1] See the discussion in *Met. Mag.*, **67–69**, 1932–4.

Colour Scale	*Examples*	
−2 blue	α of the great Dog (Sirius) . .	0·8
−1 bluish white . .	α ,, ,, Harp (Vega) . .	0·8
0 white	α ,, ,, Lion (Regulus) . .	2·1
1 yellowish white . .	α ,, ,, Little Dog (Procyon)	2·4
2 white yellow . .	α ,, ,, Eagle (Altair) . .	2·6
3 light yellow . .	α ,, ,, Great Bear	4·9
4 pure yellow . .	β ,, ,, Great Bear	2·3
5 deep yellow . .	α ,, ,, Little Bear	3·8
6 orange yellow . .	β ,, ,, Little Bear	5·8
7 orange	α ,, ,, Boötes (Arcturus) . .	4·5
8 yellowish red . .	α ,, ,, Scorpion (Antares)	7·5
9 red	Venus	3·5
	Mars	7·6
	Jupiter	3·6
	Saturn	4·8

The stars, too, naturally become more reddish the nearer they approach the horizon, but their scintillation then usually prevents us from judging their colour correctly. It is a remarkable fact that on the earth we designate a glowing body at 2,500° C. as white-hot, whereas a star of the same temperature appears to us orange-red! It is probable that this physiological phenomenon must be ascribed to the star being so much less bright, so that the red component of the light-impression on the eye is perceptible, whereas the green and blue components fall short of their threshold values.

205. *The Colour of the Clouds*

It is a pleasure to watch the beautiful summery cumulus clouds drifting past, and to try to account for the fact that certain parts are light and others darker. Where the sun illuminates these clouds they are a dazzling white, but they become grey or dark grey underneath as we watch them pass overhead. The drops of water are so closely packed that the light hardly enters the cloud, but is reflected back from the greater part by the numerous drops; the cloud resembles an almost opaque white body. If the sun is covered by cumuli, they appear dark, but their edges are light; 'every cloud has a silver lining !' Thus the distribution of light and shade provides us with interesting information concerning the various parts of the clouds, above, below, in front, behind, and the actual shapes in space of these huge masses. It is not always easy to form a correct idea of these proportions, or to realise the position of the cloud in relation to the sun. If, for instance, there are clouds in front of me, and the sun is some distance above them, I am quite surprised at seeing almost only shadow (Fig. 153, *a*). I do not realise sufficiently

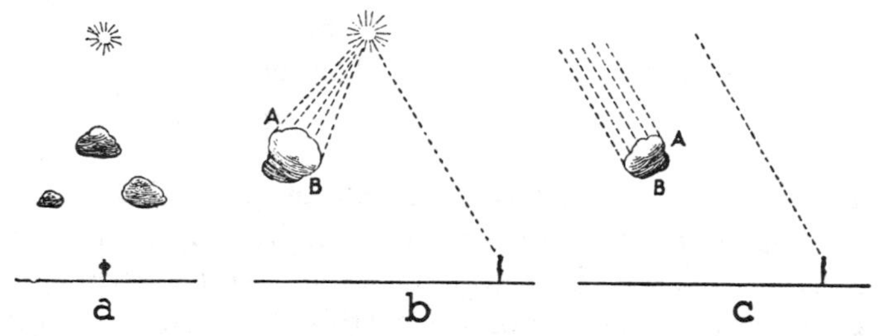

FIG. 153. Light and shadow on cumulus clouds. (*a*) The landscape and the observer seen from north to south. (*b*) Erroneous subjective conception and expectation. (*c*) True positions.
(In (*b*) and (*c*) the observer is looking from east to west.)

the enormous distance of the sun, and unconsciously I imagine it to be fairly near and expect light on AB (Fig. 153, *b*), instead of remembering that the sun's rays illuminating the cloud run parallel with the line of the sun to my eye (Fig. 153, *c*).

However capricious the play of light and shade may be on the dark masses, however complicated the shadows they throw on one another, nevertheless it seems impossible to explain all the differences in colour of the cumulus clouds by them alone. If, when the weather is clearing up after a storm, only a few small cumuli are left, brilliantly illuminated by the sun, with no possibility of one casting its shadow on another, they grow darker and darker and are finally blue-black when they are about to disappear. The impression is generally that thin parts of cumuli seen against the blue sky do not show a colour consisting of blue + white (as might be expected), but of blue + black ![1]

At other times, a cumulus cloud is grey when seen against the background formed by another big cloud that is dead white, so that there can be no question of the brightness increasing simply with the total thickness of the layers. The optics of these phenomena, though we see them day after day, has not yet been sufficiently investigated. One has, of course, to be very careful before accepting the idea that clouds really can *absorb* light; one should try first to explain everything as if they were solid white objects, and then remember that they are really scattering mists and, finally, consider the possibility of their containing dark dust-particles as well.

It is interesting to compare them with the white steam (not the smoke !) from a locomotive ! In some cases the latter appeared whiter when observed at a large angle with the incident light, and less bright when observed from the direction of the sun, so that the eye received the light-rays reflected more or less in the direction of incidence. In other cases the steam seen from all directions was much brighter than the brightest parts of the cumulus clouds, which was probably due to the great distance of the latter, and the weakening of the light by scattering in the air.

Dark cumulus clouds seen from a great distance often look bluish. This is not the colour of the cloud itself, but the scattered light reaching us from the atmosphere between the cloud and our eye. The further away such a dark cloud is from us the more its colour is bound to approach that of its background, the sky. On the other hand, bright clouds close to the horizon become yellowish in colour (§ 173).

We ought to extend our investigations to the other types of clouds as well, and try to explain why rain-clouds, for example, are so grey; why in thunder-clouds a peculiar

[1] *C.R.*, **177**, 515, 1923.

leaden colour can be seen side by side with a faded orange. Is it dust? Our knowledge, however, of all these things is so incomplete that we prefer to spur on the reader to begin investigations of his own.

The distribution of light over the celestial vault, when it is entirely and evenly overclouded, is very characteristic and forms a counterpart of the distribution of light when the sky is bright blue. Compare, for instance, by means of a small mirror, the zenith and the horizon: the latter is always the brighter of the two, the ratio being 0·50 to 0·80 (Plate XIII).

206A. *The Colour of Clouds during Sunrise and Sunset*

In our description of a sunset we began by ignoring the clouds. But now we will discuss for a moment the origin of these wonderful cloud-scenes, with their infinite wealth of colour and variety of shape and their apparent absence of all regularity. Let me begin by stating that what now follows concerns chiefly what we see *before* the sun sets, whereas the real 'twilight phenomena' themselves have been discussed in § 189. As soon as the sun has gone down the magnificence of the clouds has gone too!

Shortly before sunset the clouds are illuminated by:

1. Direct sunlight, becoming gradually yellow, orange and red coloured in succession, the lower the sun declines.

2. Light from the sky, orange-red towards the sun, blue everywhere else. This orange-red light must be attributed to the strong scattering by the large particles of dust and water-drops, which causes only a slight deviation of the rays (§§ 182, 192); the blue light arises from backward scattering by the air-molecules.

Now imagine a cloud in the neighbourhood of the sun, at first very thin, and becoming gradually denser. Its drops scatter light over small angles, so that thin veils of clouds will certainly send a great quantity of light towards us from the sun lying obliquely behind them, and the more scattering particles there are, the stronger the warm orange-pink light. But there comes an optimum when the layers become either

XIII

(*a*) The zenith is reflected in a large inclined mirror: when the sky is blue it is darker at the zenith than along the horizon

(*b*) The same experiment when the sky is uniformly clouded; the zenith is now brighter

too dense or too thick to permit light to traverse them easily. Heavy clouds transmit practically no light at all and only reflect towards us the light of the part of the sky that is still blue, which illuminates them from our side (Fig. 154). This shows that the finest sunsets are to be expected when the clouds are *thin* or when the covering of the sky is fragmentary.

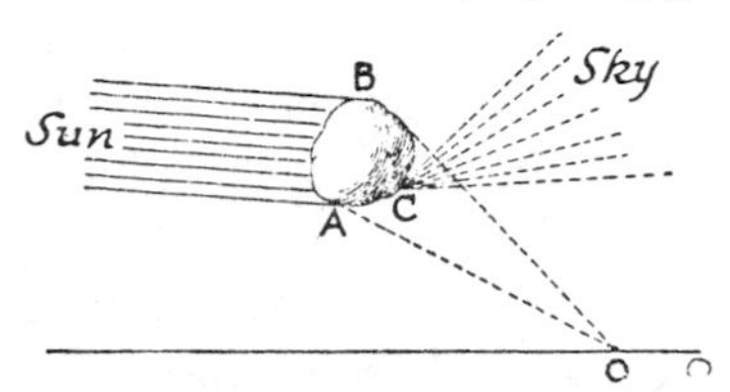

FIG. 154. Illumination of a cloud before sunset.

On the side the sun is setting we see thin clouds, illuminated from behind, thicker or denser clouds from the front, the former being bright *orange-red*, the latter a darker *grey-blue*. This contrast in colour, which often goes with differences in shape and structure, is one of the most delightful features of these cloud scenes.

The edges of heavy blue-grey clouds are often a wonderful gold. Note that the edge A, apparently nearest the sun, gives a stronger light than edge B, because (*a*) there the angle of deviation of the light-rays is smaller; (*b*) if we imagine the cloud to be perfectly round and full, it follows that on the side nearest the sun we must be able to see a little strip illuminated directly by the sun (Fig. 154).

This beautiful scattering is not shown along the edges of clouds much farther away from the sun; they are illuminated directly on one side, and on the other by blue light from the sky, so that, here too, a play of colours arises of orange and blue. As the sun sinks lower, the colours become warmer until the clouds opposite, in the east, begin to show the purple of the counter-glow.

When the sun sets completely its glow withdraws gradually from the various parts of the sky, the high clouds remaining illuminated longest of all. This develops into another lovely contrast; behind, clouds still illuminated by the sun, and before them, clouds illuminated only by light from the sky.

206B. *Illumination of Clouds by Terrestrial Sources of Light*

When we walk through the open country in the evening and the sky is uniformly clouded, we can see here and there in the distance a faint glow, low in the sky. This glow comes from a town or large village, which we can identify by the direction in which it lies. Estimate the angle α of the glow above the horizon in radians and read the distance A of the town or village from the map; the height of the clouds is then $h=\alpha A$. As an example: From Bilthoven I observed the glow of light above Utrecht at $\alpha=8{\cdot}5°$ from which $h=790$ m. (about 880 yards), above Zeist $\alpha=6°$, from which $h=780$ m. (about 870 yards). In 1884 the glow above London could be seen up to distances of about forty miles. How far would it be visible nowadays?

A closer study of this glow of light above a large town will repay you for the trouble. You will very soon notice that it differs from day to day, its variability being as great almost as that of the Northern Lights. You will discover two component parts in this light phenomenon: (1) a hazy mist of light, caused by the general illumination of the air with its particles of dust and water, which is strongest near the horizon; (2) a patch of light on the layer of cloud, the circumference of which is almost an exact replica of that of the town (that is, more or less circular), but, when seen from a distance, appears foreshortened more or less to an ellipse with fairly sharp edges, especially when the cloud-layer is smooth. If the sky is bright and cloudless, or else very misty, no light is to be seen above the town. If the sky is hazy, a mist of light is developed, not very sharply defined. If the sky is covered by a layer of clouds, the patch of light becomes clearly visible. Every sort of combination is possible and sometimes shadows are formed by isolated low-lying clouds, or irregular masses of light separate themselves from the main mass of light. The altitude of the clouds can, of course, be determined by measurements on the patch of light, the most exact values being given by the height of its boundaries. Carried out by a skilled observer this method is so precise as to enable one to ascertain whether the cloud-layer follows the unevennesses of the ground.

La Cour also succeeded in carrying out observations of this kind by day. Once after a fall of snow he observed that the cloud-layer was darker above the sea than above the snow-covered country, the line of demarcation becoming surprisingly distinct when he moved so far away that it was not higher than 20° above the horizon. Afterwards he discovered that darker regions became marked out on the clouds above woods as well; even the town of Copenhagen where the snow had already thawed on the roofs, corresponded with a darker region of this kind. From all these gradations of light, the altitude of the cloud-layer could be determined, and a set of consistent values obtained.

The easiest to observe of all these phenomena is the difference between snow-covered country and sea and is, therefore, the best one to begin with. It is nothing but the famous 'ice-blink' and 'water sky' of Arctic explorers, by which they are warned of the approach of pack-ice.

'And in the evening I saw a remarkable glow of light over the sky to the north, strongest near the horizon, though it could be followed all along the vault of the heavens, right up to the zenith, a wonderful, mysterious half-light, like the reflection of a huge fire a long way off, but in the land of ghosts, for the light was a ghostly white.'—FR. NANSEN, *Bøken om Norge*, Kristiana, 1914.

It is not so generally known that the sand of the Egyptian desert also colours the clouds with a glow, which is distinctly recognisable from a distance. A shallow place in the Indian Ocean, where the green of the sea was very pronounced, threw a pale green light on to clouds 350 to 450 yards high. And even above heath-land when the heather is in bloom and lit by the sun, the loose, drifting clouds are coloured a lovely purple underneath.

In one case a patch of light was observed on a uniform

layer of clouds, and it could be proved that this was simply the reflection of sunlight from a small lake far away.

207. *Factors determining the Colour of Water*[1]

Infinitely changeable, full of evanescent marble-patterned gradations, varying in every ripple, the fineness of its composition an everlasting joy to the eye.

Let us try to analyse it.

(*a*) Part of the light received by us from water is reflected by the surface; so long as this is smooth it acts like a mirror, and the colour of the water is blue, grey, or green, according to whether the sky is clear or heavily clouded or the gently sloping banks are covered with grass. But if the surface of the water becomes rippled, the colours of sky and banks intermingle, sparkles of the one flashing across the other. When it is very rippled, the water simply reflects a mixture of colour.

'What we commonly suppose to be a surface of uniform colour is, indeed, affected more or less by an infinite variety of hues, prolonged, like the sun image, from a great distance, and our apprehension of its lustre, purity, and even of its surface, is in no small degree dependent on our feeling of these multitudinous hues, which the continual motion of that surface prevents us from analysing, or understanding for what they are.'—RUSKIN, *Modern Painters*.

(*b*) Another part of the light has penetrated the water and is there scattered by particles of dust and by the general turbidity. These particles are as a rule so large that they scatter all rays to the same extent, the emergent light having then the same colour as the incident light; if they are particles of sand or clay, the emergent light may be of a brownish colour. In very deep, pure water, however, an

[1] Bancroft, *J. Frankl., Inst.*, **187**, 249 and 459, 1919. V. Aufsess, *Ann. d. Phys.*, **13**, 678, 1904; C. V. Raman, *Proc. R. Soc.*, **101**, 64, 1922; Shoulejkin, *Phys. Rev.*, **22**, 85, 1923; Ramanathan, *Phil. Mag.*, **46**, 543, 1925.

appreciable part of the light is scattered by the *water molecules* themselves, and has the same lovely blue colour as the sky or a thick block of glacier ice.

(*c*) Finally, in shallow water part of the light always reaches the bottom, and suffers diffuse reflection, at the same time assuming the colour of the bottom of the water.

(*d*) On their way through the water, rays of light undergo continual changes. (i) Owing to *scattering*, they lose part of their intensity; in pure water, violet and blue rays in particular are weakened. (ii) Owing to the *true absorption* of the water, which is already quite noticeable in layers a few yards deep, they lose their yellow, orange, and red rays, exactly in the same way as the light transmitted through coloured glass. (iii) In certain parts of the sea, part of the blue and violet rays is apparently transformed by *fluorescence* into green rays, the fluorescent matter being, perhaps, an organic substance sufficiently concentrated in the sea to play an important part (cf. § 230).

Scattering is invariably present, even in the purest water, for the molecules are not uniformly distributed in the water, and this causes an irregularity and a certain 'granularity'; each molecule, moreover, deviates from the spherical shape. This scattering can be compared in every respect to the scattering in the air, increasing also proportionally to $\frac{1}{\lambda^4}$, and therefore greatest for blue and violet rays. In less pure water there are floating particles; if they are extremely small their effect is added to that of the molecules and causes a blue-violet scattering. If they are larger than, say, 0.001 mm., they scatter all colours equally, and mostly in a forward direction (§ 183).

Ordinary soapy water is a good example of a liquid containing very minute scattering particles. Illuminated from the front, and seen against a dark background, it appears bluish; illuminated from behind, it appears orange (cf. § 171).

The absorption by the water of lakes and rivers must be

ascribed chiefly to the presence of chemical compounds of iron (Fe^{+++}ion) and of humic acids. For iron concentrations of 1 part in 20 million and humic acid concentrations of 1 part in 10 million (such as occur in practice) the water ought to show a much stronger colour than it actually does. Apparently, the Fe^{+++} compounds oxidise the humic acid under the influence of the light, while they themselves are changed into Fe^{++} compounds. The latter combine again with oxygen, becoming once more Fe^{+++} compounds, and so on.

We shall now give some examples, showing how these various factors combine in bringing about the colour of the water.

208. *The Colour of Puddles along the Road*

A simple case is that of puddles made in the road by the rain. If the angle at which we look at them is large, the reflection on the surface seems almost perfect and the objects reflected are rich in contrasts, the black branches very black indeed. If we draw nearer, so that we look more and more steeply, the reflection becomes much weaker (§ 52), and it seems as if it were covered all over by a sort of uniform haze; all the colours are paler, and what strikes us most is that the dark parts are no longer really dark, but grey. The haze is due to light falling from all sides on to the puddle, penetrating the water and scattered in all directions. If the water is not clear, but milky, the scattering is caused by floating particles of dust; if it is coloured with, for example, 'blue,' the scattered light will have become blue, and this colour combines with the images reflected; if the water is clear, but the bottom of it light, as in pools of sea-water on the beach, then all the reflected images become tinged with a sandy colour, and, when seen nearly perpendicularly, the bottom is visible, but only a few of the brightest reflections. When the water is clear, however, and the bottom dark, the reflected images remain, when seen nearly perpendicularly, pure and rich in contrast, albeit less luminous. In dark, quiet pools, the reflected foliage of trees shows at times a purity of colour and a distinctness greater than that of the actual objects reflected! This is a psychological effect mainly due to the fact that the surrounding scenery is less dazzling (§ 7).

Ask someone to help you by standing at different distances from the pool and watch how his reflection changes! This experiment will prove especially striking on the seashore.

We see here demonstrated on a small scale the reason why objects below sea level (such as rocks, submarines, etc.) can be seen more easily from an aeroplane than from a ship.

'Now the fact is that there is hardly a roadside pond or pool which has not as much landscape *in* it as above it. It is not the brown, muddy, dull thing we suppose it to be; it has a heart like ourselves, and in the bottom of that there are the boughs of the tall trees and the blades of the shaking grass, and all manner of hues of variable pleasant light out of the sky.'—RUSKIN, *Modern Painters*.

209. *The Colour of Inland Waterways and Canals*

Ripples on a surface cause an ever-changing variety of light and colour on every canal and every ditch (§§ 14–18). To find out whether any definite part of the surface is rippled, we should look at it from different directions. Slight ripples become visible only along the boundary lines of bright and dark reflections; in the reflection of the uniformly blue sky they cannot be seen, nor in that of the dark masses of dense woods (Plate XIV). Large ripples, however, produce a shading of shadow and light, even in fairly large uniform regions, either because they make the rays deviate so greatly or because the coefficient of reflection at the front and that at the back of the wavelets becomes appreciably different (§ 52 and Fig. 157).

Observations like these teach us that boundary lines between the rippled and the smooth parts of a surface of water are nearly always delineated with amazing sharpness.

This is not to be attributed to irregular distribution of the currents of wind, as is shown especially clearly by the fact that when it is raining and the whole surface of water is set into uniform vibrations, the boundary lines are still absolutely clear. The real cause is nothing but the presence of an extremely thin film of oil, not even a millionth of a millimetre thick (2 molecules of oil !) and yet quite enough to damp out the ripples caused by wind or rain ! This film is formed by the remains of animal and vegetable matter, of used oil left in the wake of passing boats, and of refuse in drain-water. The wind blows this greasy layer along and gathers it together at one side of the canal. You will always notice that water is rippled at the side where the wind comes from, and calm along the opposite bank. In this smooth part a lot of twigs and leaves are floating, but they barely move relative to one another, because they are kept in their places by the very thin film of oil !

In this way the striking difference between the lively sparkling surface of the water of a brook in a wood, and the leaden-coloured, syrupy waterways in the poorer districts of a big town is satisfactorily explained.

We will follow up these observations of light phenomena on the surface by studying the way in which this reflection continually competes with the light coming from below. We stand under the trees at the water's edge. Here and there we can see the reflections of dark tree-tops and between them bright patches of blue sky. In places where the clear sky is reflected we cannot see the bottom of the water, as the light coming from below is relatively too weak. In places where the dark trees are reflected we can see a dark mixture formed by the colour of their leaves, the colour of the bottom of the water, and the diffuse light scattered by the particles of dust in the water. Observe that we can only see the bottom of the water close to the bank. Looking at any distance at all across the water, it is no longer possible to see the bottom, even if the water be no deeper, for the reflected light becomes much stronger when the angle

XIV

Undulations on a sheet of water are only visible near the border of dark and light reflections

of incidence is large, and predominates over the light coming from below.

The reflection of the dark keel of a boat shows a greenish watery colour; whereas a bright white band running along the boat remains simply white.

'Under sunlight the local colour of water is commonly vigorous and active, and forcibly affects, as we have seen, all the dark reflections, commonly diminishing their depth. Under shade, the reflective power is in a high degree increased,[1] and it will be found most frequently that the forms of shadows are expressed on the surface of water not by actual shade, but by more genuine reflection of objects above.

'A very muddy river (as the Arno, for instance, at Florence) is seen during sunshine of its own yellow colour, rendering all reflections discoloured and feeble. At twilight it recovers its reflective powers to the fullest extent, and the mountains of Carrara are seen reflected in it as clearly as if it were a crystalline lake.'[2] —RUSKIN, *Modern Painters*.

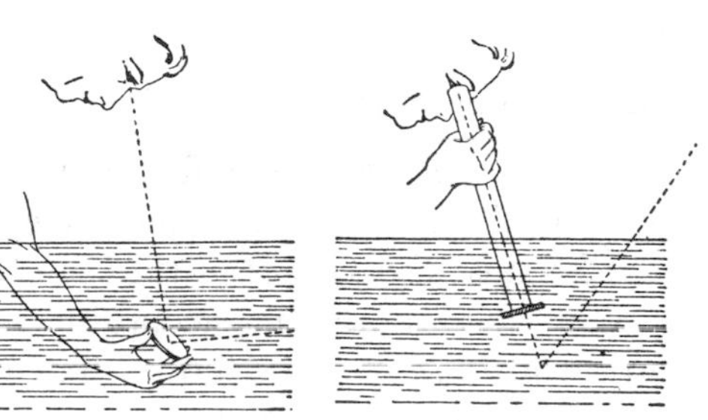

FIG. 155. Observing the colour of water without letting its reflection on the surface interfere.

There are a few simple means of eliminating the reflection on the surface:

(*a*) You can hold a black umbrella above your head.

(*b*) Take a little mirror and hold it under water (Fig. 155) at different angles, and judge in this way the colour of the light that has penetrated the water from above, after it has

[1] The physical explanation is that the reflective power is *precisely the same* in the shadow as in the sun, but the ratio $\dfrac{\text{reflected light}}{\text{scattered light from the depth}}$ is smaller in the sun, and greater in the shade.

[2] Our explanation: At twilight the light comes from one definite direction, and the general illumination has disappeared, which in the day produces the scattered light from below and was superposed on all the reflected images.

traversed the water for a certain distance. If this experiment is made in the water of any ordinary ditch you will be able to observe the yellow colour of the light due to true absorption. Where the water is very shallow a white piece of broken china, dropped on to the bed of the ditch, or a piece of white paper held under water, will serve the same purpose. At sea they use a white disc, let down to a certain depth, but this cannot be considered exactly as a *simple* experiment.

(*c*) Use a *water-telescope*, which is simply a tin tube, if possible with a piece of glass stuck to one end (Fig. 155). With this you will be able to judge the colour of the light coming from below, due to scattering from the bottom or by floating particles of dust. Use your water-telescope while bathing ! On old-fashioned ships there may still be a hole opening vertically into the water: this is a real water-telescope on a large scale !

(*d*) Look through a nicol, held in such a way as to extinguish the light reflected (§ 214).

210. *The Colour of the Sea*

Reflection is the great factor generally determining the colour of the sea. But it takes place in an endless variety of ways because the surface of the sea is a moving, living thing, rippling and undulating according to the way of the wind and the formation of the shore. The principal rule is that all far-off reflections are shifted towards the horizon, owing to the fact that our gaze falls on the slopes of the distant wavelets (§ 16). The colour of the sea in the distance is therefore about the same as that of the sky at a height of 20° to 30° and therefore darker than the sky immediately above the horizon (§ 176), and all the more so because only part of the light is reflected.

Apart from this, the sea has a 'colour of its own,' the colour of the light scattered back from below. From an optical point of view, an important characteristic of the sea is its *depth*, a depth so great that practically no light returns from the bottom of it. This 'colour of its own' is to be attributed to the combined effects of *scattering* and *absorption* in the masses of water. A sea that only scatters light would (apart from its reflection) be milky white, for all rays entering it are bound to come out again in the end. A sea that only absorbs would be as black as ink, for then the rays would return only after having reached the bottom, and the slightest absorption on this very long passage through the water is sufficient to extinguish them. However, as already stated, the colour arises from the combined effects of scattering and absorption: the kinds of light only slightly subject to scattering penetrate the water farthest before being scattered back, and during this long journey they suffer the most weakening by absorption.

Broadly speaking one can say that the quantity of light returning from below will be larger according as the fraction $\frac{\text{coefficient of scattering}}{\text{coefficient of absorption}}$ increases. A complete theory, however, is by no means simple.

The direct influence of the bottom of the sea on the colour of its expanse of water cannot be observed in the seas in our parts, at any rate when they exceed a yard in depth. Ruskin maintained that even at a depth of 100 yards the bottom contributes considerably to the colour of the sea, and further assertions of this kind may be heard from seafaring people. The truth is that a local elevation at the bottom of the sea alters the swell of the waves and the rippling of the water above it, while, naturally, more solid particles are stirred up there than where it is deeper, causing an increase in the scattering. So that the bottom of the sea has indeed some effect, though not a direct one.

211. *Light and Colour on the North Sea*

The following observations were carried out during a holiday on the flat, sandy coast of Holland which runs practically due north and south, and from where magnificent sunsets can be seen over the sea. The phenomena are, naturally, differently distributed over the day for differently orientated coasts; the essential point is the position of the sun relative to the surface of the sea.

1. *No Wind, Blue Sky.* A sea, in the calm of early morning smooth as a mirror. The sky, blue everywhere, but hazy. A tiny wave curls at our feet, on to the beach, leaving a narrow line of foam, that whispers and dies away. A hush follows. . . .

Now let us stand on a dune. The surface of the sea is spread out in front of us like a map. One part of it is so smooth that it reflects the blue-grey sky above perfectly, without any distortion, as a lake does. Other parts are blue-grey too, but darker in tone. Their boundaries are clearly marked, and their distribution is so distinct, that one can hardly refrain from sketching them. But after a comparatively short time they appear to have altered their position completely. For that reason, the lighter coloured parts cannot be 'sandbanks,' as seaside visitors usually call them; they are caused by an imperceptible, extremely thin oily film spread over the surface of the sea, similar to that on canals and ditches (§ 209), and sufficient to damp any ruffling of the water. These oily films probably arise from refuse in the wake of ships or from their used fuel oil. Where there is no film the water is slightly ruffled, as will be seen later on in the day when the sun shines above the sea, making the rippled parts sparkle like a sea of light. The colours shown by these parts now are darker, (1) because the front of each ripple reflects a higher and, therefore, a darker blue part of the sky, (2) because the reflection is less grazing and, therefore, less luminous. By using a nicol, with a vertical direction of vibration, the darker parts are seen to be much darker, and the difference between them and the lighter parts becomes more pronounced. The fact that the lines of demarcation between the various regions appear to run parallel to the coast practically everywhere is due to perspective shortening, for in reality the regions covered by oil may have all kinds of shapes (Fig. 156). A few real sandbanks are conspicuous by their rather yellower colour, but only where the sea is extremely shallow— 4 to 8 inches for instance.

If one bathes in the afternoon one is struck by the unusual clearness of the water when the sea is calm. As far as 1 yard

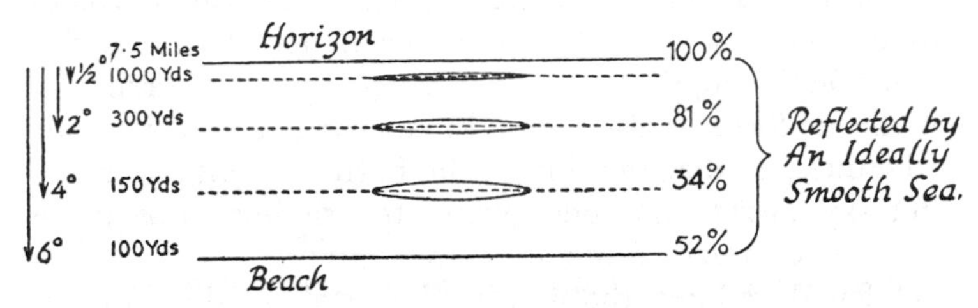

FIG. 156. The sea, seen from a dune 30 ft. high. The ellipses show the perspective shortening of a circle at different points on the surface of the sea.

down we can see every detail at the bottom, and even minute swimming creatures. There is no sand in the water or very little, and that only where a wavelet is about to break and tiny clouds of sand are whirled upwards behind it. If we look down at the water quite close to us, the reflection of the sky hardly interferes, and the yellow colour of the sand at the bottom dominates up to a depth of 8 in. At a depth of 1 to 1½ yards, the colour changes into a lovely green, and now we have to form a kind of water-telescope with our hands to avoid the reflection of the sky (§ 209). This green is the colour of the light that has penetrated the water and been scattered back. However, as soon as we look at the surface of the sea a little further off reflection predominates and the blue sky is mirrored everywhere. A wonderful interchange of sea-green and sky-blue !

In the evening the sun goes down behind a blue-grey bank of clouds a few degrees in height. Above that there is a shimmer of twilight orange and gold, merging gradually into the darker blue of evening, higher up in the sky. The sea is still as calm as ever, reflecting, without distortion, the whole scene. But as we look towards the west we begin to see very small ripples (§ 17), and in the distant parts of the sea, where the blue-grey bank of clouds is reflected, every ripple forms a little orange-yellow line (the slanting wave reflecting a higher part of the sky). And nearer to us, where the sea is orange-yellow, the wavelets produce a darker hatching by reflection of the still higher, bluer sky. Towards the north-west and south-west, where the twilight colours are disappearing and where our gaze no longer falls at right angles on the slopes of the waves, the sea is the pure reflection of the uniform bank of clouds, unchanged in colour and brightness, so that the line of the horizon vanishes and sea and sky merge into one, while the sailing ships in the distance seem to drift in a blue-grey infinity.

A few days later, the weather being about the same and the wind perhaps weaker, the parts of the sea covered by the thin film of oil were also visible in the evening, reflecting the blue-grey bank of clouds, whereas the ruffled parts, by the displacement of the images, reflected the orange-yellow sky.

2. *Slight Wind, Clear Blue Sky, with a few Isolated Clouds.* Before I even reach the top of the dune, I am struck by the strong contrast between the blue-black sea and the light sky near the horizon. Visibility is exceptionally good, horizon and distant objects stand out with clear-cut distinctness, a condition which lasts all day. There is a slight west wind blowing; breakers extend along the coast in two or three lines of foam, though in the open sea no foam is to be seen. We take up our post on the dune.

Watch the separate breakers along the shore (Fig. 157). They are dark, yellow-green grey in front, for our gaze strikes the front slope of each wave nearly at right angles, and therefore only a little reflected light reaches us, and that, moreover, from a dark part of the sky. We do, however, see yellow-green light that has either been scattered back from the depths of the sea, or has penetrated the back of a wave and emerges at the front; but since this is after all very feeble the fronts of the waves are dark. On the other hand the backs of the waves reflect the light blue sky along the horizon. In this way each wave shows a lovely contrast between its dark yellow-green front and its light blue back. These light blue backs develop between the breakers into wide flat troughs, slightly rippled and good reflectors, and likewise blue. The few rows of sandbanks running along our coast are easily distinguishable by the waves breaking on them, while the spaces in between are smoother and calmer. Further away from the beach, the shading of the waves becomes finer and finer. There are no more breakers, but the contrast between the front and back slopes remains.

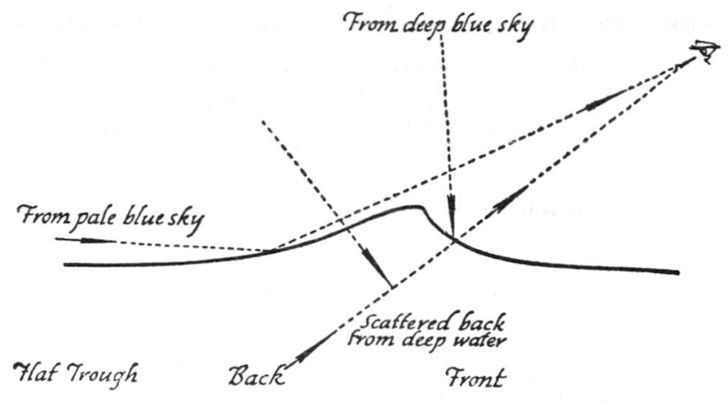

FIG. 157. How different colours are formed in a wave of the sea.

As we look more and more grazingly across the water, we can no longer see the troughs between the waves, and finally their back slopes disappear entirely. The fronts are now much less steep and reflect chiefly the sky at a height of about 25°. This 'displacement of reflected images' (§ 16) accounts for the dark blue colour of the sea, and for the contrast between sea and sky at the horizon. This contrast is at present so strong because the sky actually on the horizon is so light, and yet, at only a short distance above it, such a deep blue. Check this by projecting the reflection of the higher parts of the sky in a little mirror on to the neighbourhood of the horizon; it is amazing! Notice, at the same time, that the sea in the distance is markedly darker than the darkest parts of the sky, the reflecting power of the sea's surface being far from 100 per cent. The contrast between sea and sky is strongest in the west and becomes less to the north and the south, because the majority of the waves come from the west, whereas when we look northwards or southwards our gaze is more nearly parallel to the crests of the waves, so that their effect is less (§ 17).

We might perhaps doubt whether the strong contrast shown now between sea and sky has no other reason than the rapid increase in brightness of the blue sky near the horizon. Nature will convince us. For a moment part of the western sky is covered by veils of cirrus, so that the sky up to 30° from the horizon is practically uniformly white; the strong contrast between sea and sky in that direction disappears immediately, and the sea becomes much greener and lighter. As soon as the cirri disappear the contrast returns.

The extent to which reflection influences the colour of the sea must not lead us to neglect other factors altogether. You can see here and there the shadow of an isolated cloud. There the sea is darker; in sunlit parts it looks more sand-coloured. This is partly a contrast phenomenon however, for when you look through the hole of your half-clenched fist or through the hole in the nigrometer (§ 174) you see that in reality it is blue there too, albeit less than in the shadowy parts.[1] In any case these shadows prove clearly that the colour of the sea is not entirely determined by reflection, but that part of the light is scattered back from below. The shadow becomes visible because the light scattered back is weaker there than elsewhere, whereas the reflected light is not weakened (§ 209).

Does the sand at the bottom shine directly through the water and can the sandbanks be recognised as such from a distance? Not according to my own experience; certainly not to anyone observing them from a dune or from the beach. The sand is visible only where the water is very

[1] The sea shows the loveliest blue, when it is quite smooth, the sky bright blue, and the sun screened by a cloud so that the sea is in a shadow.

shallow, perhaps 4 to 8 in. The position of sandbanks is revealed by the formation of surf just there, and because the spaces in between are smoother (§ 210).

A striking fact is that the sea near the horizon often has a border of grey running to blue (or blue running to darker blue), its breadth not exceeding more than half a degree. It begins to disappear as soon as we leave the dune to look at it from the beach, and disappears completely if we stoop when on the beach. This shows that it is no contrast-border (§ 91). It is probably caused by the fact that the sea is comparatively dark, becoming therefore bluish in the distance, *owing to scattering by the air*[1] (§ 173). One can also imagine the sea to be less turbid at such a long distance from the coast; this would make the clearer water in the distance recognisable at once, if only we are standing high enough to see so far.

Later in the day the sun has moved, and in the afternoon we can see thousands of scintillations in the direction from which it shines. We cannot see the reflected image of the sun itself as we are looking much too closely along the surface of the water; we see only part of the immense path of light cast by it on the irregularly undulating surface. The sea in that direction becomes light grey, almost white.

After sunset, the sea in the west reflects the bright glow and the gold-coloured cirrus-veils; its undulating surface and shifting reflections show us the average colour of the western part of the sky. Towards the north and the south, the sky is less coloured and the tints of the sea are less bright. Our gaze is attracted again and again by the glory of colour in the west. Between the gold-yellow of the clouds there appears here and there a patch of blue sky, the blue wonderfully saturated by the contrast. The colours in the sky become gradually orange and the sea follows them, while the foam of the breakers seems violet by contrast. Right in the foreground there is a streak of wet sand, in

[1] This border is also clearly visible on days when they sky is a featureless grey but the wind moderate and the sea rather dark.

which reflections of certain parts of the sky are smooth and perfect (without being shifted): first a lovely clear blue, and later a tender green. Finally, the cirrus clouds in the west are no longer illuminated, their colour becomes dark violet in tone, and, in the same way, the colours of the sea are subdued; but among these peaceful evening hues the wet sand of the beach traces a warm orange-coloured stripe.

3. *Strong Wind Rising, Grey Sky.* All over the sea the gathering waves are crested with foam, four or five fringes of froth run along the coast; from the south-west the wind comes up, chasing the waves before it. The sea is grey like the clouds, a slightly greenish-grey. Close to the coast we see the waves separately and discover that the greenish component arises from their front slopes, which reflect little light, but emit grey-green light scattered from within. The water seems very turbid; stirred as it is, there will be a lot of sand floating in it. The sea is darkest towards the south-west, where the wind is coming from; towards the south, and especially towards the north, its colour becomes lighter, approaching that of the grey sky, although still remaining a little darker (we are now looking parallel to the waves). Near the horizon the sea is more bluish, the colour of the dark low clouds, which owe their bluishness to scattering over these long distances, whereas, above our heads, they are normally bright or dark grey; and, moreover, the phenomenon of the blue rim on the horizon makes the contrast still stronger (p. 321). Wherever a stray dark cloud hangs aloof in the grey sky, an indistinct, shifted reflection, dark blue-grey, is recognisable on the surface of the sea. The horizon is nowhere sharply defined; especially to the south and north a mist of water-drops is splashed into the air by the foam of the breakers, reducing our range of vision to only a few miles, and causing sea and air to intermingle in the distance.

With weather clearing and wind from the north-west, the state of affairs resembles, more or less, the one just described, but the sky is a confusion of blue patches, white clouds dazzlingly illuminated by the sun (tinged with light-yellow by aerial perspective, § 173), and dark bluish masses. At all points of the compass the sea reflects an average of the colour of the sky at a height of 20° to 30°. Only the large masses are discernible in this reflection, while the sunlit clouds are the most prominent feature, throwing a shining light over the dark, restless sea.

4. *Storm.* I am still behind the dune and the houses, but I can already hear the loud raging of the sea. From the promenade I have a full view of the foam from the breakers, more than two-thirds of the sea covered with boiling froth, white at the crests, dirty-white and frayed out to a network in the troughs between the waves. As usual, the front of the waves is darker to the west than to the south or north, and this makes the scene to the west seem wilder and richer in contrasts. In a high sea, separate crests of froth rise on all sides out of the dark water. A sunlit streak far to the south stands out sharply, a dazzling white on the foaming surface, appearing at first very narrow and long, then, as it approaches, extending over a wide area. The colour of the sand appears very distinctly in those parts where there is no foam and the sunlit sea reflects darker clouds. The light scattered back from the depth is, during such illuminations as this, as strong as possible, the more so because the raging waves stir up and keep floating great quantities of sand. The sky is very dark in some parts, lighter again in others, and there are a few patches of blue. The shifted reflections are still discernible, though only very indistinctly, in the general colour of the sea. The predominant impression is the foam.

Examine light and colour over the sea under every possible condition of wind and cloud.

Compare the tints on rocky and sandy coasts.

Examine also the colour of the sea while you are bathing ; look at the waves not only seawards, but also landwards; find the shadow of other bathers and your own. Use the water-telescope.

If you have an opportunity of walking on one of the piers of a harbour, go and compare the calm sea between the two piers with the sea outside. The condition of the sky is the same, differences arise from differences in the undulating of the sea's surface, or in its turbidity.

Examine the general brightness of the sea's surface, late in the evening and at night, this being a good time because things are not complicated by differences of colour, and the smaller details do not divert our attention.

Beware of contrast phenomena ! To compare different parts of the sky and the sea, you can benefit by the use of a little mirror (§ 176). Hold your hand, or any dark object, between the two fields A and B to be compared; in this way, both A and B will be seen bordering the same field. Use the nigrometer !

Never confuse shadows and reflections of clouds; they fall in entirely different places. When there are separate clouds in the sky, the whole of the light distribution over the sea is dependent on the combination of reflections and shadows.

212. *The Colours of the Sea, seen from a Boat*

Compared with the scene from the beach, there is one great difference, the absence of surf. This makes the whole picture round the observer much more symmetrical. This symmetry is, however, broken by the wind, which gives a definite direction to the waves, by the smoke from the ship, which has the effect of a dark cloud, by the foam in the wake of the keel, and by the sun.

Close to and behind the ship the colour of the light returning from the depths can be judged best of all, because there clouds of air-bubbles are being continually chased through the water, and then rise slowly to the top. In these places a lovely green-blue tint is clearly visible, the same tint as that seen reflected by the white bellies of porpoises frolicking round the ship, or by a white stone as it falls into the water. This tint is to be seen in every ocean, whether the sea there as a whole be indigo-blue or green. It is caused by true absorption in the water removing from the light its yellow, orange and red constituents; the violet rays are scattered away from the observer and so only the green is left, to give the characteristic colour. The parts where there is only a little froth among the foaming masses of green are mostly a kind of purple colour, the complementary colour to green, and one which we must consider as a physiological contrast colour (§ 95).

In shallow seas in the vicinity of ports or mouths of large rivers the sea-water is very muddy. This causes a comparatively large quantity of light to be scattered back from below, so that the conditions here are, to a certain extent, the same as those seen when observing the swarms of air-bubbles in the wake of a ship. The green colour predominates, probably because the water of the river brings into the sea humic acids and ferric compounds (§ 207), their yellowish absorption being superposed on the blue-green colour of the water. This is the kind of shallow green sea on which shadows of clouds stand out magnificently purple-violet on calm days (§ 216).

The 'water colour' shown by white objects at small depths is not as a rule the same as the 'proper colour' of the deep sea. In order to investigate this, the light reflected must be avoided, either by looking, for instance, at the front of a wave, or by using one of the methods mentioned in § 209. There are distinct differences in this 'proper colour' or 'self colour' of the deep sea, which depend on which sea one happens to be crossing; this can be observed splendidly during the voyage from England to Australia. The colour-distribution is, generally speaking, as follows:

Olive green ..	North of 40° Northern Latitude
Indigo	Between 40° and 30° Northern Latitude
Ultramarine ..	South of 30° Northern Latitude

It sometimes happens that olive-green regions move in patches down to lower latitudes. It might be worth while to try to ascertain whether or not this green colour changes with the seasons at a particular place, as there are already some indications in this direction. The origin of the green colour of certain deep seas has not yet been satisfactorily explained. Observations have shown that the water of these seas contains great quantities of floating particles; but though, as is shown by calculation, the ordinary absorption by water, combined with scattering by large particles, can cause every kind of transition from dark blue to light blue or grey, it can never account for the green. This has led some to ascribe it to diatoms and to the faeces of birds which feed on diatoms; and others to a yellow colour of the scattering particles, which might, for instance, consist of yellow sand. General amazement was caused a short time ago by the announcement that this green arises from *fluorescence* of some organic substance![1] Whatever may be true, observations concerning the influence of the seasons of the year point decidedly to an organic origin.

There are a few rare cases where sea-water looks milky white; evidently near the surface there must then be large numbers of floating particles which scatter light in the topmost layers, and this scattering predominates entirely over absorption.

213. *The Colour of Lakes*

The colour of lakes is a source of great beauty in mountain scenery. Their depth is usually sufficient to minimise any effect arising from the colour of the ground at the bottom, in which respect, therefore, they resemble the sea. But they differ from it in their much greater smoothness, a result in its turn of the water's surface being so much smaller and of the mountains along their banks sheltering it from the wind. The regular reflection from their surfaces for this reason plays a much more prominent part than in the case of the sea; the colours at sunset are nowhere so beautifully mirrored as in a lake, and the varying tints of mountain lakes are certainly to be attributed partly to the reflection of their shores. However, should these be high and dark, reflection on the surface is eliminated and, instead, large areas of the

[1] Ramanathan, *Phil. Mag.*, **46**, 543, 1923.

lake show the colour of the light that has penetrated the water nearly perpendicularly and is scattered back from it again. By applying the methods mentioned in § 209, some idea can be obtained of these 'individual colours.' They differ from lake to lake, and can be classified as (1) pure blue, (2) green, (3) yellow-green, (4) yellow-brown.

Closer examination in laboratories has shown that the water of *blue* lakes is almost absolutely pure, and that the colour is due to absorption by the water in the orange and red parts of the spectrum. To account for colours (2), (3) and (4), there is a constantly increasing proportion of iron-salts and humic acids, and also scattering by brown-coloured particles (§ 207).

Very often the green colour of smaller lakes is due to microscopic green algae growing there in vast quantities; often they are still a distinct green in winter, when the trees are bare and everything is covered by snow.

Red colouring can be brought about by other microscopic organisms: *Beggiatoa*, *Oscillaria rubescens*, *Stentor igneus*, *Daphnia pulex*, *Euglena sanguinea*, or *Peridinia*.

For polarisation, *see* § 214.

214. *Observations of the Colour of the Water with a Nicol*[1]

A nicol, as we may remember, only transmits those rays that vibrate parallel to its shorter diagonal. Since light reflected by water vibrates chiefly in a horizontal direction, we can weaken this reflected light by holding the nicol with its short axis vertically and this extinction is even more complete when we observe at an angle of 65° with the vertical ('angle of polarisation'). Try this in a small pool of water in the road after a shower. Stand about 5 yards away from it and hold the shorter diagonal vertical. The effect is amazing, for you can now see the bottom of the pool nearly as well as if there were no pool at all. Rotate the nicol alternately into a horizontal and vertical position, and you will see that the pool seems to become smaller and larger.

[1] E. O. Hulburt, *J.O.S.A.*, **24**, 35, 1934.

As a rule, the nicol heightens the colour of a wet beach, seaweed, granite blocks, a wet road, painted surfaces—everything, in short, that shines in a landscape. The reason is that it invariably takes away part of the surface reflection, which had mixed white with the colour of the object itself.

The contrast on a calm sea between the sunny parts and the shadows of the clouds is accentuated by a nicol with a vertical direction of vibration. The rays reflected at the surface are extinguished, causing the differences in the scattered light to appear more clearly.

It also intensifies the contrast between those parts of the sea covered by a layer of oil and the remaining parts (§ 211); perhaps because the reflection in the ripples takes place at an angle different from the one in the smooth parts, or else because the polarisation by reflection is disturbed by the layer of oil.

The effect of the nicol is striking when the wind is blowing. Look at the raging waves with the short diagonal vertical: the sea now seems much rougher than when observed with the short diagonal horizontal. For in the former case the nicol extinguishes the light reflected, thus making the surface of the sea darker, whereas the foam, retaining its brightness, is more striking.

Often, if the nicol is adjusted in the proper way, the horizon becomes more distinct. Looking in a direction at right angles to the sun we see the sea become decidedly darker and the blue sky relatively brighter when the light-vibrations are vertical (§ 211). For this reason, nicols are sometimes mounted in sextants nowadays.

The following experiments concern the polarisation of light scattered in deep tropical seas, the water of which is pure.[1] Let us assume that we can carry out the experiment when the sun is fairly high, and the surface of the water smooth. Stand with your back to the sun, look at the water more or less at the angle of polarisation, holding the shorter

[1] C. V. Raman, *Proc. R. Soc.*, **101A**, 64, 1922.

diagonal of the nicol vertically. The light reflected is extinguished and you can see the lovely blue glow of the light that comes from below after scattering. Turn the nicol so that the short diagonal is horizontal; the sea will then appear *less blue* than without the nicol.

Carry out this experiment also when the sun is moderately high, holding the short axis vertically again, and vary the azimuth. A comparison between the colour on the side towards the sun and away from the sun is particularly interesting. On the side towards the sun you see a dark indigo colour, because, looking in a direction at right angles to the sun's rays, you have not only extinguished the light reflected, but also the light scattered from the depth of the water. On the side away from the sun, the colour is bright blue, because you are looking pretty well in the direction of the sun's rays that penetrate the water, and the light scattered back in your direction is almost unpolarised. Both these experiments prove that the light scattered by the sea is to a great extent polarised like that in the air (§ 181) and the scattering is therefore by very small particles, probably the water-molecules themselves.

By using the nicol a characteristic difference has been discovered between the radiation scattered back in blue lakes, and in dark brown lakes. To see this one looks in the direction of the sun, avoiding reflection by using a water-telescope (§ 209). The nicol shows us now that in blue lakes the light scattered back vibrates horizontally, as is to be expected when the scattering particles are very small; whereas the larger particles in brown lakes scatter practically unpolarised light, in which the vertical component on emerging from the water predominates slightly (provided the water-telescope has no glass at the end).

215. *Scales for Judging the Colour of Water*

The scale generally used is that of Forel. First make a blue solution of crystals of cupric sulphate and a yellow one of potassium chromate:

0·5 gm. $CuSO_4$. Aq. in 5 c.c. ammonia made up with water to 100 c.c.

0·5 gm. K_2CrO_4 in 100 c.c. water.

Make the following mixtures:

(i)	100 blue + 0 yellow	(viii)	65 blue + 35 yellow
(ii)	98 „ + 2 „	(ix)	56 „ + 44 „
(iii)	95 „ + 5 „	(x)	46 „ + 54 „
(iv)	91 „ + 9 „	(xi)	35 „ + 65 „
(v)	86 „ + 14 „	(xii)	23 „ + 77 „
(vi)	80 „ + 20 „	(xiii)	10 „ + 90 „
(vii)	73 „ + 27 „		

Browner colours are often required, especially when judging the colour of lakes. To meet this requirement, a brown solution can be made as follows:

0·5 gr. cobalt sulphate + 5 $cm.^3$ ammonia + water up to 100 $cm.^3$.

Mix this solution with Forel's green solution (strength xi) in the following proportions:

(11)	100 green + 0 brown	(11–7)	73 green + 27 brown
(11–2)	98 „ + 2 „	(11–8)	65 „ + 35 „
(11–3)	95 „ + 5 „	(11–9)	56 „ + 44 „
(11–4)	91 „ + 9 „	(11–10)	46 „ + 56 „
(11–5)	86 „ + 14 „	(11–11)	35 „ + 65 „
(11–6)	80 „ + 20 „		

These different mixtures can be kept in test-tubes about ½in. in diameter.

The chief difficulty when applying this scale is to know which point of the water's surface should be taken as a norm of comparison. Usually one tries to judge the 'proper colour' of the water itself (§§ 209, 212).

Neither of the scales is quite satisfactory. Another way would be to try to reproduce the colour in paint to be kept for future comparisons.

216. *Shadows on the Water*

'. . . that, whenever shadow is seen on clear water and in a measure, even on foul water, it is not, as on land, a dark shade subduing the sunny general hue to a lower tone, but it is a space of an entirely different colour, subject itself, by its susceptibility of reflection, to infinite varieties of depth and hue, and liable, under certain circumstances, to disappear altogether.'—RUSKIN, *Modern Painters*.

Light sent to us from the surface of the water derives partly from that surface and partly from below, so that if we intercept the incident rays both parts can be changed.

1. *Influence of Shadow on Reflected Light*. 'When surface is rippled, every ripple, up to a certain variable distance on each side of the spectator, and at a certain angle between him and the sun, varying with the size and shape of the ripples, reflects to him a small image of the sun (cf. § 14). Hence those dazzling fields of expanding light so often seen upon the sea. Any object that comes between the sun and these ripples takes from them the power of reflecting the sun, and, in consequence, all their light; hence any intervening objects cast upon such spaces seeming shadows of intense force, and of the exact shape, and in the exact place, of real shadows.'—RUSKIN, *Modern Painters*.

The truth of Ruskin's words can be judged best of all when, on a windy night, the water of a canal, for instance, is very ruffled. As we walk along, we look at the reflection of a street lamp, spreading out into an irregularly flickering patch of light, over which there is a continual gliding of shadows—for example, from the trees between the lamps and the canal. Not until we have reached the most favourable point of view do we notice the presence of these shadows on the water, which are visible only within a small solid angle. Ruskin had long discussions with critics and other interested persons as to whether one might speak of 'shadows' at all in this sense. It is, of course, a question of words !

A rather different effect is brought about when the moon is reflected as a long path of light, and we suddenly see the dark, black silhouette of a sailing boat glide in front of this shining streak of luminosity. The boat itself is now a dark object on a background of light, but it also throws its shadow in our direction, across the rippling water, and here the above considerations apply too.

2. *Influence of Shadow on Light which is scattered Back.* Shadows are marked clearly on turbid water; the degree of distinctness of the shadow is a direct indication of the turbidity or purity of the water. Notice the shadows of bridges and trees on our waterways. When you are crossing the sea try to find your shadow on the water. You will only see it on that side where the ship has stirred the water and mixed it with air-bubbles, and not where the sea is clear and deep blue. Observe the shadows of clouds on the sea's surface.

The shadow becomes visible because the light that penetrates the water and is returned after being scattered, is less in those parts of the surface than elsewhere. On the other hand, light reflected at the surface is not weakened, and therefore becomes relatively more important. This explains why, when the sky is blue, the shadow of a cloud on the sea is often bluish, though, owing to the contrast with the surrounding green, the colour may shift a little towards purple (§§ 209, 211, 212). Apart from the clearness of the water, the direction of observation is important, too. When bathing in very clear water, you will see no shadows; when bathing in slightly muddy water you will see only your own shadow and not those of other bathers, but in very muddy water you will see those of all the bathers. Observe that a shadow cast by a post across rather turbid water of a canal can be seen properly only if you go and stand in the plane containing the sun and post, and therefore look towards that part of the sky where the sun is. You will then see the shadow loom up rather suddenly on the water. This is the same phenomenon as that described in connection with mist.

Shadows on slightly muddy water show still another phenomenon; *their edges are coloured*: the one towards us is bluish, the one away from us is orange. This phenomenon can be observed in the shadow of every post or bridge or ship. It is due to scattering by innumerable particles of dust floating in the water, many of which are so small that they show a preference for scattering blue rays. Now, we see in Fig. 158 that the particles on our side are seen to be luminous on a dark background, so that they send a bluish light to our eye; whereas on the side of the shadow away from us we can see light from the bottom of the water (or from the scattering water in the neighbourhood), deprived of its blue rays and coloured orange by the unilluminated particles in the shadow. This shows that this phenomenon is the same as that of the blue sky and the yellow setting sun (§ 172). Our eye is made particularly sensitive to it by the two contrasting colours along the edges.

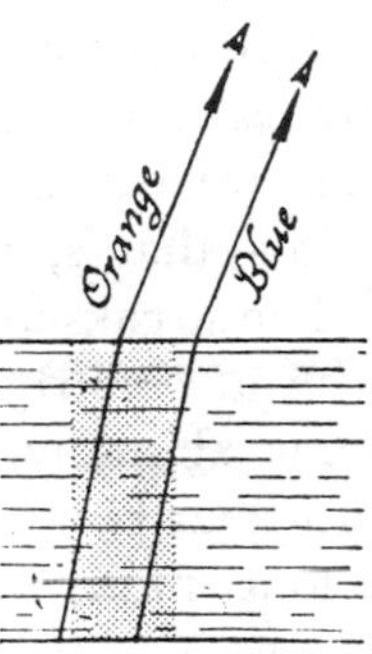

FIG. 158. How coloured edges arise on shadows on turbid water.

Examine the colour of the edges from every viewpoint for different directions of incidence of light and shadow. Note also the distinct bluish colour of narrow pencils of light which penetrate the foliage in a wood and fall on the water of a clear stream, forming an orange patch of light on the bottom.

217. *Aureole of Light about our Shadow on the Water* (Plate XV)

I looked at the fine centrifugal spokes round the shape of my head in the sunlit water. . . .
Diverge, fine spokes of light from the shape of my head, or any-one's head, in the sunlit water !

WALT WHITMAN, 'Crossing Brooklyn Ferry' (*Leaves of Grass*).

This lovely phenomenon can be seen best of all when looking from a bridge or deck of a ship at one's shadow falling on the restless dashing of the waves. Thousands of light and dark lines diverge in all directions from the shadow of our head. This aureole can be seen only around one's own head (cf. § 168). The rays do not converge precisely at one and the same point but only approximately. Another remarkable thing is the increase in the general brightness in the surroundings of the shadow.

Nothing of it is to be seen on calm water nor on water with even waves; it can only be seen well when irregular little mounds of water rise from the surface. The water must be rather turbid: the further one is away from the coast and on the open sea, the weaker the aureole becomes.[1]

The explanation is that each unevenness in the water's surface casts a streak of light or shade behind it; all these streaks run parallel to the line from the sun to the eye, so that we can see them meet perspectively in the anti-solar point—that is, in the shadow-image of our head (§ 191).

The streaks are at times so clear that they can be followed even at a fairly large angular distance from the anti-solar point. Usually, however, they are clearest of all near the anti-solar point, because in that direction our gaze traverses a long path, either through clearly illuminated, or shaded, water. The increase in the general intensity of light in the vicinity of the anti-solar point is perhaps to be attributed to the fact that the scattering of particles is stronger backwards than across the beam (§ 191).

Another aureole of this kind can be observed when we stand in the shade of a solitary tree whose spreading branches cast patches of light and shade on to the water beneath. The rays of light penetrating the liquid give the same optical effect as those caused by unevennesses on a surface.

It is interesting to realise that actually the rays of light do not run at all parallel with the line connecting sun and eye, for, as a consequence of refraction, they are deflected through a certain angle. But, on the other hand, our eye sees the traces of these pencils in their turn in the water, altered by refraction, so that, after all, the part of the pencil travelling in the water is seen as the prolongation of the part travelling through the air.

218. *The Water-line along the Sides of a Ship*

'. . . Three circumstances contribute to disguise the water-line upon the wood: where a wave is *thin*, the colour of the

[1] C. V. Raman, *loc. cit.*

(*a*) The sun casts the shadow of a heavy cumulus cloud on the hazy air below. All the beams seem to come from the same spot, although in reality they are parallel

(*b*) The shadow falls on the ruffled surface of a pond: innumerable light and dark rays diverge from the head. The camera was being held immediately in front of the eye

wood is shown a little through it; when a wave is *smooth*, the colour of the wood is a little reflected upon it; and when a wave is *broken*, its foam more or less obscures and modifies the line of junction.'—RUSKIN, *Modern Painters*.

One might, however, be equally justified in stating that the water line is made visible by these very same factors! Observe, in the case of ships, sailing or lying still, what the optical phenomena are by which we judge where the water begins, i.e. the position of the water line.

219. *Colour of Waterfalls*

When the light is favourable the green colour of water falling over rocks can often be seen very well. It is a remarkable fact that rocks emerging here and there from the water, in reality black or grey, appear now to have a reddish tinge; this must obviously be explained as a contrast colour (§ 95).

This phenomenon can be seen most distinctly where the water foams and splashes. Now it is known that in a laboratory contrast colours occur with greater intensity if the boundaries between the fields are made indistinct. In order to reproduce the case in question, we lay a strip of grey paper on a green background, with a small sheet of tissue paper over it, and then observe how beautifully the reddish contrast colour of the grey can be seen through it (*Florkontrast*). It seems not at all improbable that a similar part is played in Nature by the translucent watery mist.

220. *The Colour of Green Leaves*

Trees, meadows, fields, as well as separate leaves, show us a wealth of greens in infinite variety. In order to discover some kind of regularity in this abundance of phenomena, we will begin by examining one leaf of any 'ordinary' tree (oak, elm or beech), in order to gain an insight into the formation of groups of colour in a landscape.

A leaf on a tree is usually much more strongly illuminated

on one side than on the other, and the colour is essentially determined by whether we are looking at the side that is directly illuminated or the other side. In the former case, the light sent to us from the leaf is partly reflected at its surface, so that the colour becomes lighter but greyer. Moreover, when the leaf is illuminated from the front (relative to the observer), a bluish hue mingles with the

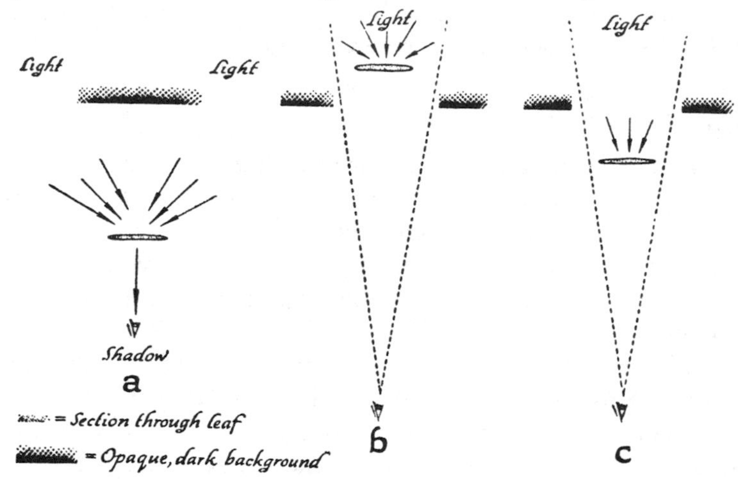

FIG. 159. Green leaves under different conditions of illumination.

green, and when from the back, a yellow hue. This reminds us of our observations concerning the scattering of light (§ 173A). And, indeed, in a leaf, though much less than 1 mm. thick, all the processes of reflection, absorption and scattering take place in the same way as in an ocean hundreds of feet deep. Absorption is caused here by the chlorophyll grains; scattering is probably brought about by innumerable grains of all kinds in which the contents of cells are so rich, or perhaps by the unevenness of the leaf's surface.

The emerald green of grass in a bright light is particularly lovely, seen from a shady spot against a dark background (Fig. 159, *a*). It seems as if each little blade were literally burning with a green inward glow. The incident light pouring on it laterally is scattered by the millions of minute grains, so that each blade casts a stream of light sideways towards our eye.

The difference in colour between grass illuminated from the front and from the back can be seen at once by standing in a meadow, and looking alternately towards and away from the sun. This difference corresponds to the difference, familiar to painters, between the green of Willem Maris in his landscapes, painted against the light, and the green of Mauve, who shows a preference for working with his back to the light.

Illumination by the sun, and by the *blue sky* differ in this, that the light of the sun is stronger, but is reflected more locally, so that a leaf seems patchy. Should the leaf reflect the sun's rays more or less at the regular angle of reflection, then the colour more nearly approaches light grey or white. If the sun is very low, flooding the landscape with a deep red light, the foliage of the trees loses its fresh green tint and becomes faded-looking: there are hardly any green rays left in the light-source to be scattered back by the leaves.

The upper sides and under sides of leaves differ in tint even if the illumination is the same. The upper side is smooth, therefore reflects better, and so is also more patchy. The under side is duller and paler and shows more pores; the cells lie much farther apart, and there are spaces filled with air, which reflect the light before it penetrates the inside of the leaf (§ 224). The upper side as a rule is at the same time the illuminated side. Note the differences in colour when we turn the leaf of a tree 180°, while the illumination, etc., remains the same! Whenever the wind is rather strong, all the trees look patchy on the light side, and the general colour becomes paler than usual; the leaves are turned about in every direction so that their upper sides are seen quite as often as their under sides.

Young leaves are fresher, of a lighter colour than old ones; this difference becomes less noticeable in the course of the summer.

The leaves on the outside of the crown of the tree differ

from those inside; they differ not only in size, thickness and hairyness, but also in colour. The colour of the shoots at the foot of the tree, and on the trunk, is usually very light.

The leaves of several plants shine under the influence of sunshine or wind as if they were varnished (e.g. monkshood). This is due to the swelling of the epidermal cells, which stretches the surface of the leaf until it is perfectly smooth.

Finally, an important part is played by the *background*! Stand under a tree and study its crown. The same leaves, first a fresh green against the background formed by other trees, change into black silhouettes as soon as one looks at them against the sky. The effect depends on the ratio of the brightness of the leaf to the brightness of the sky as a background. It is, therefore, weak if the leaf is illuminated from all sides, especially if the sun shines on it (Fig. 159, *b*), and strongest if the leaf receives only light from a limited part of the sky, as generally occurs when a tree is surrounded by others (Fig. 159, *c*) or during the one-sided illumination of twilight. The difference between the ordinary green and the black silhouette is then so great, that one can hardly believe it to be merely an optical illusion! And yet it is nothing but a matter of contrasts; the bright sky is so enormously brighter than terrestrial objects.

221. *Vegetation in the Landscape*[1]

1. *Separate Trees.* Trees, among all the elements that go to form a landscape, are practically the only ones to show, when illuminated laterally, the wonderful beauty of contrast between their sunlit and their shaded sides. For this reason, they give us the impression of being solid things, 'showing again and again that space with its three dimensions is a visible reality.' This contrast is toned down by the roundness of the tree-tops but accentuated again by the contrast in colour.

[1] *See* Vaughan Cornish, *Geogr. Journ.*, **67**, 506, 1926, for the first part of this section.

Trees seen against the light stand out darkly against the distant background, and make us feel its distance, its remoteness, more sharply; this is due to the stereoscopic effect quite as much as to the difference in the shade of colour. This explains why a tree is so often depicted in the foreground of stereoscopic prints and of landscape paintings. The effect is to a certain extent comparable to that of a landscape seen through an open window, or from under the vault of an archway. Seen from under the high trees of our avenues, the buildings of a town seem larger and more stately.

The most striking contrast with its background is shown by a tree when delineated against the orange glow of an evening sky. The silhouette of the fantastically distorted juniper on a solitary sandhill, or of the solemn firs, with their dense growth of needles, is black, its outline very sharp. The other trees are more open; the thinnest is the birch, with its graceful arabesques, its colour forming, especially against the light, a lovely contrast to the colour of the sky.

'On a sunny morning at the end of February I will show you the colour of the birch-twigs against the azure vault. All their slender sprays seem to glow with a purple fire, while, across that delicate glow, the sky looks down on you with wonderful tenderness. Wait, watch attentively, and do not depart before you have understood. Such a store of happiness is to be gained that one can wait patiently until the winter for this wonderful light to appear again.'—DUHAMEL, *La Possession du Monde*, p. 126.

2. *Woods*. The silhouette of a wood seen close to us and against the light, is certainly very irregular, but the wood itself is too transparent and its light-effects too varied, to give an impression of strength and massiveness. The effect of unity is more pronounced at a greater distance, when the tops of trees shine gold and green against the deepening blue of the hills behind, or when the sunlit foliage of groups of leafy trees stand out against the tall dark fir trees. A distant wood in a flat country can actually be compared

to rows of hills, its shade being at least as dark, its colour, owing to scattering in the atmosphere, being just such a lovely misty blue; it is arranged in successive rows, made individually distinct by aerial-perspective (§ 91).

The scenery in the interior of the wood is unique of its sort, showing, as it does, neither horizon, nor a definite outline. In the spring we see above our heads young green leaves everywhere, glistening in the yellow-green transmitted light. In summer-time, after the tiring glare of the white sky, so trying to look at, our eyes can rest here, once more free to look about in every direction.

The wood is illuminated most of all at midday, when the sun shines from such a height as to penetrate the tops of the trees. The play of light and shade differs in every plane; its charm disappears as soon as we focus our eyes on a certain distance, but it reappears again when we no longer consciously try to find it, but surrender spontaneously and naturally to the influence of our surroundings. On autumn mornings the sun's rays strike here and there between the trunks, and their path can be followed through the slightly hazy air, especially when looking in a direction near the sun (§ 183); in this way the fascination of aerial-perspective is brought very close to us.

3. *Flowers.* Heather is about the only one of our flowers to cover extensive surfaces of land. In August, when it is in full bloom, a curious harmony of colour arises, of purple country and dark blue sky, which some do not admire, but which to others, in the abundant light and freedom of Nature, is exceptionally impressive. Grey clouds covering the sky soften the harmony of colours, but at the same time weaken the contrasts between light and shade.

Fruit-trees in blossom owe their splendour largely to the meagre development of foliage at that time of the year. The white and the pale soft pink against the blue sky are at their very best only when the sun shines on them, or when seen from the top of a dyke or hill with meadows in the background.

4. *Meadows.* One level expanse of a single colour, giving an impression of smoothness and open space, and yet, thanks to its many details, sufficiently varied to suggest buoyancy and softness. Why else should its aspect be so different from that of an expanse of sand? Seen at a distance the green runs to blue-green, and, further off, it approaches more and more the atmospheric blue of the sky.

222. *Shadows and Dark Patches*

Look around and see where there are dark patches in the scenery!

(*a*) In woods and shrubberies: between the trunks and stems. (*b*) In the town: open windows in the distance.

Both these cases are excellent examples of the 'black body,' as it is termed by physicists. They are spaces in which we can only look through a narrow aperture; the light-rays which enter it can emerge only after being reflected a number of times, becoming considerably feebler each time. A body of this kind absorbs nearly all the rays—dark woods re-emit only 4 per cent. of the incident light. On the other hand, one must bear in mind that the darkness of a wood is only relative, and if we draw near, the eye having once become adapted to the light, we can see that everything in it shows light and colour. Similarly, every detail in a room, seen from inside, can be distinguished, whereas the same room seen from outside through the open window looks pitch dark.

Delicate objects outlined against the bright sky usually look black, but this is only due to contrast (§ 220).

Examine systematically the colours of the shadows!

'All ordinary shadows must be coloured somehow, never black or nearly black. They are evidently of a luminous nature. . . . It is a fact that shadows are colours just as much as light parts are.'—RUSKIN.

Where the sun shines its brilliant yellowish rays predominate over the light radiated from the sky; but in the shade light only falls from the blue or grey sky. Shadows are, therefore, generally bluer than their surroundings and this difference becomes accentuated by contrast.

'From my window, I see the shadows of people walking along the shore; the sand in itself is violet, but the sun makes it golden; the shadows of those people are so violet that the ground seems yellow.'—DELACROIX.

If the sun is half-hidden by clouds, shadows become hazier; if it disappears altogether, there are no longer any cast shadows, but there still remain parts that are darker and lighter.

'Observe in the streets at twilight, when the day is cloudy, the loveliness and tenderness spread on the faces of men and women.'—LEONARDO DA VINCI.

This remark has made me feel reconciled, more than once, to a dreary, dull, grey day !

At night in the light of an arc-lamp (more or less a point-source of light), dominating every other source in its vicinity, the shadows cast are very sharp, and, by exaggerating the wrinkles in faces, make them look old.

All kinds of transitions exist between the sharp shadows of the sun and the diffuse light on all sides from the clouded sky. An open space in a wood, for instance, is lit by a limited part of the sky, the effect produced by it altering according as it is large or small.

In flat or gently undulating country shadows accentuate the relief only when the sun is very low. Its rays then almost graze the surface of the ground, causing very curious variations of light and shade. This can be seen on a small scale, though to an exaggerated extent, on a smooth expanse of sand towards sundown; every pebble, every unevenness, casts a lengthy shadow; the ground resembles a photo of a lunar landscape, and gives the impression of being unreal.

223. *Illumination of a Landscape, towards and away from the Sun*

There is a remarkable difference in colour and structure to be seen in almost any landscape, depending on whether we look at it towards the sun, or away from it. The entire aspect of the scenery changes! Make use of a mirror in order to view the scene in two directions simultaneously (Plate XVI).

1. A field of young corn, a meadow, a field of lupine, are yellow-green towards the sun, but away from it they are bluish; and the reason? Look 'microscopically' at one little leaf in particular. Pick it, and hold it towards the sun and then away from the sun. In the former case, you will see chiefly light falling through the leaf, in the latter case, light reflected by its surface (§ 220). The colours and lights are sometimes influenced by the direction of the wind.

2. The waves in a field of ripe rye are chiefly caused by the changing aspect of the ears. Suppose that the wind is blowing towards the sun: *facing* the sun, we see practically only bright light waves; these arise when the ears incline towards the sun to such an extent as to reflect the sunlight towards our eye; *looking away* from the sun we see a few bright but more dark waves. The latter arise when the ears bend in such a way as to cast shadows on those near to them.

These phenomena change with every direction of wind, of gaze, and with the altitude of the sun.

3. A lawn mown with a lawnmower, looks much lighter in colour when the direction of mowing is away from us than when it is towards us; in the former direction we see much more reflected light (Plate XVI; cf. § 220). The contrast on a stubble-field is very strong; there the successive rows are alternately light and dark because the mowing-machine has passed alternately up and down. If you turn round, you will see the shades the other way about. Freshly ploughed land glistens if one looks at right angles to the direction of the still damp furrows.

4. Duckweed on the water of a ditch behaves in exactly the opposite way to grass. Away from the sun it is yellow-green, but towards the sun pale grey-green. 'Microscopic' observation shows that in the latter case irregular reflection on the surface is much stronger. We do not see through the leaves of this plant.

5. Heath-land, when the heather is over, is darker seen towards the sun, away from the sun more glistening, silky, light brown-grey, evidently owing to reflections (Plate XVI).

6. Fruit-trees in full bloom are white only when seen away from the sun. Seen towards the sun, the blossoms stand out black against the sky (§§ 220, 221).

7. Similarly the branches and twigs of trees are grey and brown seen away from the sun; black, without detail, towards the sun.

8. A brick-paved road is brown-red towards the sun, white-grey away from the sun.

9. A gravel road is white-grey towards the sun, brown-grey away from the sun.

10. Sea-foam is a pure white away from the sun; towards the sun, however, it is rather darker than its surroundings, among the myriads of reflections and flickerings of the playful water.

11. An uneven road, covered with snow, seen towards the sun, looks, as a whole, darker than the smooth snow at the side; away from the sun it is the other way round.

12. Waves on a lake, the wind blowing towards the sun: if we look away from the sun, the water appears a sombre blue, with here and there blue-black streaks radiating from the point of observation, corresponding to the blue parts of the sky; each one of the many waves stands out separately. Looking towards the sun, everything is a laughing bright blue, waves can be seen only in the distance, in endless numbers (§ 211).

13. Observe that when you look towards the sun all objects having their shaded side towards you appear dark, but with lovely light edges. This is the charm of photos taken against the light.

These and many other instances give altogether an inexhaustible opportunity for observation. Always try to find the explanation by first observing things as a whole and then individually.

XVI

A patch of heath-land seen with the sun behind and, by reflection in a mirror, with the sun in front

The tracks of a mowing machine on a lawn. The light and dark strips disappear when viewed at right angles

224. *How Colours are affected by Humidity*

'It is true that the dusky atmosphere "obscures all objects," but it is also true that Nature, never intending the eye of man to be without delight, has provided a rich compensation for this shading of the tints with *darkness*, in their brightening by *moisture*. Every colour, wet, is twice as brilliant as it is when dry; and when distances are obscured by mist, and bright colours vanish from the sky, and gleams of sunshine from the earth, the foreground assumes all its loveliest hues, the grass and foliage revive into their perfect green, and every sunburnt rock glows into an agate.'—RUSKIN, *Modern Painters*.

Humidity alone is not sufficient to explain this enlivening of colour. One must also take into account that as soon as objects are covered by a thin film of water, their surface becomes smoother, they no longer scatter white light on all sides, and, therefore, their own colour predominates and becomes more saturated.

Rain alters the colour of the ground altogether. Street cobbles reflect more strongly the farther they are away from us and the more steeply our gaze falls on them. It is surprising how splendidly not only asphalt, but also very uneven paved roads, can reflect at large angles. The colours of roads of sand, soil and gravel grow darker and warmer; the first drops of rain stand out as dark spots. Why? The water penetrates every interstice between the grains of sand. A ray of light, which otherwise would have been scattered by the topmost layers, can now penetrate much deeper before being sent back to our eye; and is almost entirely absorbed over this longer path.

A pool of water on an asphalt road reveals beautiful shades of colour:

(*a*) the surface of water reflecting the blue sky;
(*b*) a black edge where the ground is still damp;
(*c*) the grey surroundings.

Algae in a ditch form a dark green, fibrous mass; a portion protruding from the water looks a much paler green on account of the air between the fibres. But hold those paler parts under water, shake them and press them together, and air-bubbles will issue from them, whilst at the same time they become darker.

225. *Human Figures in the Landscape*

'From my window I see a man stripped to the waist, working at the floor of the gallery. When I compare the colour of his skin with that of the wall outside, I notice how coloured the half-tints of the flesh are compared with those of the inanimate material. I noticed the same yesterday in the Place St. Sulpice, where a young urchin had clambered on to one of the statues of a fountain, standing in the sun. Dull orange was his flesh, bright violet the gradations of the shadows and golden the reflections in shaded parts turned towards the ground. Orange and violet predominated in turn, or became intermingled. The golden colour was slightly tinged with green. The true colour of the flesh can be seen only in the sun and in the open air. If a man puts his head out of a window its colouring is quite different from what it is indoors. Which shows the absurdity of studies done in a studio, where each one does his best to reproduce the wrong colour.'—DELACROIX, *Journal.*

XIII

Luminous Plants, Animals and Stones

226. *Glow-worms*

'Tell B. that I have crossed the Alps and the Apennines, that I have visited the "Jardin des Plantes," the museum arranged by Buffon, the Louvre with its masterpieces of sculpture and painting, the Luxembourg with the works of Rubens, that I have seen a *Glow-worm* ! ! !'—Letter from FARADAY to his mother, *Life and Letters*.

As a matter of fact glow-worms are not worms at all but beetles. The female glow-worms are wingless and creep about, the males fly. The common glow-worm (*Lampyris noctiluca*) is abundant in some of the southern counties of England and is found in Scotland south of the Tay, but not in Ireland. The luminous organs occupy the last two segments of the hinder abdomen, and contain a substance which when oxidised becomes luminous by chemi-luminescence. The colour of the rays emitted is precisely the one to which our eyes are most sensitive, and contains no infra-red, so that this beetle might be called a really ideal source of light—if only it would shine somewhat brighter !

227. *Phosphorescence of the Sea*

Phosphorescence of the sea is caused mainly, in our parts, by millions of microscopic marine animals of the species (*Noctiluca miliaris*). These are protozoa belonging to the group known as 'flagellates,' about 0·2 mm. in size, that is, just large enough to be seen with the naked eye as tiny separate dots. They emit light only when oxygen is dissolved in the water, as by stirring or by the breaking of the waves. This causes a certain substance to become oxidised, but not noticeably heated; nor does its light show the same composition as that of a glowing body; it is not a case of temperature-radiation but of chemi-luminescence:[1] it contains neither ultra-violet rays nor infra-red rays, but only those colours that convey a strong impression of light to our eye, such as, in particular, yellow and green.

If you immerse your fingers in the sea when phosphorescent organisms are present in great numbers, you may feel a slight pricking. By this you can foretell by day whether the beautiful phosphorescence will be visible at night.

Phosphorescence on the sea can often be seen splendidly on thundery evenings in summer after a hot day. The glow of lamps along a promenade or from hotels can always make one doubt whether what one sees is really phosphorescence or the white foam on the crests of waves; for this reason the beauty of this phenomenon is only perfect on an absolutely dark night. If, however, the conditions for observing are not so ideal, the next best thing is to take off your shoes and stockings, go into the water and stir it with your hand below the surface.

If phosphorescence cannot be seen clearly, one will nevertheless see, while stirring, many a small stray spark here and there emitting light for a second and then going out. Fill a small bucket with sea-water and put it somewhere where it is absolutely dark. Even on less favourable days you will see signs of phosphorescence by pouring the water into a basin, or by exciting the microscopic beings by adding alcohol, formol or some acid to the water. Pour the phosphorescent water into a glass and the little animals collect on its surface. Tap the glass, the mechanical vibration will cause them to emit light, and if you do this repeatedly the emission of light becomes gradually feebler.

Occasionally sea-water is phosphorescent without our being able to distinguish the sparks. This is accounted for by the presence of bacteria (*Micrococcus phosphoreus*).

[1] Strictly speaking, the word 'phosphorescence' has quite a different meaning, and ought never to be used in connection with the luminescence of the sea!

Draw up a scale for the phosphorescence on the sea!

Practise on cold evenings when phosphorescence is certain not to be present, and examine the appearance of foaming crests; on favourable evenings you will then be able to observe the difference.

If you are on a sea-voyage, especially in the tropics, you should go and stand, on a dark night, on the bow or stern of the ship out of the way of lamplight. You will see an almost continual display of sparks of light shooting past; this is made up of all kinds of luminous marine animals.

In the Indian Ocean the entire sea seems at times to be luminous, while a system of enormous bands of light seems to rotate like the spokes of a wheel over its surface: these are wind-waves and bow-waves of the ship, which, as they pass along, make the water turbulent and therefore luminous.

228. *Luminous Wood. Luminous Leaves*

Sometimes on warm summer nights in a damp wood one can see how decaying wood emits a feeble light. This is caused by the fibres of the honey fungus (*Armillaria mellea*), which lies embedded in it everywhere.

In spring or winter try to find tree-trunks from which the bark can easily be peeled, and on which are dark, bifurcated fibres. Lay pieces of these trunks in damp moss and take them home. Keep them in a shady place under a bell-glass. In a few days' time the fibres of the fungus covering the wood will begin to give out light. Occasionally decaying branches, too, emit light; this is caused by bacteria.

The dry leaves of the beech and oak, gathered together in thick layers and half-decomposed, clearly emit light at a certain stage of decay. Try to find layers 4 to 12 in. deep; do not take the loose leaves at the very top, but those lying underneath, close together, with yellow-white spots, and carry a handful of them into a perfectly dark room. The luminosity in this case is ascribed to fungus fibres of a species not yet determined.

229A. *Cats' Eyes at Night*[1]

We all know the fierce light cats' eyes seem to radiate. Yet in reality this light is only reflected, but *directed reflected light*, like that of a reflector on a bicycle, or of the *Heiligenschein* on dewy grass (§ 168). The rays penetrating the cornea form a very clear image on the background of the eye, and this image reflects its rays through the same cornea, the pencil of light returning practically along the same path as that along which it entered. To see the phenomenon most clearly, the lamp, the eye of the cat and the eye of the observer must lie in one straight line. This can be achieved by holding an electric torch level with your eyes; the shining of the cat's eyes will still be visible at a distance of 90 yards.

The light reflected by dogs' eyes is reddish. Sheep, rabbits and horses have also luminous eyes, but human beings have not.

229B. *Reflection of Light on Mosses*

It is a beautiful clear morning, with dew everywhere on the grass. Luxuriant clumps of moss of the kind *Mnium* grow in a fairly dark ditch, two rows of leaflets on their delicate little stems, these leaflets giving the impression of being strewn with little shining stars. Each star radiates a goldish-green light, much steadier than the light of a sparkling dew-drop. On closer inspection, we discover that there are little drops hanging everywhere under the leaflets and we come to the conclusion that sunlight penetrates the edges of a leaf, whereupon it suffers total reflection in the drop and emerges after passing once more through the leaf, the gold-green colour arising during this process.

Schistostega osmundacea, the famous luminous moss in the caverns and crevices of the Fichtelgebirge in Bavaria, shows light-reflections, which are still more beautiful. In this moss the spherical cells themselves play the part of reflecting drops.

[1] *Nat.*, **88**, 377, 1912.

230. *Fluorescence of Plant Juices*

In the spring, cut some pieces of the bark or leaves of horse chestnut and put them in a glass of water. The plant juice mixes with the water, and this begins to exhibit a peculiar blue luminescence, which can be seen best if you cast a cone of sun's rays through the fluid by means of a convex lens (eye-glass or magnifying glass). The phenomenon is accounted for by the liquid absorbing the violet and the, to us invisible, ultra-violet rays of the sun, and emitting blue rays in their stead. A transformation of this kind is called 'fluorescence.'

The bark of the largely cultivated manna ash tree (*Fraxinus Ornus*) is said to show this phenomenon, too.

231. *Phosphorescent Ice and Snow*

An old legend tells us that ice-fields, after being illuminated for a long time by the sun, give a feeble light at night. Snow, several degrees below zero is also said to give light when brought into a dark room after the sun has cast its rays on it. Hailstones, especially those that fall first in a hailstorm, are said to show a kind of electric luminescence.

Who will put this to the test?

232. *Scintillation from Stones*

Occasionally we see how the hooves of a horse strike the cobbles of a street with such force as to cause sparks.

Look for flints or quartzites (i.e. ordinary pebbles) along the roadside. The latter are brownish stones slightly transparent along the edges, usually softly rounded without crystalline structure. Knock two stones of this kind together in as dark a place as possible; sparks will arise, and there is a peculiar smell. This can be observed with other stones as well. The sparks are caused by particles being knocked off and becoming heated by the collision so that they glow. Certain gases are set free, which cause the peculiar odour.

233. *Will-o'-the-Wisps*[1]

Folk-lore tells us of will-o'-the-wisps dancing like tiny flames over the churchyard, or enticing travellers into the morass. Their existence, however, is by no means a fairy tale! They have been seen and described by the famous astronomer Bessel and other excellent observers; the difficulty is that this phenomenon can occur in many different shapes.

Will-o'-the-wisps are found in bogs, or places where peat is dug, and along dykes; they have been seen occasionally on the damp, freshly-manured ground of a nursery garden whenever one stamped on the soil, and in muddy ditches or in drains when one stirred the water in them. They occur more during summer and on rainy, warm autumn nights than during cold seasons. They resemble tiny flames, about ½ in. to 5 in. high and not more than 2 in. broad. Sometimes they are right on the ground, at other times they float about 4 in. above it. That they dance about is apparently not true. What really happens is that they go out suddenly while another flame arises quite near, and this probably accounts for the impression of rapid movement. Occasionally they are blown along by the wind a few feet before they become extinguished. Many other cases have been observed where a will-o'-the-wisp has burnt steadily for hours at a stretch, a whole night long and even in the daytime. When a new flame arises one hears at times the pop of a little explosion. The colours are said to be sometimes yellow, sometimes red or blue. In many cases when one puts one's hand in the little flame, no heat is to be felt; a walking-stick with a copper ferrule, after being held in the flame for a quarter of an hour, was practically at the same temperature as before; even dry reeds did not catch fire. In other cases, paper and cotton-waste could be lighted by the flame. Generally there is no smell, occasionally a faint smell of sulphur.

[1] W. Müllen-Erzbach, *Abh. Naturw. Verein Bremen*, **14**, 217, 1897; *Wetter*, **20**, 46, 1903, and **33**, 18 and 71, 1916.

What do these mysterious flames consist of? Nobody has yet succeeded in collecting the gas that takes fire. It has been suggested that it might be hydrogen phosphide, which is capable of spontaneous combustion in air; apparently a mixture of PH_3 and H_2S catches fire without smoke and without smell, reproducing rather closely the actual phenomenon. Such gases can arise by the decomposition of rotting substances. The flame is a form of chemiluminescence and its low temperature is a peculiar feature, occurring frequently in reactions of this kind.

Appendix

234. *A Few Suggestions for photographing Natural Phenomena*

With each optical phenomenon described in this book, the question arises as to whether it would not be possible to take photographs of it. It is amazing how much there is to be done in this line, and how little has been done up to now! An ordinary camera is generally good enough. If a stand is used, it should be fitted with a ball-joint (to be had for a few shillings), which makes it possible to incline the camera in any direction. Phenomena like the rainbow and the halo will require the use of a large aperture objective. When taking photos of the coronae and distortions of the setting sun, the focal length of the camera must be at least 12 in.

Always use plates or films with an anti-halation backing, and preferably ortho- or panchromatic. For landscapes showing snow, hoar-frost, trees in bloom, clouds, distant horizons, use a yellow filter with ortho- or panchromatic films or plates. Prevent the sun from shining on your objective by fixing a cylinder in front of the lens.

Take landscape photographs preferably when the sun is not too high. Distinguish the effect of illumination obliquely from the front, from behind, or from above (cf. § 223).

The time of exposure varies from $\frac{1}{100}$ second for photos taken from an aeroplane to one hour for photos taken by moonlight.

Develop with a metol-hydroquinone developer.

235. *How to measure Angles in the Field*

(*a*) Try to estimate, without any auxiliary means at all, the altitude of stars. To this end, try to fix the position of the zenith first, then turn round and judge whether you would again locate it at the same place. After this try to determine an altitude of 45°, then of 22·5°, and 67·5°. You will find that you have a tendency not to bend your head sufficiently backwards (cf. § 109). The errors made by a *good* observer never exceed 3°.

(*b*) Stick three pins A, B and C in a piece of wood or a post card in such a way that the angle to be measured is exactly included between the sighting directions BA and

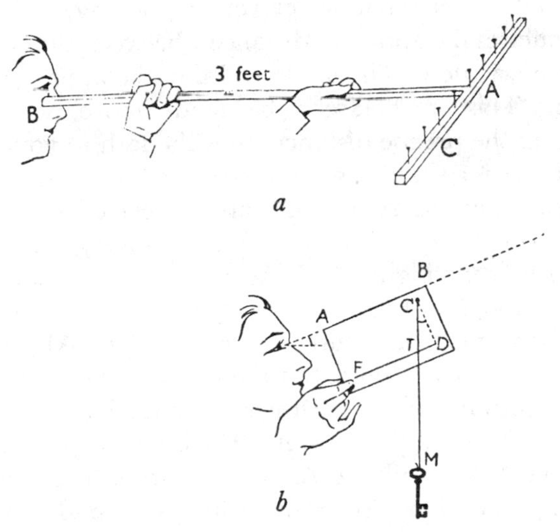

FIG. 160. Simple means of the estimating of angles.

BC. The piece of wood must be properly fixed, either flat on a table or nailed to a tree. Then draw the lines BA and BC and read the required angle on a graduated arc (Fig. 106).

(*c*) Fasten the middle of a lath, provided with nails or pins at equal distances from each other, to the end of another lath (3 ft. long) and at right angles to it (Fig. 160, *a*). Hold the end of the rake thus produced with its stick B pressed against your cheek-bone, then, when the nails A and C appear to cover the points in question, $\frac{AC}{BA}$ will be equal to the required angle expressed in 'radians,' one radian being

57°. If we have, for example, AC=3 in., then $\frac{AC}{BA}=$ 0·08 rad.=4·7°. For angles exceeding 20° the calculation is slightly less simple.

(*d*) Stretch your arm right in front of you and spread your fingers as widely apart as possible. The angle subtended by the tops of thumb and little finger will be about 20°. Or, again, with your arm outstretched, hold a short lath at right angles to the direction in which you are looking, and measure in centimetres the apparent distance *a* between the two points under observation. The angle will then be approximately *a* degrees. This method can be made more accurate by measuring the precise distance from the lath to your eye.

(*e*) There is a simple apparatus for measuring angles above the horizon, giving results that are correct to 0·5°. Take a rectangle of cardboard pierced at C, and hang a thread CM through C, carrying a weight to make it serve as a plumb-line (Fig. 160, *b*). The observer sights, say, a tree top whose height he wants to measure, accurately along AB, turns the cardboard ever so little out of the vertical plane, so that the thread hangs free and then turns it slowly back, so that the thread presses gently against it. The lines CD perpendicular to AB and DT parallel to AB are drawn on the cardboard. The length of CD is preferably 4 in. The angle DCM will then be equal to the angle between AB and the horizontal plane, and can be measured with a graduated arc, or computed from its tan $\frac{TD}{CD}$; for small angles $\frac{\text{TD (in inches)}}{4}$ is equal to the angle in radians.

Cf. §§ 1, 120.

Index

A CATALOG OF SELECTED

DOVER BOOKS

IN ALL FIELDS OF INTEREST

THE CLARINET AND CLARINET PLAYING, David Pino. Lively, comprehensive work features suggestions about technique, musicianship, and musical interpretation, as well as guidelines for teaching, making your own reeds, and preparing for public performance. Includes an intriguing look at clarinet history. "A godsend," *The Clarinet,* Journal of the International Clarinet Society. Appendixes. 7 illus. 320pp. $5\frac{3}{8}$ x $8\frac{1}{2}$. 0-486-40270-3

HOLLYWOOD GLAMOR PORTRAITS, John Kobal (ed.). 145 photos from 1926-49. Harlow, Gable, Bogart, Bacall; 94 stars in all. Full background on photographers, technical aspects. 160pp. $8\frac{3}{8}$ x $11\frac{1}{4}$. 0-486-23352-9

THE RAVEN AND OTHER FAVORITE POEMS, Edgar Allan Poe. Over 40 of the author's most memorable poems: "The Bells," "Ulalume," "Israfel," "To Helen," "The Conqueror Worm," "Eldorado," "Annabel Lee," many more. Alphabetic lists of titles and first lines. 64pp. $5\frac{3}{16}$ x $8\frac{1}{4}$. 0-486-26685-0

PERSONAL MEMOIRS OF U. S. GRANT, Ulysses Simpson Grant. Intelligent, deeply moving firsthand account of Civil War campaigns, considered by many the finest military memoirs ever written. Includes letters, historic photographs, maps and more. 528pp. $6\frac{1}{8}$ x $9\frac{1}{4}$. 0-486-28587-1

POE ILLUSTRATED: Art by Doré, Dulac, Rackham and Others, selected and edited by Jeff A. Menges. More than 100 compelling illustrations, in brilliant color and crisp black-and-white, include scenes from "The Raven," "The Pit and the Pendulum," "The Gold-Bug," and other stories and poems. 96pp. $8\frac{3}{8}$ x 11. 0-486-45746-X

RUSSIAN STORIES/RUSSKIE RASSKAZY: A Dual-Language Book, edited by Gleb Struve. Twelve tales by such masters as Chekhov, Tolstoy, Dostoevsky, Pushkin, others. Excellent word-for-word English translations on facing pages, plus teaching and study aids, Russian/English vocabulary, biographical/critical introductions, more. 416pp. $5\frac{3}{8}$ x $8\frac{1}{2}$. 0-486-26244-8

PHILADELPHIA THEN AND NOW: 60 Sites Photographed in the Past and Present, Kenneth Finkel and Susan Oyama. Rare photographs of City Hall, Logan Square, Independence Hall, Betsy Ross House, other landmarks juxtaposed with contemporary views. Captures changing face of historic city. Introduction. Captions. 128pp. $8\frac{1}{4}$ x 11. 0-486-25790-8

NORTH AMERICAN INDIAN LIFE: Customs and Traditions of 23 Tribes, Elsie Clews Parsons (ed.). 27 fictionalized essays by noted anthropologists examine religion, customs, government, additional facets of life among the Winnebago, Crow, Zuni, Eskimo, other tribes. 480pp. $6\frac{1}{8}$ x $9\frac{1}{4}$. 0-486-27377-6

TECHNICAL MANUAL AND DICTIONARY OF CLASSICAL BALLET, Gail Grant. Defines, explains, comments on steps, movements, poses and concepts. 15-page pictorial section. Basic book for student, viewer. 127pp. $5\frac{3}{8}$ x $8\frac{1}{2}$. 0-486-21843-0

THE MALE AND FEMALE FIGURE IN MOTION: 60 Classic Photographic Sequences, Eadweard Muybridge. 60 true-action photographs of men and women walking, running, climbing, bending, turning, etc., reproduced from a rare 19th-century masterpiece. vi + 121pp. 9 x 12. 0-486-24745-7

CATALOG OF DOVER BOOKS

LIGHT AND SHADE: A Classic Approach to Three-Dimensional Drawing, Mrs. Mary P. Merrifield. Handy reference clearly demonstrates principles of light and shade by revealing effects of common daylight, sunshine, and candle or artificial light on geometrical solids. 13 plates. 64pp. $5\frac{3}{8}$ x $8\frac{1}{2}$. 0-486-44143-1

ASTROLOGY AND ASTRONOMY: A Pictorial Archive of Signs and Symbols, Ernst and Johanna Lehner. Treasure trove of stories, lore, and myth, accompanied by more than 300 rare illustrations of planets, the Milky Way, signs of the zodiac, comets, meteors, and other astronomical phenomena. 192pp. $8\frac{3}{8}$ x 11. 0-486-43981-X

JEWELRY MAKING: Techniques for Metal, Tim McCreight. Easy-to-follow instructions and carefully executed illustrations describe tools and techniques, use of gems and enamels, wire inlay, casting, and other topics. 72 line illustrations and diagrams. 176pp. $8\frac{1}{4}$ x $10\frac{7}{8}$. 0-486-44043-5

MAKING BIRDHOUSES: Easy and Advanced Projects, Gladstone Califf. Easy-to-follow instructions include diagrams for everything from a one-room house for bluebirds to a forty-two-room structure for purple martins. 56 plates; 4 figures. 80pp. $8\frac{3}{4}$ x $6\frac{3}{8}$. 0-486-44183-0

LITTLE BOOK OF LOG CABINS: How to Build and Furnish Them, William S. Wicks. Handy how-to manual, with instructions and illustrations for building cabins in the Adirondack style, fireplaces, stairways, furniture, beamed ceilings, and more. 102 line drawings. 96pp. $8\frac{3}{4}$ x $6\frac{3}{8}$. 0-486-44259-4

THE SEASONS OF AMERICA PAST, Eric Sloane. From "sugaring time" and strawberry picking to Indian summer and fall harvest, a whole year's activities described in charming prose and enhanced with 79 of the author's own illustrations. 160pp. $8\frac{1}{4}$ x 11. 0-486-44220-9

THE METROPOLIS OF TOMORROW, Hugh Ferriss. Generous, prophetic vision of the metropolis of the future, as perceived in 1929. Powerful illustrations of towering structures, wide avenues, and rooftop parks–all features in many of today's modern cities. 59 illustrations. 144pp. $8\frac{1}{4}$ x 11. 0-486-43727-2

THE PATH TO ROME, Hilaire Belloc. This 1902 memoir abounds in lively vignettes from a vanished time, recounting a pilgrimage on foot across the Alps and Apennines in order to "see all Europe which the Christian Faith has saved." 77 of the author's original line drawings complement his sparkling prose. 272pp. $5\frac{3}{8}$ x $8\frac{1}{2}$. 0-486-44001-X

THE HISTORY OF RASSELAS: Prince of Abissinia, Samuel Johnson. Distinguished English writer attacks eighteenth-century optimism and man's unrealistic estimates of what life has to offer. 112pp. $5\frac{3}{8}$ x $8\frac{1}{2}$. 0-486-44094-X

A VOYAGE TO ARCTURUS, David Lindsay. A brilliant flight of pure fancy, where wild creatures crowd the fantastic landscape and demented torturers dominate victims with their bizarre mental powers. 272pp. $5\frac{3}{8}$ x $8\frac{1}{2}$. 0-486-44198-9

Paperbound unless otherwise indicated. Available at your book dealer, online at **www.doverpublications.com**, or by writing to Dept. GI, Dover Publications, Inc., 31 East 2nd Street, Mineola, NY 11501. For current price information or for free catalogs (please indicate field of interest), write to Dover Publications or log on to **www.doverpublications.com** and see every Dover book in print. Dover publishes more than 400 books each year on science, elementary and advanced mathematics, biology, music, art, literary history, social sciences, and other areas.